Advances in Intelligent Systems and Computing

Volume 875

Series editor

Janusz Kacprzyk, Polish Academy of Sciences, Warsaw, Poland
e-mail: kacprzyk@ibspan.waw.pl

The series "Advances in Intelligent Systems and Computing" contains publications on theory, applications, and design methods of Intelligent Systems and Intelligent Computing. Virtually all disciplines such as engineering, natural sciences, computer and information science, ICT, economics, business, e-commerce, environment, healthcare, life science are covered. The list of topics spans all the areas of modern intelligent systems and computing such as: computational intelligence, soft computing including neural networks, fuzzy systems, evolutionary computing and the fusion of these paradigms, social intelligence, ambient intelligence, computational neuroscience, artificial life, virtual worlds and society, cognitive science and systems, Perception and Vision, DNA and immune based systems, self-organizing and adaptive systems, e-Learning and teaching, human-centered and human-centric computing, recommender systems, intelligent control, robotics and mechatronics including human-machine teaming, knowledge-based paradigms, learning paradigms, machine ethics, intelligent data analysis, knowledge management, intelligent agents, intelligent decision making and support, intelligent network security, trust management, interactive entertainment, Web intelligence and multimedia.

The publications within "Advances in Intelligent Systems and Computing" are primarily proceedings of important conferences, symposia and congresses. They cover significant recent developments in the field, both of a foundational and applicable character. An important characteristic feature of the series is the short publication time and world-wide distribution. This permits a rapid and broad dissemination of research results.

Advisory Board

Chairman

Nikhil R. Pal, Indian Statistical Institute, Kolkata, India
e-mail: nikhil@isical.ac.in

Members

Rafael Bello Perez, Universidad Central "Marta Abreu" de Las Villas, Santa Clara, Cuba
e-mail: rbellop@uclv.edu.cu

Emilio S. Corchado, University of Salamanca, Salamanca, Spain
e-mail: escorchado@usal.es

Hani Hagras, University of Essex, Colchester, UK
e-mail: hani@essex.ac.uk

László T. Kóczy, Széchenyi István University, Győr, Hungary
e-mail: koczy@sze.hu

Vladik Kreinovich, University of Texas at El Paso, El Paso, USA
e-mail: vladik@utep.edu

Chin-Teng Lin, National Chiao Tung University, Hsinchu, Taiwan
e-mail: ctlin@mail.nctu.edu.tw

Jie Lu, University of Technology, Sydney, Australia
e-mail: Jie.Lu@uts.edu.au

Patricia Melin, Tijuana Institute of Technology, Tijuana, Mexico
e-mail: epmelin@hafsamx.org

Nadia Nedjah, State University of Rio de Janeiro, Rio de Janeiro, Brazil
e-mail: nadia@eng.uerj.br

Ngoc Thanh Nguyen, Wroclaw University of Technology, Wroclaw, Poland
e-mail: Ngoc-Thanh.Nguyen@pwr.edu.pl

Jun Wang, The Chinese University of Hong Kong, Shatin, Hong Kong
e-mail: jwang@mae.cuhk.edu.hk

More information about this series at http://www.springer.com/series/11156

Ajith Abraham · Sergey Kovalev
Valery Tarassov · Vaclav Snasel
Andrey Sukhanov
Editors

Proceedings of the Third International Scientific Conference "Intelligent Information Technologies for Industry" (IITI'18)

Volume 2

 Springer

Editors
Ajith Abraham
Scientific Network for Innovation
and Research Excellence
Machine Intelligence Research Labs
(MIR Labs)
Auburn, WA, USA

Sergey Kovalev
Rostov State Transport University
Rostov-on-Don, Russia

Valery Tarassov
Bauman Moscow State Technical University
Moscow, Russia

Vaclav Snasel
VSB-Technical University of Ostrava
Ostrava, Czech Republic

Andrey Sukhanov
Rostov State Transport University
Rostov-on-Don, Russia

ISSN 2194-5357   ISSN 2194-5365   (electronic)
Advances in Intelligent Systems and Computing
ISBN 978-3-030-01820-7   ISBN 978-3-030-01821-4   (eBook)
https://doi.org/10.1007/978-3-030-01821-4

Library of Congress Control Number: 2018958808

This Springer imprint is published by the registered company Springer Nature Switzerland AG
The registered company address is: Gewerbestrasse 11, 6330 Cham, Switzerland

Preface

This volume of Advances in Intelligent Systems and Computing contains papers presented in the main track of IITI 2018, the Third International Scientific Conference on Intelligent Information Technologies for Industry held in September 17–21 in Sochi, Russia. The conference was jointly co-organized by Rostov State Transport University (Russia) and VŠB-Technical University of Ostrava (Czech Republic) with the participation of Russian Association for Artificial Intelligence (RAAI).

IITI 2018 is devoted to practical models and industrial applications related to intelligent information systems. It is considered as a meeting point for researchers and practitioners to enable the implementation of advanced information technologies into various industries. Nevertheless, some theoretical talks concerning the state of the art in intelligent systems and soft computing were also included in proceedings.

There were 160 paper submissions from 11 countries. Each submission was reviewed by at least three Chairs or PC members. We accepted 94 regular papers (58%). Unfortunately, due to the limitations of conference topics and edited volumes the Program Committee was forced to reject some interesting papers, which did not satisfy these topics or publisher requirements. We would like to thank all the authors and reviewers for their work and valuable contributions. The friendly and welcoming attitude of conference supporters and contributors made this event a success!

The conference was supported by Russian Fund for Basic Research (grant no. 18-07-20024 G).

September 2018

Ajith Abraham
Sergey M. Kovalev
Valery B. Tarassov
Václav Snášel
Andrey V. Sukhanov

Organization

Organizing Institutes

Rostov State Transport University, Russia
VŠB-Technical University of Ostrava, Czech Republic
Russian Association for Artificial Intelligence, Russia

Conference Chairs

Sergey M. Kovalev	Rostov State Transport University, Russia
Alexander N. Guda	Rostov State Transport University, Russia

Conference Vice-chair

Valery B. Tarassov	Bauman Moscow State Technical University, Russia

International Program Committee

Alexander I. Dolgiy	JSC "NIIAS", Rostov branch, Russia
Alexander L. Tulupyev	St. Petersburg Institute for Informatics and Automation of the Russian Academy of Sciences, Russia
Alexander N. Shabelnikov	JSC "NIIAS", Russia
Alexander N. Tselykh	Southern Federal University, Russia
Alexander P. Eremeev	Moscow Power Engineering Institute, Russia
Alexander V. Smirnov	St. Petersburg Institute for Informatics and Automation of the Russian Academy of Sciences, Russia

Alexey B. Petrovsky	Institute for Systems Analysis of Russian Academy of Sciences, Russia
Alexey N. Averkin	Dorodnitsyn Computing Centre of Russian Academy of Sciences
Alla V. Zaboleeva-Zotova	Volgograd State Technical University, Russia
Anton Beláň	Slovak University of Technology in Bratislava, Slovakia
Dusan Husek	Institute of Computer Science, Academy of Sciences of the Czech Republic
Eid Emary	Cairo University, Egypt
Eliska Ochodkova	VSB-Technical University of Ostrava, Czech Republic
František Janíček	Slovak University of Technology in Bratislava, Slovakia
Gennady S. Osipov	Institute for Systems Analysis of Russian Academy of Sciences, Russia
Georgy B. Burdo	Tver State Technical University, Russia
Habib M. Kammoun	University of Sfax, Tunisia
Hussein Soori	VSB-Technical University of Ostrava, Czech Republic
Igor B. Fominykh	Moscow Power Engineering Institute, Russia
Igor D. Dolgiy	Rostov State Transport University, Russia
Igor N. Rozenberg	JSC "NIIAS", Russia
Igor V. Kotenko	St. Petersburg Institute for Informatics and Automation of the Russian Academy of Sciences, Russia
Ildar Batyrshin	National Polytechnic Institute, Mexico
Ivan Zelinka	VSB-Technical University of Ostrava, Czech Republic
Jana Nowakova	VSB-Technical University of Ostrava, Czech Republic
Jaroslav Kultan	University of Economics in Bratislava, Slovakia
Jiří Bouchala	VŠB-Technical University of Ostrava, Czech Republic
Jiří Hammerbauer	University of West Bohemia, Czech Republic
Josef Paleček	VŠB-Technical University of Ostrava, Czech Republic
Juan Velasquez	University of Chile, Chile
Konrad Jackowski	Wrocław University of Technology, Poland
Leszek Pawlaczk	Wrocław University of Technology, Poland
Marcin Paprzycki	IBS PAN and WSM, Poland
Michal Wozniak	Wroclaw University of Technology, Poland
Milan Dado	University of Žilina, Slovakia
Mohamed Mostafa	Arab Academy for Science, Technology, and Maritime Transport, Egypt

Nadezhda G. Yarushkina	Ulyanovsk State Technical University, Russia
Nashwa El-Bendary	Scientific Research Group in Egypt (SRGE), Egypt
Nour Oweis	VSB-Technical University of Ostrava, Czech Republic
Oleg P. Kuznetsov	Institute of Control Sciences of Russian Academy of Sciences
Pavol Špánik	University of Žilina, Slovakia
Petr I. Sosnin	Ulyanovsk State Technical University, Russia
Petr Saloun	VSB-Technical University of Ostrava, Czech Republic
Santosh Nanda	Eastern Academy of Science and Technology, Bhubaneswar, Odisha, India
Sergey D. Makhortov	Voronezh State University, Russia
Stanislav Kocman	VŠB-Technical University of Ostrava, Czech Republic
Stanislav Rusek	VŠB-Technical University of Ostrava, Czech Republic
Svatopluk Stolfa	VSB-Technical University of Ostrava, Czech Republic
Tarek Gaber	VSB-Technical University of Ostrava, Czech Republic
Teresa Orłowska-Kowalska	Wrocław University of Technology, Poland
Vadim L. Stefanuk	Institute for Information Transmission Problems, Russia
Vadim N. Vagin	Moscow Power Engineering Institute, Russia
Vladimir V. Golenkov	Belarus State University of Informatics and Radioelectronics, Belarus
Vladimír Vašinek	VŠB-Technical University of Ostrava, Czech Republic
Yuri I. Rogozov	Southern Federal University, Russia
Zdeněk Peroutka	University of West Bohemia, Czech Republic

Organizing Committee Chair

| Alexander N. Guda | Rostov State Transport University, Russia |

Organizing Vice-chair

| Andrey V. Sukhanov | Rostov State Transport University, Russia |

Local Organizing Committee

Andrey V. Chernov	Rostov State Transport University, Russia
Anna E. Kolodenkova	Samara State Technical University, Russia
Ivan A. Yaitskov	Rostov State Transport University, Russia
Jan Platoš	VSB-Technical University of Ostrava, Czech Republic
Maria A. Butakova	Rostov State Transport University, Russia
Maya V. Sukhanova	Azov-Black Sea State Engineering Institute, Russia
Pavel Krömer	VSB-Technical University of Ostrava, Czech Republic
Vitezslav Styskala	VSB-Technical University of Ostrava, Czech Republic
Vladislav S. Kovalev	JSC "NIIAS", Russia

Contents

Probabilistic Models, Algebraic Bayesian Networks and Information Protection

Computer-Aided Event Tree Synthesis on the Basis of Case-Based Reasoning

Aleksandr F. Berman[1], Olga A. Nikolaychuk[1],
and Aleksandr Yu. Yurin[1,2(✉)]

[1] Matrosov Institute for System Dynamics and Control Theory,
Siberian Branch of the Russian Academy of Sciences,
134, Lermontov st., Irkutsk 664033, Russia
iskander@icc.ru
[2] Irkutsk National Research Technical University,
83, Lermontov st., Irkutsk 664074, Russia

Abstract. Emergency analysis and technogenic risk assessment are the key concepts of the technogenic safety research. There are some methods used for their investigation, for example, the event tree analysis method. But the high complexity of building and verification of event trees of complex technogenic systems significantly weakens the effectiveness of the practical application of this method and requires development of special software and modification of the standard methodology. This paper describes a new algorithm for computer-aided event tree synthesis for technical systems in petrochemistry. The proposed algorithm is based on the original model of the object technical state dynamics which describes cause-effect relationships between the parameters in different time intervals and a case-based reasoning approach. The model of the technical state dynamics is formalized in the form of the technical states matrix. The case-based reasoning approach is used for implementation of the algorithm proposed. The elements of software, including functions, the architecture and the information process of event tress synthesis are described.

Keywords: Case-based reasoning · Event tree · Synthesis

1 Introduction

Technogenic safety analysis and provision are one of the most significant aspects of sustainable development of a country and its regions. In this context technogenic safety is considered as a state of protection of the population, industrial personnel, objects of economy and environment against hazardous technogenic events.

The key concepts of the research of technogenic safety are emergency (as a result of a safety violation) and its risk analysis. The last one includes the identification of emergency causes and losses as well as substantiation of actions for emergency risk reduction. Risk analysis of emergency involves the following tasks: to define a sequence of possible events (a base scenario) which can violate the safety of a hazardous industrial object; to define their probability (or rate) and calculate their effects (consequences).

© Springer Nature Switzerland AG 2019
A. Abraham et al. (Eds.): IITI 2018, AISC 875, pp. 3–12, 2019.
https://doi.org/10.1007/978-3-030-01821-4_1

A hazardous industrial object is meant an industrial object, which can be used for production, processing, storage or transportation of radioactive, inflammable, hazardous chemical and biological substances. The hazardous industrial objects in this paper are presented as complex technical systems (CTSs) along with their components which are characterized by the following main factors: the presence of extreme pressures and temperatures during operation; the risk for people and the environment.

Decision errors and delayed actions result in incorrect risk assessment, and therefore in unreasonable actions (measures) and more significant losses. Decision-making efficiency and accuracy in this case can be improved by combining contemporary risk assessment methods (for example, fault tree and event tree methods, fault type and effects analysis, a check list method and etc.) [1–3], their automation [3–7] and intellectualization [8, 9] by means of artificial intelligence methods, in particular, expert systems.

The most accurate and one of the basic scientific methods used for technogenic risk assessment [1, 2, 10, 11] is the event tree (ET) analysis method [1–3]. However the high complexity of building and verification of ETs of complex systems significantly weakens the effectiveness of the practical application of this method and requires development of special software [3] and modification of the standard methodology [5, 12].

There is special software for these purposes: Risk Spectrum [13], FaultTree+ [14], SAPHIRE [15], REY, PC ASM SZMA [16], TREE-EXPERT and etc. Most of them don't provide a way to synthesize (generate) ETs and only support drawing trees without taking into account their semantic content. In turn the semantic content can be provided only by specialists who have experience in describing the simulated processes. Thus, the formation of adequate and consistent content result in not computational, but semantic complexity of the trees construction, which can be reduced. So, the problem of the synthesis of semantically significant ETs is relevant.

In this paper we propose to automate the ET synthesis on the basis of the original model of object technical state dynamics [17–19] and special software that implements a case-based reasoning approach. The main result of the paper is an approach that reduces the semantic complexity of ETs building and supports expert decision-making on the basis of the accumulated information about the technical diagnosis and maintenance of a hazardous object.

2   The Technical States Dynamics Model

The technical states dynamics model [17–20] is a basis for development of the proposed algorithm. This model provides: the complete and detailed description of degradation (aging) of technical systems (resulting in catastrophic faults); taking into account effecting factors; and definition (specifying) of regularities of the technical state changing at submicro-, micro-, and macro-levels.

The processes of degradation are the objective physical-chemical processes conditioned by both different technological processes and structural, manufacturing, and maintenance irregularities which cause damages, failures and emergencies. Each process of degradation is characterized by a mechanism and kinetics. A mechanism of degradation is a set of properties of the technological object and effecting factors.

A kinetics is a set of micro-and (or) macroscopic phenomena resulting from accumulating elementary movement acts. A description of kinetics includes: events; event parameters; functional relations (if it is possible) for definition of event parameters at a specified moment of time. In the field of safety the degradation processes is called as hazardous processes.

The model of dynamics of technical states is based on the following structure: part (P) $\rightarrow$ assembly unit (AU) $\rightarrow$ mechanical system (MS) $\rightarrow$ complex technical system (CTS).

There are two main states of CTSs and their structural elements: a safe state and a hazardous state.

It is proposed to represent the dynamics of technical states in the form the following sequence [17]: initial defectiveness (Def), damage (Dam), destruction (Des), failure (F) – in the case of MSs; failure (F), pre-emergency (PE), emergency (E), technogenic catastrophe (TC) – in the case of CTSs.

Then, the transitions between states are represented in the form of a state matrix (Fig. 1) where the "reliability" block describes states of MSs, the "safety" block – the states of CTSs where $S_k^j\left(s_{k1}^j, \ldots, s_{kN_k}^j\right)$ – k is a class of a state an object at j-level of a CTS hierarchy in some phase space, $s_{k1}^j, \ldots, s_{kN_k}^j$ are parameters describing states.

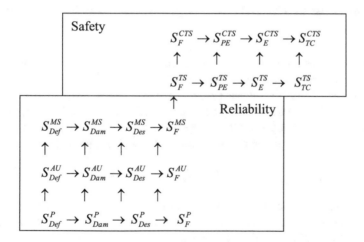

Fig. 1. The technical states matrix.

This model proposes the form for conceptualization of the cause-effect structure of technical states dynamics and improves the efficiency of its formalization due to the clear description of states relationships.

The standard procedure (algorithm) of the ET construction includes the following main steps: building a subtree based on the structure of the technical object (e.g. P, AU, MS, and etc.); building a sequence of events causing a failure. The additional steps expanding standard ET construction procedure are proposed (Fig. 2). These steps reflect mechanisms and kinetics of events (inspired by the degradation processes). The additional symbol ($\triangleleft$) denotes a mechanism.

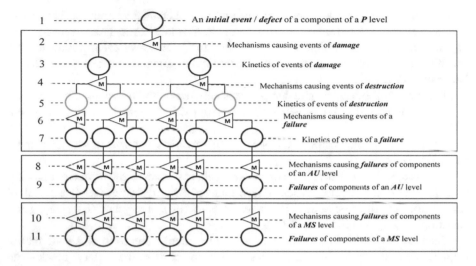

Fig. 2. The proposed ET construction procedure based on the technical states matrix (a fragment).

3 Application of Case-Based Approach for the Event Tree Synthesis

The ET synthesis is a creative problem requiring appropriate knowledge and experience that in turn can be represented as cases.

The automation of ET synthesis based upon the technical states matrix is carried out by means of artificial intelligence methods, in particular, a case-based approach [21].

A case (as a structured pattern) can be represented as a combination of two basic components [21–23]: a problem description and a problem solution:

$$Case = \langle \text{Description, Solution} \rangle.$$

In our case the problem description contains a list of diagnostic properties of technical states. The solution is a list of actions (measures) for prevention, localization and liquidation of failures and emergencies.

The properties are divided into groups (classes) related to possible states: initial defectiveness, damage, destruction, etc. [23]. Thus, a case is represented as follows: $c^i = \left\{ \langle Spr_k^i, Sd_k^i \rangle \right\}_{k=1}^{k_i}$ where Spr_k^i is a description of the k-class of a state, Sd_k^i is a description of solution for the k-class of a state. The problem is described by a set of tuples: $Spr_k^i = \left\{ \langle p_{kj}^i, v_{kj}^i, r_{kj}^i \rangle \right\}_{j=1}^{J}$ where p_{kj}^i is a name of a j-property describing the state, v_{kj}^i is a value of a j-property, r_{kj}^i is a restriction for a value on a j-property; a description of solution is represented as a set of actions (measures) $Sd_k^i = \left\{ sd_{km}^i \right\}_{m=1}^{M}$ where sd_{km}^i is a description of a m-action that is represented in the form of a sequence of control decisions for prevention, localization and liquidation of failures and emergencies.

A case base has a structure (with vertical and horizontal links) in accordance of the technical state matrix: a case for a part connected with a case of an assembly unit, which, in turn, connected with the case of MS, and so on (Fig. 3).

A generalized algorithm of the ET synthesis on the basis of the case-based reasoning approach is the following:

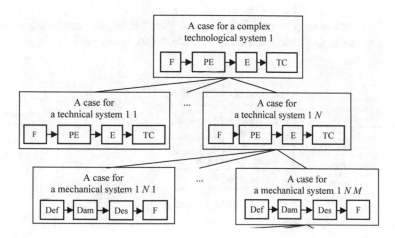

Fig. 3. A hierarchy of cases (vertical links) and chains of states (horizontal links).

- Step 1. Selection of an object and its technical state.
- Step 2. Analysis of information on the object involving description of its structure and observed characteristics.
- Step 3. Retrieving similar cases.
- Step 4. Analysis of retrieved cases and creation of event scenarios and subtrees on the basis of their vertical and horizontal links.
- Step 5. Integration of event scenarios into ETs, according to the following rule: i-node of j-scenario can be merged with i-node of k-scenario, if i-nodes (and all previous) are similar cases (Fig. 4).
- Step 6. Integration of event scenarios into ETs if i-node of j-scenario is similar to i-node of k-scenario with a specified assessment (0.5, for example).

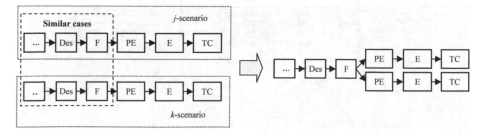

Fig. 4. Integration of event scenarios into the event tree.

It is possible to calculate the probabilities (or frequencies) of events based on the number of their occurrences in a scenarios description. These probabilities can be replaced by expert appraisals if a representative case base is absent.

The proposed algorithm is implemented in the software ESRA (Expert System for Risk Analysis).

The architecture of the software includes the following modules: an ETs synthesis module; an editor of ETs; a database of hazardous industrial objects; a library of computation modules (for calculating event parameters for undesired processes, assessing the most probable scenario of undesired processes, etc.).

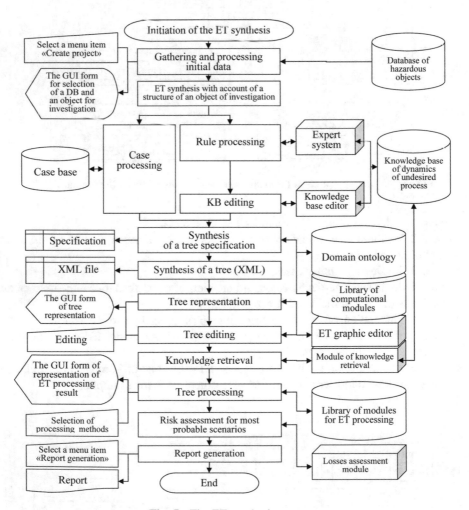

Fig. 5. The ET synthesis process.

The main functions of the software are the following:

- the automated ET synthesis using logical rules and cases;
- the ET editing including;
- event editing, including: input of/editing initial data on ET events; calculation of event parameters (for example, a volume of fluid outflow, destruction range, and some others); probability calculation for events;
- textual and graphic ETs representation;
- generation of reports.

Let's describe the ET synthesis process (Fig. 5) supported by ESRA. The ET synthesis module builds (on the basis of initial data) a subtree representing an object structure and possible transitions of states. Next, a case-based expert system retrieves the similar cases in the case-base and forms event sequences (chains). Then the results are processed by the ET specification creation module.

The results of interpretation are represented in the ET graphic editor, at that specifications are translated into XML format.

4 Discussion

Efficiency and quality of the results obtained (the algorithm and software) were evaluated by comparing ETs built in different ways:

- A1: using the algorithm and software proposed, we used a database of equipment failures in the petrochemical industry [20] that contains detailed information on 200 cases;
- A2: using a standard procedure based on information from safety reports for similar equipment.

At the same time, efficiency is understood as the economic effect obtained from the meaningful interpretation of the constructed trees. Identification of a new possible cause of failure will prevent it or reduce its consequences, thus avoiding unplanned downtime. In particular, the unplanned downtime of the technological line for high-pressure polyethylene production for 1 day results in losses of about 105 000 dollars.

Quality is understood according to ISO 8402-86. The main property of the algorithm and software proposed characterizing their quality is the adequacy of the results obtained, i.e. the correspondence of the created event trees to the real undesirable processes occurring at the studied hazardous objects.

So, the main objective of the case study was to assess the adequacy of the ETs built.

The details of mechanical systems with description of critical failures, the causes of which were processes of degradation presented in the knowledge base, namely, corrosion cracking, corrosion fatigue and hydrogen embrittlement, were selected for the experiments. Seven experiments (Table 1) were carried out, during which ETs were synthesized according to the algorithm proposed. Then the synthesized trees were compared with the standard ones.

Table 1. Results of the case study (DP – a process of degradation, AD – an angular diameter).

Exp. N	Detail	Count of DP	A1: count of events/scenarios	A1: count of cases	A2: count of events/scenarios	Difference in the events/ scenarios
1	Bend 180 AD 60	2	36/3	8	12/2	24/1
2	Bend AD 24	1	26/2	6	13/2	13/0
3	Lens outlet	2	24/2	6	11/2	13/0
4	T-piece	1	20/1	3	8/1	12/0
5	Bent tube AD 15	3	35/3	2	9/2	24/1
6	Bent tube AD 16	1	27/2	18	15/2	12/0
7	Bent tube AD 25	2	39/3	8	24/2	15/1

The results of the case study showed that the synthesized ETs provide more complete information about the genesis of processes of degradation (more events) and in some cases (Exp. N: 1, 5, 7) allow one to identify new processes and develop adequate preventive measures.

5 Conclusions

The automation of the event trees synthesis (generation) is actual problem. In this paper we propose to automate this process on the basis of the original model of object technical state dynamics formalized in the form of the technical states matrix, and special software (ESRA), that implements the case-base approach.

The approach allows to reduce the experts workload when creating high dimensional ETs; however, it will require more efficient preparation of initial data when filling knowledge bases.

The ETs construction is a complex and time-consuming process, the complexity of which is associated with the need for experience and detailed information to describe hazardous processes, which is not always the case.

The algorithm proposed can be utilized to generate the first draft of the event tree and to increase the overall efficiency of the process by increasing the adequacy of the resulting trees.

In addition, the algorithm proposed is a way to support the expert decision-making on the basis of accumulated information about the technical diagnosis and maintenance of a hazardous object and provides the ability to automate the synthesis of trees based on the experience of the organization. Further, the trees can be supplemented by experts on the basis of their own knowledge.

The reported study was partially supported by project IV.38.1.2 (AAAA-A17-117032210079-1) and RFBR projects 18-07-01164, 18-08-00560.

References

1. Faisal, I., Abbasi, S.: Analytical simulation and PROFAT II: a new methodology and a computer automated tool for fault tree analysis in chemical process industries. J. Hazard. Mater. **87**(1–3), 23–56 (2001). https://doi.org/10.1016/S0304-3894(00)00169-2
2. Henley, E., Kumamoto, H.: Reliability Engineering and Risk Assessment. Prentice-Hall Inc., Englewood Cliffs (1981)
3. Kumamoto, H.: Chapter 7 – Fault tree analysis. In: Fundamental Studies in Engineering, vol. 16, pp. 249–311. Elsevier, New York (1993)
4. Papazoglou, I.: Functional block diagrams and automated construction of event trees. Reliab. Eng. Syst. Saf. **61**(3), 185–214 (1998). https://doi.org/10.1016/S0951-8320(98)00011-8
5. Walker, M., Papadopoulos, Y.: Synthesis and analysis of temporal fault trees with PANDORA: the time of Priority AND gates. Nonlinear Anal. Hybrid Syst. **2**, 368–382 (2008). https://doi.org/10.1016/j.nahs.2006.05.003
6. Wang, Y., Teague, T., West, H., Mannan, S., O'Connor, M.: A new algorithm for computer-aided fault tree synthesis. J. Loss Prev. Process. Ind. **15**(4), 265–277 (2002). https://doi.org/10.1016/S0950-4230(02)00011-6
7. Yang, Z.-X., Zheng, Y.-Y., Xue, J.-X.: Development of automatic fault tree synthesis system using decision matrix. Int. J. Prod. Econ. **121**(1), 49–56 (2009). https://doi.org/10.1016/j.ijpe.2008.02.021
8. Bobbio, A., Ciancamerla, E., Franceschinis, G., Gaeta, R., Minichino, M., Portinale, L.: Sequential application of heterogeneous models for the safety analysis of a control system: a case study. Reliab. Eng. Syst. Saf. **81**, 269–280 (2003). https://doi.org/10.1016/S0951-8320(03)00091-7
9. Wang, Y., Li, Q., Chang, M., Chen, H., Zang, G.: Research on fault diagnosis expert system based on the neural network and the fault tree technology. Procedia Eng. **31**, 1206–1210 (2012). https://doi.org/10.1016/j.proeng.2012.01.1164
10. ISO/IEC 31010:2009 Risk management - Risk assessment techniques
11. Majdara, A., Wakabayashi, T.: Component-based modeling of systems for automated fault tree generation. Reliab. Eng. Syst. Saf. **94**(6), 1076–1086 (2002). https://doi.org/10.1016/j.ress.2008.12.003
12. Ferdous, R., Khan, F., Veitch, B., Amyotte, P.: Methodology for computer-aided fault tree analysis. Process. Saf. Environ. Prot. **85**(1), 70–80 (2007). https://doi.org/10.1205/psep06002
13. Risk spectrum homepage. http://www.riskspectrum.com/en/risk/. Accessed 16 Mar 2018
14. FaultTree+ homepage. https://www.isograph.com/software/reliability-workbench/fault-tree-analysis-software/. Accessed 16 Mar 2018
15. SAPHIRE homepage. http://www.princeton.edu/~achaney/tmve/wiki100k/docs/SAPHIRE.html. Accessed 16 Mar 2018
16. SZMA homepage. http://www.szma.com/pkasm.shtml. Accessed 16 Mar 2018
17. Berman, A., Nikolaichuk, O.: Technical state space of unique mechanical systems. J. Mach. Manuf. Reliab. **36**, 10–16 (2007). https://doi.org/10.3103/S105261880
18. Berman, A.: Formalization of formation processes of failure for unique mechanical systems. J. Mach. Manuf. Reliab. **3**, 89–95 (1994)
19. Berman, A., Nikolaychuk, O., Pavlov, A., Yurin, A.: A methodology for the investigation of the reliability and safety of unique technical systems. Proc. Inst. Mech. Eng., Part O: J. Risk Reliab. **228**(1), 29–38 (2014). https://doi.org/10.1177/1748006X13494820

20. Berman, A., Khramova, V.: Automated data base for failures in pipelines and tubular high-pressure apparatus. Chem. Pet. Eng. **29**, 63–66 (1993)
21. Aamodt, A., Plaza, E.: Case-based reasoning: foundational issues, methodological variations, and system approaches. AI Commun. **7**, 39–59 (1994). https://doi.org/10.3233/aic-1994-7104
22. Bergmann, R.: Experience Management. LNAI, vol. 2432. Springer, Heidelberg (2002)
23. Nikolaychuk, O., Yurin, A.: Computer-aided identification of mechanical system's technical state with the aid of case-based reasoning. Expert. Syst. Appl. **34**, 635–642 (2008). https://doi.org/10.1016/j.eswa.2006.10.001

Security of Information Processes
in Supply Chains

Yury Iskanderov[1](✉) and Mikhail Pautov[2]

[1] The St. Petersburg Institute for Informatics and Automation of RAS,
39, 14-th Line, St. Petersburg, Russia
iskanderov_y_m@mail.ru
[2] TPF Forwarding Network, Avd. Paralell 15, 2, 08004 Barcelona, Spain
management@tpfnetwork.org

Abstract. This report reviews and summarizes legal, technical and psychological methods of data and information protection considered as critical elements of a comprehensive information security management system for global supply chains. An intelligent system to automate the functions of such information security management system is suggested.

Keywords: Supply chains · Corporate information · Data protection
Information security management system · Intelligent system

1 Introduction

Supply chain management is a coordinated control of material, financial and information traffics throughout a supply chain. Novel IT are designed and implemented in order to increase the efficiency of supply chains. The basic role of these IT is to facilitate document and information interchange, at both in-house and inter-organizational levels. Modern supply chains generate and transmit significant information flows containing confidential data. Thus, activities of 3PL, 4PL, LLP and other major operators in supply chains, as well as coordination of subcontractors require the use of complex information systems that enable rapid and safe flow of a wide range of data and information. The role of information processes in supply chains is reflected in the principles of supply chain management formulated by APICS (American Production and Inventory Control Society, now renamed into The Educational Society for Resource Management). These principles (velocity, variability, vocalize, visualize, value) also known as the 5V principles of supply chain management are described in detail in [1]. The third of these 5V principles (vocalize) concerns ensuring the flow of information between cooperating units, appropriate in format, place and time. Realization of this principle requires a thoroughly designed system of information security for supply chains [6]. Modern research in the field of information security in supply chains is mainly focused on either legal or technical or psychological aspects separately. Single attempts were made to synthesize legal and technical (e.g. in [2]) or psychological and technical (e.g. in [7]) methods. In this paper we suggest a concept of comprehensive security management system for global supply chains which includes legal, technical and psychological elements.

© Springer Nature Switzerland AG 2019
A. Abraham et al. (Eds.): IITI 2018, AISC 875, pp. 13–22, 2019.
https://doi.org/10.1007/978-3-030-01821-4_2

2 Information Related Risks and Data Protection in Supply Chains

Identification and classification of supply chain data and information in terms of their value and confidentiality is fundamental for any system of information security for supply chains. Sensitive data and information that come across the supply chains may include commercial information and documents, financial data, personal data, intellectual property. Data loss, leak, damage and counterfeit may happen due to purposeful malicious human activities, negligence, errors and omissions, hard- or software damages or malfunctions. The channels of data breach in supply chains normally include: e-communication channels, phone/fax, postal and courier channels, direct physical contact with data sources. The supply chain breach threats can be people, process and technology related. The "anatomy" of a typical supply chain breach was suggested by Shackleford [8]. When sensitive data is breached and exposed, the impact to organizations and consumers is far-reaching. Businesses may experience financial penalties, legal costs, loss of consumer confidence, drops in stock price and overall hits to their reputation [8]. In this paper we overview the three classes of data protection strategies which in our opinion must be integrated into a comprehensive information security management system for supply chains.

3 Legal Approach

As it is put forward in [2] the problem of information security in supply chains can be analyzed through the transaction cost theory [3], agency theory and incomplete contract theory [4]. According to transaction costs theory, bounded rationality (associated with asymmetry in access to information), opportunism of parties to transaction and the specificity of the assets used in transaction were recognized by O. Williamson as main sources of transaction costs [3]. Using agency theory assumptions it should also be noted that agents typically act for their own benefit, and they represent the opportunistic attitude [3, 4]. Considering information security, the use of information needed for the transaction by one party for its own purpose, as well as limiting access of other parties to information may be examples of opportunistic behavior. Limiting access to data and information, as well as the use of data and information for individual purpose of one party are also important factors of incompleteness of contract [2]. For the sake of minimization of risks associated with limited access to data and information important for cooperation within a supply chain or with improper use of data and information it is worth referring to the role of contractual regulations concerning informational support of cooperation (creation of information, use, transfer, access etc.). ISO/IEC 27001 guidelines help formalize and standardize such regulations [5].

Data breaches in global supply chains also become a major concern of governmental and international organizations. In 2017 cybersecurity regulations introduced on the state level more than doubled. While the federal governments have not passed security regulations as fast, pressure is increasing for businesses due to global and state level regulatory changes [9]. With the number of data breaches in 2017 and the coming implementation of the General Data Protection Regulation (GDPR) [10], data security has become a core focus for governments for a good reason. Warehouses have traditionally been a place not just to store goods, but also to store sensitive data for greater insights. In the past any business in the supply chain simply had to be compliant and pass an occasional audit if regulations were in place. Things have changed just in the last few years [9]. According to a Thales Security report [11], 67% of enterprises were breached in 2017. Countries are moving very quickly to implement penalties and reign in poor cybersecurity practices in organizations. The GDPR represents the most significant paradigm shift in decades regarding data security [9].

Despite the growing threat and evidence surrounding the supply chain attack vector, there are few specific compliance mandates addressing third parties, although third party risk is usually implicated in a number of other areas, e.g. vendor due diligence, risk management and contract requirements (as mentioned above). However, some compliance and regulatory bodies have issued guidance explicitly dealing with vendor management and third-party risk management. As vendor management and supply chain security are increasingly linked to data breaches and security incidents, it is likely that more explicit requirements will become commonplace in most major compliance mandates. Organizations should stay aware of all mandates and standards and make necessary revisions in policy or procedure to accommodate those changes [8].

Best practices have only increased in the last few years due to the increase in data security regulations. The legal practices listed below also take into account what will be required to be in compliance with the GDPR [9]:

- Integrate Data Security into Supplier Governance;
- Define and Classify Suppliers and Data;
- Appoint a Data Protection Officer;
- Audit Suppliers.

4 Technical Approach

However even when the value and confidentiality hierarchy of the supply chain data and information is well determined and adopted, and measures are taken to protect most sensitive data and information through contracts and internal instructions in compliance with the GDPR and other relevant regulations and recommendations, other risks still persist (even if they are contemplated by contracts and regulations). Supply chains account for roughly 80% of all cyberattacks around the globe according to the SANS Institute [8]. Suppliers, distributors, recipients, logistics providers and subcontractors in a supply chain often use heterogenic hard- and software, belonging to different generations and designed by different developers. Data and information may also be heterogenic and/or incompatible when they are represented by different data- and knowledge bases [6]. As the result information processes in supply chains are exposed to risks, especially in the most vulnerable "bottlenecks" like, for instance, in the Supply Chain Collaboration (SCC) module of mySAP SCM software using Internet to gain visibility to suppliers and manage the replenishment process, also enabling to gain data and information on the status of manufacturing process at all sources and to receive alerts regarding too low levels of inventory, as well as to quickly respond [2]. The most serious threats for the IT infrastructure of supply chains nowadays come from various viruses (malware of trojan or worm types), spyware, adware, spam and phishing attacks of the denial-of-service type and other cyberattacks, homepage counterfeiting and social engineering. These threats may come from both external and internal sources. Internal risks are normally associated with deliberate or arbitrary behavior of staff. Cyberattacks may be detrimental for the supply chain data and information since on the one hand they may lead to system rifts and on the other hand they may result in data losses, counterfeiting or leaks, intellectual property theft [6]. Data Loss Prevention (DLP) solution suggests an advanced technical approach to tackle the majority of the above risks. DLP is still an adolescent technology that may provide significant value for supply chains. Basically, DLP concept encompasses the following main features [12]:

- Data Loss Prevention/Protection
- Data Leak Prevention/Protection
- Information Loss Prevention/Protection
- Information Leak Prevention/Protection
- Extrusion Prevention
- Content Monitoring and Filtering
- Content Monitoring and Protection

The DLP core concept can be briefly formulated as follows: products that, based on central policies, identify, monitor, and protect data at rest, in motion, and in use, through deep content analysis. Thus the key defining characteristics of DLP are [12]:

- Deep content analysis
- Central policy management
- Broad content coverage across multiple platforms and locations

One of the defining characteristics of DLP solutions is their content awareness. This is the ability of products to analyse deep content using a variety of techniques, e.g. file cracking. File cracking is the technique used to read and understand a file, even if the content is buried multiple levels down (e.g. a PDF file embedded in a CAD file). Many of the products on the market today support around 300 file types, embedded content, multiple languages, double byte character sets for Asian languages, and pulling plain text from unidentified file types. Quite a few use the Autonomy or Verity content engines to help with file cracking, but all the serious tools have quite a bit of proprietary capability, in addition to the embedded content engine. Some tools support analysis of encrypted data if enterprise encryption is used with recovery keys, and most tools can identify standard encryption and use that as a contextual rule to block/quarantine content. Seven major content analysis techniques forming the basis for most of the DLP products on the market and used to find policy violations are discussed in detail in [12], each technique having its own strengths and weaknesses. The goal of DLP is to protect content throughout its lifecycle. In terms of DLP, this includes three major aspects [12]:

- DATA AT REST: scanning of storage and other content repositories to identify where sensitive content is located.
- DATA IN MOTION: "sniffing" of traffic on the network (passively or inline via proxy) to identify content being sent across specific communication channels.
- DATA IN USE: addressed by endpoint solutions that monitor data as the user interacts with them.

Another solution is known under the name of Security Information and Event Management (SIEM). With some subtle differences, there are four major functions of SIEM solutions [13]:

- Log Consolidation: centralized logging to a server.
- Threat Correlation: the artificial intelligence used to sort through multiple logs and log entries to identify attackers.
- Incident Management: workflow from identification to containment and eradication.
- Reporting:
  - Operational Efficiency/Effectiveness
  - Compliance /SOX, HIPPA, FISMA etc.
  - Ad Hoc /Forensic Investigations

A good SIEM tool can provide the analytics and knowledge of a good security engineer and process a mountain of events from a range of devices. Instead of 1,000 events per day an engineer with a SIEM tool can handle 100,000 events (or more). Besides, SIEM works uninterruptedly on 24/7 basis [13].

An alternative to SIEM is the Unified Threat Management (UTM) solution – a single do-all-correlate-all security device. However it may represent a "good enough" solution that affects the greatest increase in overall security in a small or medium size business only. In a larger organization (supply chain) with significant security assets already in place, the replacement of those devices may be cost prohibitive. This is not to say that a UTM may be able to augment an existing security infrastructure by adding one or more functions needed without replacing existing systems. In some cases if a UTM can fill two or more needs it may be justified by those features alone [13].

Protagonists of decentralized Blockchain approach to supply chains pay special attention to information security through implementation of verified and immutable transactions. They claim all Blockchain transactions are verified because the members sign them using public-private key cryptography before sharing them with the network. Therefore, only the owner of the private key can initiate them. However, the members can stay anonymous because the keys are not linked to real-world identities [14].

5 Psychological Approach

Apart from the aforementioned legal and technical methods of data protection, the methods of applied psychology known as profiling techniques aiming at detection of potential insider attacks must become an integral part of comprehensive information security management system for supply chains. Supply chains always face challenges when employees become malicious giving away confidential information. These threats happen seemingly without advance notice and cause severe consequences. However, in retrospect, there is often a pattern or trail before the fact that could be traced and uncovered. The novel technologies for detecting malicious insider behavior include, but are not limited to [7]:

- Machine learning to recognize malicious intent in information gathering commands;
- Detection of anomalies in document accesses and queries with respect to a Hidden Markov Model of text content;
- User processes modeling and detection of deviations from the model.

There are also many commercial tools for detecting malicious insider behavior through monitoring network activity and the use of enterprise applications. While these tools can accurately identify known attacks, they are necessarily reactive in their enforcement, and may be eluded by previously unseen, adversarial behaviors. Modern supply chains seek capability to proactively identify malicious intent before this intent is carried out. To tackle this problem the authors of [7] suggest a combination of two research areas: (a) graph learning from large enterprise behavior data and (b) psychological modeling of adversarial insiders. The first area, Structural Anomaly Detection (SA), extracts information from large-scale information network data (social networks, messages, internet visits, etc.). SA defines notions of similarity between individuals, normal patterns, and anomalies, and uses technologies including graph analysis, dynamic tracking, and machine learning to detect structural anomalies in the data. The second area, Psychological Profiling (PP), builds a dynamic psychological model from behavioral patterns. PP constructs psychological profiles, detects psychological anomalies and hence provides semantics for information network data. The authors of [7] suggest a synergic interplay between the two threads. PP provides focal points for SA by filtering out a large portion that is considered irrelevant based on psychological semantics, and hence improves scalability. At the same time, PP reduces SA's false alarm rate, making the threat detection more usable and actionable.

6   Comprehensive Information Security Management System (CISMS) for Supply Chains

A Comprehensive Information Security Management System (CISMS) for supply chains is sought to be a solution to the aforementioned problems. Its scope must comprise the development of security policy at a strategic level, assessment of risks, determination and implementation of security controls aiming to eliminate threats, monitor the system through internal audits and management review [2]. A CISMS for a supply chain is considered an integrator for thoroughly designed special tools like risk management program, DLP, SIEM, personnel profiling and others. A model of CISMS architecture suggested by us is shown on Fig. 1. The three blocks of the system closely interact with each other. The Legal block is designed not only to adopt international and national information security regulations and recommendations through contracts and internal security policies, but also to document, legalize and govern security control strategies and tactics generated in the Technical and Psychological blocks, verify their compliance with the existing information security policies of the supply chain and upgrade these policies if necessary.

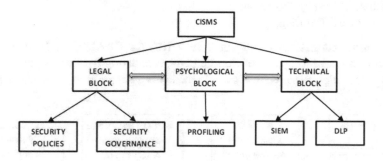

Fig. 1. Model of CISMS architecture.

Inter alia, a CISMS for supply chains should be accountable for [2, 6–9]:

– Proactive identification of malicious intent before this intent is carried out.
– Access to data for authorized users.
– Access to information services, data sourcing and data mining functions.
– Guaranteed authenticity and integrity of information, its protection from unautho-rized alteration or destruction.
– Defense of confidentiality.
– Submission and explanation of every doubt concerning regulations on informational support of cooperation.
– Regulations influencing motivation of parties and discouraging opportunistic behavior (mutual benefits resulting from improvements of long-term cooperation, further joint investments, ability to extend the range of cooperation etc.).
– Specification of required compatible, reliable hard- and software.
– Ability to renegotiate the terms of contract if [specified] changes influencing information management occur.
– Requirement of agreement on any change in terms of cooperation introduced by each party involved.
– Procedure of notification of the changes applying to those responsible for cooper-ation in each cooperating organization within a supply chain.
– Emergency actions plans (actions of parties adjusted to new conditions).

For implementation of a CISMS for a supply chain the following data protection methods are suggested [2, 6, 7, 9]:

– Risk management policy (best practices) complying with GDPR regulation and ISO/IEC 27001 guidelines;
– Registration;
– Protocoling;
– Authentication;

- Identification;
- Access Control;
- Firewalls;
- Cryptography;
- Structural Anomaly Detection;
- Psychological Profiling.

We further suggest an Intelligent System (IS) for CISMS [15, 16] allowing for selection of a relevant method of riposting and eliminating the threats pulled out of a specific Knowledge Base (KB). Figure 2 represents a generalized structure of such IS:

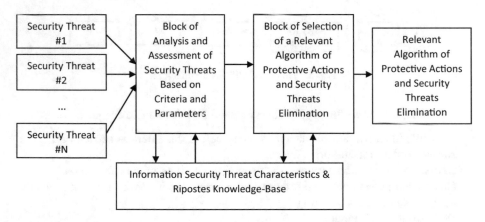

Fig. 2. Generalized structure of an IS designed to fulfil the functions of a CISMS for supply chains.

The IS for CISMS suggested by us will provide the following important features [15, 16]:

- Determined steady sequence and speed of information security control operations;
- Reasonability and certitude of the final solutions;
- Smooth and confidential interactions between the members of a supply chain;
- Minimization of various resource expenditures needed for efficient interaction;
- Uniformity and simplicity of the user interface;
- Prevention of possible conflicts between supply chain members.

Analysis of evolution of the transportation and logistics systems demonstrates that software based on the state-of-the-art technologies, including multi-agent technologies and cloud computing, is still far from total implementation since it requires elevated level of personnel training. Development of IS for CISMS will inevitably provide for considerable increase in efficiency of information and telecommunication systems used in supply chains, augmented competitiveness and performance quality of enterprises due to process optimization and facilitation based on workplace and business process automation (BPA), advanced software protection and security for users [15, 16].

7   Conclusions

The problem of security of information processes in global supply chains is particularly important given the risks produced by ever changing economic and political environment, administrative impedances and corrupt governmental practices affecting supply chains at the local, regional, national and international levels, competition, technological evolution, succession of human resources, cultural diversity and other humanitarian and technological factors. An intelligent Comprehensive Information Security Management System (CISMS) for supply chains suggested by the authors of this report to minimize information related risks can be a response to the above challenges and should obligatorily unite legal, technical and psychological data and information protection solutions under one umbrella.

References

1. Walker, W.T.: Supply Chain Architecture: A Blueprint for Networking the Flow of Material, Information and Cash. CRC Press LLC, Boca Raton, London, New York, Washington, D.C. (2005)
2. Malkus, T., Wawak, S.: Information security in logistics cooperation. Acta Logist. **2**(1), 9–14 (2015)
3. Williamson, O.: The Economic Institutions of Capitalism. The Free Press, A Division of Macmillan Inc., Collier Macmillan Publishers, New York, London (1985)
4. Hart, O.: Firms, Contracts and Financial Structure. Clarendon Press, Oxford (1995)
5. ISO/IEC 27001: Information Technology - Security Techniques - Information Security Management Systems - Requirements. ISO, Geneva (2013)
6. Iskanderov, Y., Pautov, M.: Protection of corporate information in transport and logistics networks. Collection of articles X of the St. Petersburg Interregional Conference Information Security of the Regions of Russia (ISRR-2017). SPb., 1–3 November 2017. SPOISU. - SPb., ISBN 978–5–906931–64–1 (2017)
7. Brdiczka, O., Liu, J., Price, B., Shen, J., Patil, A., Chow, R., Bart, E., Ducheneaut, N.: Proactive insider threat detection through graph learning and psychological context. In: IEEE Symposium on Security and Privacy Workshops, San Francisco (2012)
8. Shackleford, D.: Combatting Cyber Risks in the Supply Chain. A SANS Institute Whitepaper (2015). https://www.sans.org/reading-room/whitepapers
9. Kohen, I.: Data Security Best Practices for Mitigating Supply Chain Risk. Supply & Demand Chain Executive Magazine, 8 March 2018
10. GDPR Portal. www.eugdpr.org
11. Thales Data Threat Report. Thales eSecurity, Inc. https://dtr.thalesesecurity.com
12. Mogull, R.: Understanding and Selecting a Data Loss Prevention Solution. A SANS Institute Whitepaper, Securosis L.L.C. © (2016). http://creativecommons.org/licenses/by-nc-nd/3.0/us
13. Swift, D.: A Practical Application of SIM/SEM/SIEM Automating Threat Identification. A SANS Institute Whitepaper (2007). https://www.sans.org/reading-room/whitepapers
14. Hackius, N., Petersen, M.: Blockchain in logistics and supply chain: trick or treat? In: Proceedings of the Hamburg International Conference of Logistics (HICL), vol. 23 (2017)

15. Iskanderov, Y., Doroshenko, V.I.: Organization of transport-technological processes on the basis of integrated information systems. In: Collection of Articles of the International Scientific and Practical Conference on the New Economy and the Main Directions of its Formation, pp. 53–62. St. Petersburg Polytechnic University of Peter the Great (2016). ISBN: 978-5-7422-5215-3
16. Iskanderov, Y.: Design of models of an integrated information system of transport logistics based on multi-agent technologies. In: Collection of Articles of the International Scientific and Practical Conference on the New Economy and the Main Directions of its Formation, pp. 62–69. St. Petersburg Polytechnic University of Peter the Great (2016). ISBN: 978-5-7422-5215-3

External Consistency Maintenance Algorithm for Chain and Stellate Structures of Algebraic Bayesian Networks: Statistical Experiments for Running Time Analysis

Nikita Kharitonov[1]([⊠]), Ekaterina Malchevskaia[1], Andrey Zolotin[1],
and Maksim Abramov[1,2]

[1] St. Petersburg Institute for Informatics and Automation of the Russian Academy
of Sciences, 39, 14-th line Vasilyevsky Ostrov, St. Petersburg 199178, Russia
nikita.kharitonov95@yandex.ru, katerina.malch@gmail.com,
andrey.zolotin@gmail.com
[2] St. Petersburg State University, St. Petersburg, Russia
mva16@list.ru

Abstract. This article describes an experiment demonstrating the running time of the algorithm for maintainance of external consistency in algebraic Bayesian networks. In the experiment, the stellate and chain structures of algebraic Bayesian networks are compared. The results of the experiment demonstrate the dependency of the algorithm complexity on the number of atoms in the network, as well as on the intersections and in the fragments of knowledge.

Keywords: Probabilistic graphical models
Algebraic bayesian networks · Knowledge patterns
Global probabilistic-logic inference
Maintenance of external consistency · Statistical experiments

1 Introduction

Probabilistic graphical models (PGM) are widely distributed and used in many areas [3,4]. One of the classes of PGM are the algebraic Bayesian networks considered in this paper. Related Belief Bayesian networks, which have found application, in particular, in such areas as analysis of protection against socio-engineering attacks [1,2] and medicine [10]. But unlike Belief Bayesian networks [7], which could work only with scalar probability estimates, algebraic Bayesian networks can also process interval estimates.

Algebraic Bayesian networks (ABN) are represented as an undirected graphs with ideals of conjuncts in their nodes. Conjuncts are assigned to scalar or interval probability estimates.

© Springer Nature Switzerland AG 2019
A. Abraham et al. (Eds.): IITI 2018, AISC 875, pp. 23–30, 2019.
https://doi.org/10.1007/978-3-030-01821-4_3

ABNs are the representations of the bases of knowledge patterns that are set over the selected set of assertion variables. Further, this set is decomposed into intersecting sets of vertices (from 2 to 4), over which knowledge patterns are built. KP is an ideal of conjuncts with scalar or interval probability estimates attributed to its elements. Such structure with intersecting fragments of knowledge is ABN.

After creation or changes of an algebraic Bayesian network, it is necessary to check whether the system of given to its elements probability estimates satisfies the axiomatics of probability theory. This process is called maintenance of consistency, and an algebraic Bayesian network that has been "tested" is consistent. There are four degrees of consistency:

1. *Local.* ABN is locally consistent, if each knowledge pattern that is considered is consistent.
2. *External.* ABN is externally consistent if it is locally consistent and probability estimates of the elements located at the intersections of the knowledge patterns of the network coincide.
3. *Internal.* ABN internally consistent, if it is locally consistent and for each conjunct and for any probability estimate from the interval of its validity can be taken such probability estimates of other conjuncts that the resulting ABN estimates with the scalar estimates will be consistent.
4. *Global.* ABN is globally consistent, if it can be placed in a single, comprehensive knowledge pattern that will be locally consistent.

It is worth noting that from the external consistency for an algebraic Bayesian network with scalar estimates follows the internal and global ones. More details with the aspects of each degree of consistency can be found in the [5,6,9].

In this paper a study of external consistency maintainance of algebraic Bayesian networks of chain and stellate structures [8] with interval probability estimates of their elements was carried out.

An algebraic Bayesian network has a *chain* structure if knowledge pattern have one or two intersections with other patterns, and the intersection can be common only for two KPs. An example of such network is shown in Fig. 1.

An algebraic Bayesian network has a *stellate* structure, if knowledge pattern have one common intersection for all KPs. An example of such network is shown in Fig. 2.

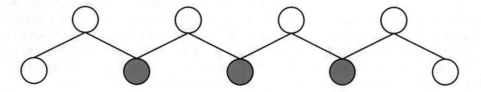

Fig. 1. ABN with chain structure

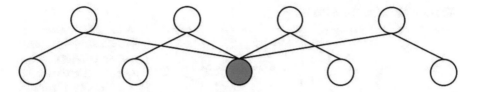

Fig. 2. ABN with stellate structure

2 Description of the Experiment

To study the time of the external consistency maintainance, algebraic Bayesian networks [11] with chain Fig. 1 and the stellate Fig. 2 structure were considered. Knowledge patterns were built over a different number of atoms n - from 2 to 4, and the number of atoms at the intersections ranged from 1 to $n - 1$, depending on the size of the KP.

For each pair of these values (the size of the knowledge pattern - the number of atoms at the intersection) a solution to maintain the external consistency was performed. This solution was repeated 100 times. In order to obtain correct data, an algebraic Bayesian network was created anew each time. Next, the mathematical expectation, variance, and standard deviation of the obtained sets of values were calculated. For each type of structure the results are presented in the table and the dependences are shown on the graphs.

2.1 Algebraic Bayesian Network with Chain Structure

The values of the mathematical expectation of the time for the external consistency maintenance obtained from the experiment carried out for ABN having a chain structure are given in Table 1.

In addition, dispersion and root-mean-square deviation were calculated. The standard deviation did not exceed 5% of the average for each of the calculations, which indicates the correctness of the experiment. The time is specified in milliseconds.

Table 1. Results of the first experiment. ABN with chain structure

Num. of atoms in ABN	2 at. in KP, 1 at. at intersection	3 at. in KP, 1 at. at intersection	3 at. in KP, 2 at. at intersection	4 at. in KP, 1 at. at intersection	4 at. in KP, 2 at. at intersection	4 at. in KP, 3 at. at intersection
3	221	243	213	0	0	0
4	405	357	427	465	491	462
5	490	471	656	635,333	723,5	983
6	585	591	931	805,667	956	1539
7	690	711	1157	976	1220	1879
8	790	823,5	1505	1263,667	1484	2322
9	940	936	1650	1551,333	1700,5	2819
10	1002	0	1976	1839	1917	3312

ABN with chain structure

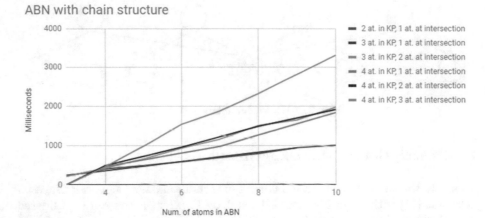

Fig. 3. Results of the first experiment. ABN with chain structure

ABN with chain structure: same number of atoms at intersection.

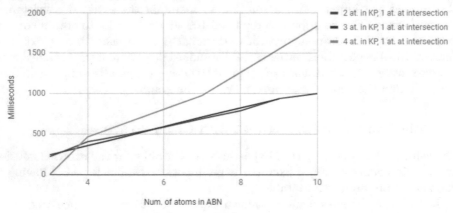

Fig. 4. Results of the first experiment. ABN with chain structure: same number of atoms at intersection

The Table 1 contains missing values, which is explained as follows: an algebraic Bayesian network with the specified number of atoms can not consist of knowledge pattern's data with the number of atoms at the intersections.

Figure 3 shows a graph illustrating the experimental results presented in the Table 1. The number of atoms in the algebraic Bayesian network is shown horizontally, and the time for consistency maintenance (in milliseconds) is shown vertically. Different colors correspond to different variations of the number of atoms in the knowledge pattern and at the intersections in the considered networks.

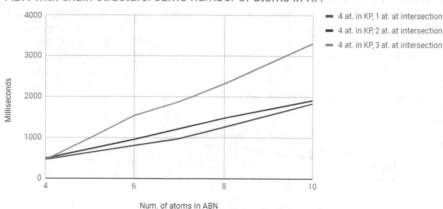

Fig. 5. Results of the first experiment. ABN with chain structure: same number of atoms in KP

Figure 4 shows the dependency of the time on the number of atoms in the knowledge pattern for consistency maintenance. The number of atoms at the intersections is the same.

Figure 5 shows the dependency on the number of atoms at the intersection of the time for consistency maintenance. The number of atoms in the fragments of knowledge is the same.

2.2 Algebraic Bayesian Network with Stellate Structure

For algebraic Bayesian network, which has a stellate structure, a similar experiment was carried out. Its results are shown in Table 2 and in Figs. 6, 7 and 8.

Figure 6 shows a graph illustrating the experimental results presented in the Table 2. The number of atoms in the algebraic Bayesian network with stellate structure is shown horizontally, and the time for consistency maintenance (in

Table 2. Results of the second experiment. ABN with stellate structure

Num. of atoms in ABN	2 at. in KP, 1 at. at intersection	3 at. in KP, 1 at. at intersection	3 at. in KP, 2 at. at intersection	4 at. in KP, 1 at. at intersection	4 at. in KP, 2 at. at intersection	4 at. in KP, 3 at. at intersection
3	312	255	233	0	0	0
4	373	0	466	506	518	488
5	537	474	689	0	0	979
6	624	0	923	0	984	1492
7	749	702	1151	1027	0	2078
8	835	0	1376	0	1490	2497
9	936	928	1610	0	0	3059
10	1062	0	1901	1500	2027	3437

ABN with stellate structure

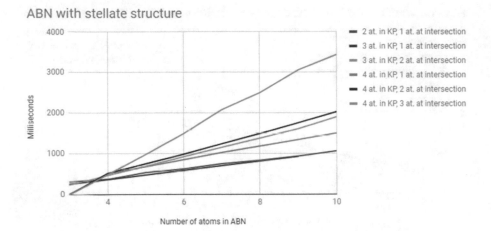

Fig. 6. Results of the second experiment. ABN with stellate structure

milliseconds) is shown vertically. Different colors correspond to different variations of the number of atoms in the knowledge pattern and at the intersections
in the considered networks.

Figure 7 shows the dependency of the time for consistency maintenance on
the number of atoms in the knowledge pattern. The number of atoms at the
intersections is the same.

Figure 8 shows the dependency on the number of atoms at the intersection
of the time for consistency maintenance. The number of atoms in the fragments
of knowledge is the same.

ABN with stellate structure: same number of atoms at intersection.

Fig. 7. Results of the second experiment. ABN with stellate structure: same number
of atoms at intersection

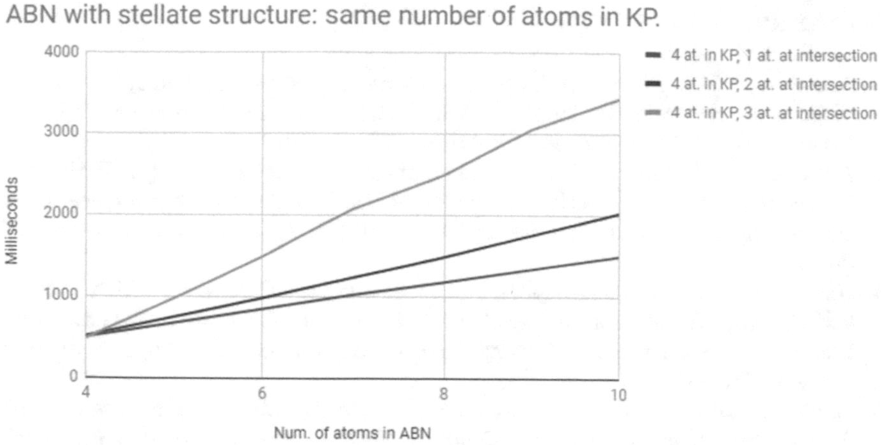

ABN with stellate structure: same number of atoms in KP.

Fig. 8. Results of the second experiment. ABN with stellate structure: same number of atoms in KP

The results of the first and second experiments are quite close, which is explained by the fundamental similarity of the studied structures of the algebraic Bayesian networks. Based on the obtained graphs, we can say that for both types of structures the time dependency on the number of atoms in the ABN is linear. As the number of atoms increases at the intersection, the time for consistency maintenance increases, as shown in Figs. 4 and 7. Also, the time increases with the increase of the number of atoms at the intersection, as can be seen in Figs. 5 and 8.

3 Conclusion

The experiment showed that the time for maintainance of the external consistency linearly depends on the number of atoms at the intersection and the number of atoms in the ABN if the network has a stellate or chain structure. In addition, there is no fundamental difference and no gain in computational time when choosing a particular structure. These points should be taken into account when creating an ABN structure for carrying out computational experiments on data.

The obtained results enable us to study externally consistent algebraic Bayesian networks and a global inference on them. Also at the moment, work is underway to create algebraic Bayesian networks with an arbitrary structure.

Acknowledgments. The research was carried out in the framework of the project on state assignment SPIIRAS No. 0073-2018-0001, with the financial support of the RFBR (project № 18-01-00626; project № 18-37-00323).

References

1. Abramov, M.V., Azarov, A.A.: Social engineering attack modeling with the use of Bayesian networks. In: 2016 XIX IEEE International Conference on IEEE Soft Computing and Measurements (SCM), pp. 58–60 (2016)
2. Abramov, M.V., Azarov, A.A.: Social engineering attack modeling with the use of Bayesian networks. Collection of reports of the International Conference on Soft Computing and Measurement (SCM-2016), vol. 1-2. T. 1, St. Petersburg, pp. 71–74 (2016)
3. Herrera-Vega, J., Orihuela-Espina, F., Ibarguengoytia, P.H., Garcia, U.A., Morales, E.F., Sucar, L.E.: A local multiscale probabilistic graphical model for data validation and reconstruction, and its application in industry. Eng. Appl. Artif. Intell. **70**, 1–15 (2018)
4. Jiang, J., Xie, J., Zhao, C., Su, J., Guan, Y., Yu, Q.: Max-margin weight learning for medical knowledge network. Comput. Methods Prog. Biomed. **156**, 179–190 (2018)
5. Kharitonov, N.A., Zolotin, A.A., Tulupyev, A.L.: Software implementation of algebraic Bayesian networks consistency algorithms. In: 2017 XX IEEE International Conference on Soft Computing and Measurements (SCM), 24–26 May 2017, pp. 8–10 (2017)
6. Tulupyev, A.L., Nikolenko, S.I., Sirotkin, A.V.: Bayesian networks: a probabilistic-logic approach. SPb.: Nauka, p. 607 (2006). (in Russian)
7. Tulupyev, A.L., Sirotkin, A.V., Nikolenko, S.I.: Bayesian belief networks. SPb.: SPbSU Press, p. 400 (2009). (in Russian)
8. Tulupyev, A.L., Stolyarov, D.M., Mentyukov, M.V.: A representation for local and global structures of an algebraical Bayesian network in java applications. Tr SPIIRAN **5**, 71–99 (2007). (in Russian)
9. Tulupyev, A.L., Tulupyeva, T.V., Suvorova, A.V., Abramov, M.V., Zolotin, A.A., Zotov, M.A., Azarov, A.A., Malchevskaya, E.A., Levenets, D., Toropova, A.V., Kharitonov, N.A., Birillo, A.I., Solntsev, R.I., Mikoni, S.V., Orlov, S.P., Tolstov, A.V.: Soft calculations and measurements. Models and methods: monograph. In: Prokopchina, S.V. (ed.), vol. III, Moscow, ID. "SCIENTIFIC LIBRARY", p. 300 (2017)
10. Xing, F.W., Chen, J.L., Zhao, B.L., Jiang, J.Z., Tang, A.L., Chen, Y.L.: Real role of beta-blockers in regression of left ventricular mass in hypertension patients Bayesian network meta-analysis. Medicine **96**(10), e6290 (2017)
11. Zolotin, A.A., Levenets, D.G., Malchevskaya, E.A., Zotov, M.A., Birillo, A.I., Berezin, A.I., Ivanova, A.V., Tulupyev, A.L.: Algorithms for processing and visualization of algebraic Bayesian networks. Educ. Tech. Soc. **1**, 446–457 (2017)

Perspectives of Fast Clustering Techniques

Ilias K. Savvas$^{(\boxtimes)}$ and Georgia Garani

Department of Computer Science and Engineering,
T.E.I. of Thessaly, Larissa, Greece
{savvas,garani}@teilar.gr

Abstract. Nowadays, when the data size grows exponentially, it becomes more and more difficult to extract useful information in reasonable time. One very important technique to exploit data is clustering and many algorithms have been proposed like k-means and its variations (k-medians, kernel k-means etc.), DBSCAN, OPTICS and others. The time complexity of all these methods is prohibitive (NP hard) in order to make decisions on time and the solution is either new faster algorithms to be invented, or increase the performance of the old well tested ones. Distributed, parallel, and multi-core GPU computing or even combination of these platforms consist a very promising method to speed up clustering techniques. In this paper, parallel versions of the above mentioned algorithms were used and implemented in order to increase their performance and consequently, their perspectives in several fields like industry, political/social sciences, telecommunications businesses, and intrusion detection in big networks. The parallel versions of clustering techniques are presented here and two different cases of their applications on different fields are illustrated. The results obtained are very promising concerning their quality and performance and therefore, the perspective of using clustering techniques in industry and sciences is increased.

Keywords: Clustering · k-means · DBSCAN · Parallel computing

1 Introduction

Big Data is a key issue these days. Data is not produced solely by computers; almost all electronic instruments produce huge amounts of raw data, to mention a few, radio telescopes, biology and biochemistry labs, IP-networks, and several kinds of different sensors [1]. In addition, human activity adds data using not only social networks or mobile phones but in many cases, human behavior by recording it. Ordinary people are not any more information consumers from media etc. but also, producers of information. By mining from social networks key words, it is easy to identify if a geographical area suffers from flu epidemic or if the rainfall was extremely heavy without looking the medical news or meteorological forecasts respectively. Any company, in order to survive in a global environment needs to know, analyze, and use appropriate information. For example, when a new business started its operation a few decades ago, the required information came from the competitors which were activated in the same geographical area. Now, e-commerce and WEB offer to customers the ability to shop from any place in the Globe and therefore the same business today has to use the knowledge of their competitors all around the world.

© Springer Nature Switzerland AG 2019
A. Abraham et al. (Eds.): IITI 2018, AISC 875, pp. 31–40, 2019.
https://doi.org/10.1007/978-3-030-01821-4_4

Clustering is a very useful technique to extract information of data sets. K-means [2], and DBSCAN [3] algorithms (and their variations) are heavily studied and used for this purpose. The two techniques form clusters of "similar" objects by using different approaches. The number of clusters has to be predefined for k-means, while DBSCAN needs the minimum distance between objects to be considered "similar" and in addition, the minimum number of objects that could be characterize them as cluster. Both algorithms suffer from their computational complexity making them inefficient to apply to big data sets. In order to overcome this problem, either new techniques have to be invented or old algorithms need to be modified.

The last decades, distributed and parallel systems offer cheap platforms for increasing the efficiency of algorithms without changing them dramatically. MPI [4], OpenMP [5], and General Purpose Graphics Processing Units (GPGPUs) [6] can be used to speed up the above mentioned techniques, making them efficient to apply to big data sets and producing results in reasonable computational time.

Hybrid computing, these days can be also easily used. In a distributed system, where the tasks can be distributed along their computational nodes, each node can in addition divide its workload to the available physical cores and GPU cores. Combining MPI, OpenMP, and CUDA is the ultimate step in order to take advantage of all available hardware resources of a computational cluster adding value to the perspectives of clustering techniques.

The rest of the paper is organized as follows. Applications of clustering in industry and science are presented in Sect. 2. In Sect. 3, two clustering techniques are presented namely k-means and DBSCAN, while some case studies of the application of clustering are given in Sect. 4, followed by discussion in Sect. 5. Finally, Sect. 6 concludes the paper.

2   Applications of Clustering Techniques

The most popular tool in statistical data analysis is cluster analysis. It is used widely in many different scientific fields, including data mining, information retrieval, machine learning, image analysis, computer graphics, bioinformatics, medicine, business, marketing and insurance. The main techniques used for clustering are K-means and DBSCAN algorithms which are the most popular methods, used currently for companies' clustering.

Many different clustering methods are available, and those are grouped into the following main classifications, partitioning methods, hierarchical methods, model-based methods, methods for high-dimensional data, density-based methods, constraint-based clustering, and grid-based methods.

The benefits of using clustering techniques have been proved by the testing of the K-means algorithm to a number of business and industrial applications. In this section, a number of applications are presented where different variations of the K-means algorithm have been applied successfully to an industrial or business environment with promising outcomes.

The water quality in the Haihe River in China has been analyzed by the varying weights K-means clustering algorithm. For the identification of the quality of water sample a modified version of the clustering algorithm has been proposed in [7] where the margin of the iteration not being calculated is avoided and consequently, the efficiency of data processing is improved. The results show that variations of the K-means algorithm can be implemented successfully for the confrontation of water pollution.

The problem of manual detection of infected fruits has been tried to be solved by applying the K-means algorithm in [8]. An image segmentation approach based on color features has been implemented to the K-means clustering unsupervised algorithm. It takes place in two phases. Initially, the pixels of the images are clustered based on their color and spatial features followed by combining the clustered blocks to a specific number of regions. K-means clustering algorithm has been used successfully to detect defected fruit accurately and timely.

Another area where the K-means clustering algorithm has been applied with positive results is insurance industry. Insurance data are perfect candidates for data analytics due to their diverse nature. A new technique of selecting initial centroids using Multiple Random method in K-means algorithm has been implemented in [9] particularly for the identification of new customers for insurance products and claiming in general insurance for testing the presented approach.

Insurance companies are investing a lot of money for the detection of fraudulent claims since they lose annually too much money in these kinds of claims. A wide diversity of types of insurance frauds occurs in all areas of insurance industry. In [10], K-means clustering technique has been used to detect fraud in automobile insurance specifically. The approach combines also data mining methods for the extraction of hidden knowledge and patterns on huge volume of data. The results indicate considerable relations among efficient factors in relevant fraud cases.

The clustering of high-performance companies is very important for stock exchange markets. Many people can take advantage of the results including shareholders, stockholders and creditors for planning and decision making. Data are collected from financial information databases. The k-means clustering algorithm is used in [11] for the classification of three different types of industries, cements industry, metal and automotive and parts industries, in Tehran Stock Exchange.

3 Clustering Techniques

3.1 *K*-Means Algorithm

K-means algorithm is a very popular and straightforward clustering technique (Algorithm 1) [2]. Besides the data set, k-means requires as input the number of the desired cluster (k). The measure of distance and consequently the similarity of two objects used in this work is the Euclidean distance.

Algorithm 1. Original *k*-means

Input: number of clusters *k*
Choose *k* points from the data set *D* // these represent the initial centroids
REPEAT
    Assign each point to its closest centroid
    Recalculate the centroids
UNTIL (no centroid changes)
Output: List of centroids

The main disadvantages of *k*-means are twofold. Firstly, it requires the number of clusters beforehand and secondly, its computational complexity which makes the technique not appropriate to be applied on big data sets on time. The time complexity is $T_s = O(kNM)$ [2], where *k* is the number of clusters, *N* stands for the size of data set, and *M* is the number of iterations needed to finalize the algorithm. *K*-means produces the *k* predefined clusters where all data points are assigned to one of them no matter their distance from the centroid and does not points any outliers.

In [12–18] distributed/parallel, multi-core and Hadoop/MapReduce versions of k-means are proposed which reduce its computational complexity T_P according to the number of participating nodes as in *Eq.* 1.

$$T_P = O\left(n\frac{kNM_P}{P-1} + C\right) \tag{1}$$

where P represents the number of participating computational nodes, M_P stands for the iterations needed in the parallel version (considering that the data set has been divided equally to the participating nodes, then $M_P \ll M$), C is for the communication overhead produced and finally, *n* denotes the dimensionality of data points. It is obvious that the parallel version outperforms the original sequential one.

3.2   DBSCAN Algorithm

DBSCAN (Density-Based Spatial Clustering of Applications with Noise) [3] input requirements are the minimum distance between two points in order to be considered as neighbors (*eps*) and the smaller number of points to form a cluster (*MinPts*). DBSCAN produces high-quality clusters and additionally, characterizes the points of the data set as core, border, or noise (outliers) which, in many cases, is very useful information. A brief description of the original sequential DBSCAN is presented in Algorithm 2. So, *DBSCAN* seems a suitable algorithm to find anomalies to the data set it is applied. Once again, its computational complexity makes it inefficient to apply in big data sets giving results on the fly.

Algorithm 2. DBSCAN (brief description)

Input: minimum distance between points to be considered as neighbors (*eps*) and minimum number of points in order to form a cluster (*MinPts*)

For each point of data set:

Identify its neighbor points with respect to *eps*

Identify if it is a core point (according *MinPts*)

Find if there is a connection with other core points to create larger clusters ignoring the non-core ones

For all the remaining points:

Identify if they belong to any cluster or not (if not these are noise/outliers)

Output: List of clusters and point characterized as core, border, outliers

---

In [19] a variation of DBSCAN is proposed which operates in parallel computational environments and reduces substantially its complexity. The complexities of the sequential versus parallel version of DBSCAN are presented in Eq. 2. Is it is obvious again that the parallel version outperforms the original sequential one and scales efficiently [19].

$$T_S = O(N^2) > T_P = \left(\frac{N^2}{P-1} + C \right) \tag{2}$$

where P represents the number of participating nodes, N stands for the data size, and C denotes the communicational overhead.

Therefore, by parallelizing two well known and heavily used clustering techniques, their efficiency has been increased and can be used to real time, real world applications which use huge amounts of data and demand fast responses by the used techniques. As an added value, the proposed methods can apply to variations of these algorithms, like Kernel k-means or OPTICS, for increasing also their efficiency.

4   Case Studies

For all the experiments of this work, 20 computational nodes were used (AMD@ FX (tm)-4100 Quad-Core Processorx4, with UBUNTU 14.04 LTS operating system, MPICH 3.04, gcc 4.84, and Ethernet 100 Mbit/s). The data points used are provided from real applications and the distance measure is the Euclidean distance.

4.1   Clustering the Behavior of Telecommunication Companies' Customers

Telecommunication companies nowadays provide customers a lot of different services as mobile network, internet and many others. On the other hand, customers have different needs and so, different politics from the companies must be provided to them. A well tested technique to analyze and use properly customers' needs is to cluster them according to their characteristics (for example use of mobile phone, SMSs', internet services and so on). Depending on the company's policy, k-means or a variation of it, can produce the desired number of discrete clusters making contract's policies easier. Thus, contracts could be more or less personalized, covering the majority of their needs for keeping them happy, and simultaneously increasing the incomes of the companies.

Another very important issue is the knowledge that some of the customers potentially are going to change provider or not to pay on time. A clustering technique like DBSCAN or OPTICS can easily identify this type of customers, according to their characteristics, as outliers or noise, giving to the company the ability to predict them and treat them properly.

An improvement of the K-means clustering algorithm has been also proposed for customer segmentation of telecommunication companies in [20]. Customer segmentation is the classification of customers into different groups according to one or more attributes. It is essential for the precisely analysis of customer composition and the promotion of the quality of service and marketing. The results can be used for example, for phone packages recommendation.

So, both k-means and DBSCAN (and/or variations of them) are an extremely useful tool for a telecommunication company in order to categorize and analyze their customers' needs. Once again, the computational complexity of these techniques is a crucial issue that has to be taken under consideration. To overcome this, we proposed distributed/parallel/multi-core techniques which reduce their computational complexity and as a result their performance is increased, making them effective to be used in real time problems [13, 19]. In Fig. 1, the speedup and scalability of both k-means and DBSCAN is presented using their distributed/parallel versions proposed in [21].

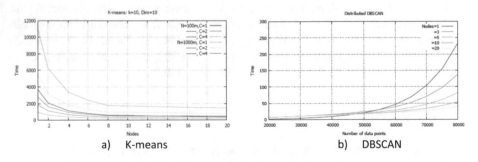

a)   K-means   b)   DBSCAN

Fig. 1. Performance and scalability of (a) k-means, (b) DBSCAN

4.2   Political/Social Sciences

According to Mazis theoretical basis [22] Systemic Geopolitical Analysis provides a geographical method to analyze power redistribution. Each country can be characterized as a set of indicators (attributes or dimensionality of the space considering each country as a point of the n-dimensional Euclidean space). The indicators used in this work are from the World Bank [23] and can be distinguished in 5 main categories, namely Geography (17 indicators), Culture (51 ind.), Defense (11 ind.), Economy (35 ind.),

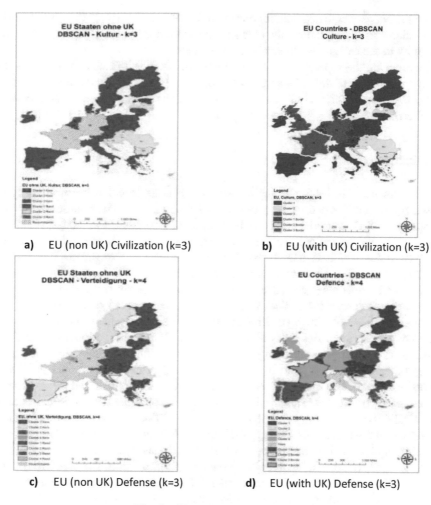

Fig. 2. Clustering of EU countries

and Politics (6 ind.). In order to clarify the categories (or pillars according Mazis's theory), pillar Economy is used as an example which includes indices as national debt, energy production and supply, trade, synthesis of labor force and so on. Pillar Defense includes indicators such as military expenditure, arms imports and exports. The total number of indicators is 120 and in this study was used either for each category as stand alone or combination of them (for example Geography + Politics + Defense). The countries on which the clustering techniques were applied were the European Union countries (considering U.K. as EU and non-EU country). We applied k-means and DBSCAN and the obtained results proved that DBSCAN behaves better. This is mainly because DBSCAN does not need to know beforehand the number of clusters and in addition, it can discover outliers which in this case represent countries with many differences compared to the rest. Another very important issue of DBSCAN is the distinction between core and border points which in this case can lead to very interesting observations. In Fig. 2, the results obtained are presented, where DBSCAN is applied separately to the categories Civilization and Defense.

Once again, the problem of the computational complexity makes very difficult to follow the changes of the indicators in real time and especially if all indicators will be used (more than 1,000) and all the countries of the world participate. Thus, increasing the efficiency of DBSCAN is a necessity [24].

5   Discussion

As it is shown in Sect. 4.2, the application of clustering techniques can improve our knowledge in many aspects and with different outcomes. In Political/Social science, clustering the E.U. countries offers to the researcher or the political analyst a different view than the theoretical political one. For example, in many cases the dipole North – South countries which is heavily used by the politicians changed to West – East countries when DBSCAN is applied to their characteristics. Of course, this is an early approach and further research is needed on this direction. A very important issue in this study is the weight of the indices (attributes – characteristics) of the countries. For example, has the Gross Domestic Product (GDP) of a country the same importance as its costal length? Questions like that have to be analyzed and explored by the Politicians but on the contrary, clustering even without this kind of information, can give a very good and interesting point of view. Considering a country as a data point in n-dimensional Euclidean space, the dimensionality which is represented by the characteristics of each country, and the total number of countries in the Globe increase dramatically the complexity of clustering techniques. Taking also under consideration the velocity of data, makes this type of approach even more infeasible; thus, fast clustering algorithms become the best option in such cases.

Computer networks are another big producer of information. However, they suffer of malfunctions or intrusions. To detect such situations in real time is a difficult task. Once again, fast clustering techniques (Sect. 2) can become a useful tool if they can produce results also in real time.

6   Conclusions

In this work, applications of fast clustering techniques in different cases are presented. As it is shown, these applied methods can be a powerful tool for analyzing data and extracting useful information from huge even raw data sets. Companies and more general, industry can use this information to understand customers' view and improve or modify their products beneficially for both business and customers.

There are two important issues that still have to be explored. Each product, service, customer, network signal, country etc. is characterized of a variety of attributes or indices. Firstly, to each one of these characteristics a weight according to its importance has been given. Unfortunately, this is not an easy job. In Political sciences the politicians have to decide and agree on that, which is an extremely difficult procedure.

In the contrary, the services used by a mobile phone user are much easier to be identified and consequently, to be given a weight. Secondly, even a fast clustering algorithm will suffer of the computational complexity when the dimensionality (characteristics) of data is extremely large (curse of dimensionality). In such a case, several techniques have been suggested, like the Principal Component Analysis (PCA) [25]. Decreasing the dimensionality, the performance of clustering is increasing, but the quality of the results must be very carefully examined before using them in real applications. On the other hand, in some cases *the curse of dimensionality* is a blessing. For example, trying to identify unreliable bits of a transmission, since the attributes are very close to each other, any clustering technique will have no success. So, projecting data points to a higher dimension could be a good approach and algorithms like kernel k-means can operate successfully.

Data heavy bombing is something new. The last two decades are distinguished by an explosion of data, changing human behavior and as a result industry and science. Extracting useful information from large data sets of high diversity and velocity is a difficult, time consuming but important and necessary task. Clustering techniques can apply beneficially, but their computational complexity makes them insufficient. Software platforms and libraries which can offer capabilities of distributed, parallel, or multi core programming (and combination of them) provide the ideal mean to increase the performance and make clustering a useful tool in the hands of scientists, ordinary people, and businesses.

References

1. Emani, C.K., Cullot, N., Nicolle, C.: Understandable big data: a survey. Comput. Sci. Rev. **17**, 70–81 (2015). https://doi.org/10.1016/j.cosrev.2015.05.002
2. MacQueen, J.: Some methods for classification and analysis of multivariate observations. In: Proceedings of the Fifth Berkeley Symposium on Mathematical Statistics and Probability, vol. 1: Statistics, pp. 281–297. University of California Press, Berkeley (1967). https://projecteuclid.org/euclid.bsmsp/1200512992
3. Ester, M., Kriegel, H.-P., Sander, J., Xu, X.: A density based algorithm for discovering clusters in large spatial databases with noise. In: KDD-96 Proceedings, pp. 226–231 (1996). https://www.aaai.org/Papers/KDD/1996/KDD96-037.pdf
4. MPICH: High-Performance Portable Message Passing Interface (2018). https://www.mpich.org/
5. OpenMP: The OpenMP API Specification for Parallel Programming (2018). https://www.openmp.org/
6. CUDA Zone: NVDIA Accelerated Computing (2018). https://developer.nvidia.com/cuda-zone
7. Zou, H., Zou, Z., Wang, X.: An enhanced K-means algorithm for water quality analysis of the Haihe River in China. Int. J. Environ. Res. Public Health **12**(11), 14400–14413 (2015). https://doi.org/10.3390/ijerph121114400
8. Dubey, S.R., Dixit, P., Singh, N., Gupta, J.P.: Infected fruit part detection using k-means clustering segmentation technique international. J. Artif. Intell. Interact. Multimed. **2**(2), 65–72. https://doi.org/10.9781/ijimai.2013.229

9. NallamReddy, S., Behera, S., Karadagi, S., Desik, A.: Application of multiple random centroid (MRC) based k-means clustering algorithm in insurance-a review article. Oper. Res. Appl. Int. J. **1**(1), 15–21 (2014)
10. Ghorbani, A., Farzai, S.: Fraud detection in automobile insurance using a data mining based approach. Int. J. Mechatron. Electr. Comput. Technol. **8**(27), 3764–3771 (2018). https://doi.org/IJMEC/10.225163
11. Momeni, M., Mohseni, M., Soofi, M.: Clustering stock market companies via k-means algorithm. Kuwait Chapter Arab. J. Bus. Manag. Rev. **4**(5), 1–10 (2015). https://doi.org/10.12816/0018959
12. Zhao, J., Zhang, W., Liu, Y.: Improved k-means cluster algorithm in telecommunications enterprises customer segmentation. In: 2010 Information IEEE International Conference on Theory and Information Security (ICITIS), Beijing, pp. 167–169 (2010). https://doi.org/10.1109/ICITIS.2010.5688749
13. Savvas, I.K., Tselios, D., Garani, G.: Distributed and multi-core version of k-means algorithm. Int. J. Grid Util. Comput. (2018, accepted). http://www.inderscience.com/info/ingeneral/forthcoming.php?jcode=ijguc
14. Savvas, I.K., Tselios, D.: Combining distributed and multi-core programming techniques to increase the performance of k-means algorithm. In: 26th IEEE International WETICE Conference, pp. 96–100 (2017)
15. Savvas, I.K., Sofianidou, G.N.: A novel near-parallel version of k-means algorithm for n-dimensional data objects using MPI. Int. J. Grid Util. Comput. **7**(2), 80–91 (2016)
16. Savvas, I.K., Sofianidou, G.N.: Parallelizing k-means algorithm for 1-d data using MPI. In: 2014 IEEE 23rd International Conference on Enabling Technologies: Infrastructure for Collaborative Enterprises (WETICE), Milano, pp. 179–184 (2016). https://doi.org/10.1109/wetice.2014.13
17. Savvas, I.K., Sofianidou, G.N., Kechadi, M.: Applying the k-means algorithm in big raw data sets with Hadoop and MapReduce. In: Big Data Management, Technologies, and Applications, pp. 23–46. IGI Global (2014). https://doi.org/10.4018/978-1-4666-4699-5, ISBN13: 9781466646995, ISBN10: 1466646993
18. Savvas, I.K., Kechadi, M.: Mining on the cloud: k-means with MapReduce. In: 2nd International Conference on Cloud Computing and Services Science, CLOSER, pp. 413–418 (2012)
19. Savvas, I.K., Tselios, D.: Parallelizing DBSCAN algorithm using MPI. In: 2016 IEEE 25th International Conference on Enabling Technologies: Infrastructure for Collaborative Enterprises (WETICE), Paris, pp. 77–82 (2016). https://doi.org/10.1109/wetice.2016.26
20. Ye, L., Qiuru, C., Haixu, X., Guangping, Z.: Customer segmentation for telecom with the k-means clustering method. Inf. Technol. J. **12**, 409–413 (2013)
21. Savvas, I.K., Chaikalis, C., Messina, F., Tselios, D.: Understanding customers' behaviour of telecommunication companies increasing the efficiency of clustering techniques. In: 25th IEEE Telecommunications Forum TELFOR, Serbia (2017)
22. Mazis, I.T.: Dissertationes academicae geopoliticae. Papazisis Publications, Athens (2015)
23. World Bank: Countries and Economies, January 2015. http://data.worldbank.org/country
24. Savvas, I.K., Stogiannos, A., Mizis, I.T.: A study of comparative clustering of EU-countries using the DBSCAN and k-means techniques within the theoretical framework of systemic geopolitical analysis. Int. J. Grid Util. Comput. **8**(2), 94–108 (2017)
25. Jolliffe, I.T.: Principal Component Analysis, Series: Springer Series in Statistics, 2nd edn., XXIX, 487, p. 28 illus. Springer, New York (2002). ISBN 978-0-387-95442-4

Impact of Security Aspects at the IOTA Protocol

Tomáš Janečko[✉] and Ivan Zelinka

Department of Computer Science, VŠB - Technical University of Ostrava,
17. listopadu 15, 708 33 Ostrava - Poruba, Czech Republic
{tomas.janecko,ivan.zelinka}@vsb.cz

Abstract. This paper presents an impact of security aspects at the IOTA protocol. Based on the different usage of computational resources and the difficulty of the selected Proof of Work (PoW) algorithms have been explored the consequences in the final behavior of the IOTA network and the throughput of the protocol implementation. The main feature of the IOTA is the *tangle* which is the name for the directed acyclic graph (DAG). This graph is highly responsible for persisting transactions in the network. The goal of this network is to provide peer-to-peer communication between machines, humans and as well for the Internet of Things (IoT) industry.

Keywords: IOTA · Proof of work · Performance · Security · IoT

1 Introduction

With the advent of distributed ledger technologies, we are now able to distribute and synchronize ledgers of data and money in secure, distributed, decentralized and permissionless environments. By removing the need for trusted third-parties as the gatekeepers and arbiters of truth, enormous efficiency gains, innovation opportunities and new value propositions emerge [1].

In the increasing adoption of IoT technologies and industry is necessary to provide any system which will be able to handle massive load of microtransactions and data integrity for machines. For that was created cryptocurrency called IOTA which is the main token in the current network. To master this the [1] IOTA's distributed ledger does not consist of transactions grouped into blocks and stored in sequential chains, but as a stream of individual transactions entangled together.

2 The Tangle

The Tangle, which is the data structure behind IOTA, is a particular kind of directed graph, which holds transactions. Each transaction is represented as a vertex in the graph. When a new transaction joins the tangle, it chooses two

A. Abraham et al. (Eds.): IITI 2018, AISC 875, pp. 41–48, 2019.
https://doi.org/10.1007/978-3-030-01821-4_5

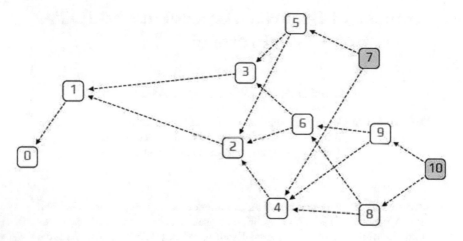

Fig. 1. The Tangle overview

previous transactions to approve, adding two new edges to the graph [1]. The example of the graph is presented in Fig. 1.

In the example above transaction number 6 approves transactions number 2 and 3. The greyed out transactions number 7 and 10 are called *tips*. The tips are the unconfirmed transactions in the tangle graph because no one approved it yet. Each new incoming transaction needs to choose two tips to approve. The key point for the correct behavior of the network is to choose the correct algorithm for tips selection.

3   Experiment Design

In our testnet environment we have been observing how exactly the transactions are behaving in the IOTA network depending on the different Proof of Work (PoW) algorithm implementations and their selected difficulties. Because as we know from the real world scenarios, each node uses different strategy and that various implementations have direct impact on the whole Tangle and the organic growth of the network.

Making a transaction in the IOTA network is 3 step process which needs to be completed before the transaction is propagated into the network [4]:

- Signing - Your node creates a transaction and sign it with your private key
- Tip Selection - Your node chooses two other unconfirmed transactions (tips) using the Random Walk Monte Carlo (RWMC) algorithm
- Proof of Work - Your node checks if the two transactions are not conflicting. Next, the node must do some Proof of Work (PoW) by solving a cryptographic puzzle (hashcash)

Hashcash works by repeatedly hashing the same data with a tiny variation until a hash is found with a certain number of leading zero bits. This PoW is to prevent spam and Sybil attacks. A Sybil attack is based on the assumption, that half of all hash power is coming from malicious nodes [4].

An important property of the hashcash puzzle (and all proof-of-work puzzles) is that they are very expensive to solve, but it is comparatively cheap to verify the solution [3].

3.1   Trinary Numeral System

For the computational purposes and outputs IOTA uses the trinary numeral system. This system can have two types:

– Balanced trinary system - Trit can have values -1, 0, 1
– Unbalanced trinary system - Trit can have values 0, 1, 2

Trit unit is analogous to the bit and means Trinary Digit. Tryte means Trinary Byte and is anologous to byte where Tryte consists of 3 bits.

A tryte has 3 trits, so the maximum value will be $(3^3 - 1)/2 = 13$ and it has $3^3 = 27$ combinations. This is caused because the values in the trinary system are balanced around the zero.

For the IOTA purposes and the better human readability have been created IOTA tryte alphabet. The tryte alphabet consists of 26 letters of the latin alphabet plus the number 9 and the tryte alphabet has a total of 27 characters. Because 1 tryte has $3^3 = 27$ combinations, each tryte can be represented by a character in the tryte alphabet: 9ABCDEFGHIJKLMNOPQRSTUVWXYZ [4]. The alphabet is visible in the Table 1.

Table 1. IOTA tryte alphabet

Tryte	Decimal	Char	Tryte	Decimal	Char
0, 0, 0	0	9			
1, 0, 0	1	A	$-1, -1, -1$	-13	N
$-1, 1, 0$	2	B	$0, -1, -1$	-12	O
0, 1, 0	3	C	$1, -1, -1$	-11	P
1, 1, 0	4	D	$-1, 0, -1$	-10	Q
$-1, -1, 1$	5	E	$0, 0, -1$	-9	R
$0, -1, 1$	6	F	$1, 0, -1$	-8	S
$1, -1, 1$	7	G	$-1, 1, -1$	-7	T
$-1, 0, 1$	8	H	$0, 1, -1$	-6	U
0, 0, 1	9	I	$1, 1, -1$	-5	V
1, 0, 1	10	J	$-1, -1, 0$	-4	W
$-1, 1, 1$	11	K	$0, -1, 0$	-3	X
0, 1, 1	12	L	$1, -1, 0$	-2	Y
1, 1, 1	13	M	$-1, 0, 0$	-1	Z

3.2   Minimum Weight Magnitude

The difficulty of the PoW is set by a variable called Minimum Weight Magnitude (MWM). This refers to the number of trailing zeros (in trits) in transaction hash. MWM is proportional to the difficulty of the Proof of Work [5].

The device which does the PoW will bruteforce the transaction hash to find a nonce that, hashed together with the transaction's trits, will result in a transaction hash that has the correct number of trailing 0's. Every extra trailing zero to be found will increase the difficulty of PoW by 3 times [5].

The currently applied parameters are as follows for the IOTA reference implementation (IRI) [4]:

- On the mainnet the minWeightMagnitude = 14 (Applies to IRI release: v1.4.1.2)
- On the testnet the minWeightMagnitude = 9 (Applies to IRI release: testnet-v1.4.1.2)

Higher minWeightMagnitude values should be no problem but will just cause the Proof of Work to take longer unnecessarily. The other side effect is that this longer time will make transactions temporarily invisible for the rest of the transactions and the network could end in the single chain output.

A simplified explanation how hashcash works is as follows. Lets assume that MWM = 3:

- hash(transaction data + counter) = ...704c19cddf95 (PoW is not completed)
- hash(transaction data + counter) = ...721b564b9000 (PoW is completed)

3.3   Quantum Computers Resistance Transactions

As of 2018, the development of actual quantum computers is still in its infancy, but experiments have been carried out in which quantum computational operations were executed on a very small number of quantum bits [6].

Large-scale quantum computers would theoretically be able to solve certain problems much more quickly than any classical computers that use even the best currently known algorithms [7].

The goal of the IOTA network and the experiment is to be quantum resistant. To achieve that, the solution for the PoW and hash computations, need to be done by One Time Signatures (OTS) cryptography.

The term implies that a single public/private key pair must only be used once. Otherwise, an attacker is able to reveal more parts of the private key and spoof signatures [4].

For the IOTA network was choosed the Winternitz One Time Signature (WOTS) scheme. The WOTS is most suitable for combining it with Merkle's tree authentication scheme because of the small verification key size and the flexible trade-off between signature size and signature generation time. Further

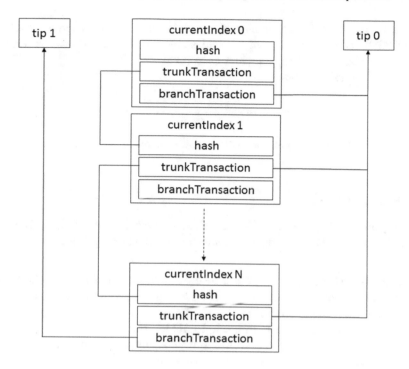

Fig. 2. Bundle of transactions

it is possible to compute the corresponding verification key given a W-OTS signature. So a Merkle signature scheme does not need to contain the verification key [8].

For the performance optimization purposes was choosed the solution where the transactions are bundled together, because upon that the signature needs to be passed into each transaction. The structure of the bundle is shown at Fig. 2 [4].

All transactions in the same bundle should be treated as an atomic unit. It means that either all transactions of a bundle are confirmed, or none of them are confirmed and every transaction in the bundle requires it own PoW [4], so they have the different nonces in the transactions.

For the transactions we have used 3 different security levels that impacted the final size of the transaction.

– Security level 1 - The signature is stored in 1 transaction
– Security level 2 - The signature is stored and partitioned into 2 transactions
– Security level 3 - The signature is stored and partitioned into 3 transactions

By increasing the security level we increased the signature size and thus the number of transactions needed to store the signature and that leads to that IOTA signatures are larger than Bitcoin signatures due to IOTA's use of Winternitz one-time signatures to gain quantum resistance.

As an output we have seen that each single transaction inside a bundle consists of 2673 trytes and much of it is taken by the signature message fragment which has a size of 2187 trytes which is approx. 82% [4].

Convert trytes to bytes:

$$bytes = \frac{trytes \times 3 \times ln(3)}{ln(2)}/8 \tag{1}$$

The final size of a single transaction inside a bundle requires 2673 trytes or 1.55 kB.

3.4 Proof of Work

The key point in the building of transactions is the PoW mechanism. To be fully comply with the IOTA protocol, the transaction consists of 3 steps:

Constructing the Bundle. Constructing the bundle and signing the transaction inputs with your private keys. IOTA uses a bundle which consists of multiple transactions containing credits to the receiving addresses (outputs) and debits from the spending addresses (inputs). In IOTA there are two types of transactions: one where you transfer value and thus, have to sign inputs, and ones where you simply send a transaction to an address with no value transfer (e.g. a message) [4].

Tip Selection. The tip selection is a process whereby you traverse the tangle in a random walk to randomly chose two transactions which will be validated by your transaction. Your transaction checks for example if the descendants of that transaction is valid. If these transactions are valid they will be added to your bundle construct and are called branchTransaction and trunkTransaction [4].

Computing Hashcash. Once the bundle is constructed, signed and the tips are added to the bundle, the PoW has to be done for each transaction in the bundle. Every transaction in a bundle requires a nonce (this is the result of the PoW) in order to be accepted by the tangle network [4].

To meet these conditions have been implemented 2 different versions of the hash algorithm of the PoW:

- Webgl 2 Curl implemention - WebGL uses the system Graphics Processing Unit (GPU)
- CCurl implementation - CCurl means C port of the Curl library, which uses the system Central Processing Unit (CPU) (Fig. 3)

During the gathering of the metrics was observed that Webgl 2 Curl implementation executes PoW faster, but method will not work for all users due to the incompatibility by the GPUs. Otherwise CCurl implementation will always work for all users.

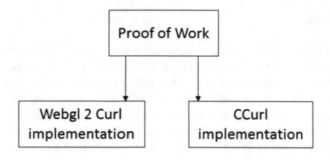

Fig. 3. Proof of work

These results leads to that we use the CCurl implementation as a fallback function whether the more efficient Webgl 2 Curl implementation isn't sufficient or compatible.

For the testing purposes as the single node was used machine with specification Intel Core i5-7300 2.7 GHz with 16GB RAM together with Intel HD Graphics 620.

4   Results

To uncover the nature of the security in the network, it was necessary to analyze the differences between GPU and CPU implementation of PoW.

Upon closer measurement, it was found that the nodes which are responsible for hashcash calculation and transactions propagation into the network are network creators and determines the direction of the organic growth of the IOTA which has big impact on the whole network.

We distinguish between two regimes, single transaction and bundle. There is only one hash calculation in the single transaction regime and a single transaction can obviously contain multiple inputs and outputs [1].

In the bundle regime IOTA uses an account-like scheme. This means that we have inputs (addresses) which you have to spend in order to transfer tokens. Addresses are generated from private keys, which in turn are derived from a trytes-encoded seed. A transfer in IOTA is a bundle consisting of outputs and inputs. Bundles are atomic transfers, meaning that either all transactions inside the bundle will be accepted by the network, or none. A typical transfer in IOTA is a bundle consisting of 4 transactions [1].

A unique feature of bundles is that the transactions are identified via the bundle hash, but also via the trunkTransaction. What this means is that the tail transaction (currentIndex: 0), references in the trunkTransaction the transaction hash at index: 1, currentIndex 1 transaction references (and approves) index 2 and so on. This makes it possible to get the full bundle of transactions from just a tail transaction by traversing down the trunk transaction [1].

5   Conclusion

In this paper, IOTA network was used for investigation of the impact of the security features at the IOTA protocol. From the above findings, we came to the conclusion that the main advantage over other networks is the preparation for resistance to the quantum computers. But the same drawback for that feature is that for each transaction we need to generate new pair of private/public key.

This leads to the weak point of the whole network from the users perspective, because any incorrect usage of the transactions could lead to loss of control over own address balance.

For further research we would like to focus how to apply evolution algorithms which could help with more stable transactions distribution and propagation into the tangle.

The next path for research leads towards to improve nodes scalability to be more agile and be able quickly respond to the changes in the nonce (minimum weight magnitude) setup of the particular network.

Acknowledgement. The following grants are acknowledged for the financial support provided for this research by Grant of SGS No. 2018/177, VSB-Technical University of Ostrava and under the support of NAVY and MERLIN research lab.

References

1. IOTA Foundation. https://www.iota.org
2. Popov, S.: The Tangle. https://assets.ctfassets.net/r1dr6vzfxhev/2t4uxvsIqk0EUau6g2sw0g/45eae33637ca92f85dd9f4a3a218e1ec/iota1_4_3.pdf. Accessed 26 June 2018
3. Hashcash.org: Hashcash proof-of-work paper. http://www.hashcash.org/papers/proof-work.pdf. Accessed 2 July 2018
4. Mobilefish. https://www.mobilefish.com. Accessed 2 July 2018
5. IOTA documentation. https://docs.iota.org. Accessed 2 July 2018
6. Gershon, E.: New Qubit control bodes well for future of quantum computing. Phys.org. Accessed 26 Oct 2014
7. Simon, D.R.: On the power of quantum computation. In: Proceedings of 35th Annual Symposium on Foundations of Computer Science, pp. 116–123. CiteSeerX 10.1.1.655.4355 Freely accessible (1994). https://doi.org/10.1109/SFCS.1994.365701. ISBN 0-8186-6580-7
8. Buchmann, J., Dahmen, E., Ereth, S., Hülsing, A., Rückert, M.: On the Security of the Winternitz one-time signature scheme. In: Nitaj, A., Pointcheval, D. (eds.) Progress in Cryptology - AFRICACRYPT 2011. AFRICACRYPT 2011. Lecture Notes in Computer Science, vol. 6737. Springer, Heidelberg (2011)

Cryptographic Protocol Security Verification of the Electronic Voting System Based on Blinded Intermediaries

Liudmila Babenko[✉] and Ilya Pisarev

Information Security Department, Southern Federal University, Taganrog, Russian Federation
lkbabenko@sfedu.ru, ilua.pisar@gmail.com

Abstract. During developing secure systems the security analysis of the main algorithms is a priority goal. This paper considers the analysis of the improved cryptographic voting protocol, which is used in the electronic voting system based on blind intermediaries. The protocol of voting is described, the messages transmitted between the parties are shown and their contents are explained. The Dolev-Yao threat model is used during protocols modeling. The Avispa tool is used for analyzing the security of the selected protocol. The protocol is described in CAS+ and subsequently translated into the HLPSL (High-Level Protocol Specification Language) special language with which Avispa work. The description of the protocol includes roles, data, encryption keys, the order of transmitted messages between parties, parties' knowledge include attacker, the purpose of verification. The verification goals of the cryptographic protocol for resistance to attacks on authentication, secrecy and replay attacks are set. The data that a potential attacker may possess is detected. The security analysis of the voting protocol was made. The analysis showed that the objectives of the audit were put forward. A detailed diagram of the messages transmission and their contents is displayed in the presence of an attacker who performs a MITM-attack (Man in the middle). The effectiveness of protocol protection from the attacker actions is shown.

Keywords: e-voting · Cryptographic protocols · Cryptographic security
Cryptographic protocols security verification

1 Introduction

The creation of e-voting systems is a serious problem. There are a number of ready-made systems [1, 2] that are used in practice, but they are far from a sufficient level of reliability and the presence of necessary mechanisms, such as complete anonymity of the voter or vote checking opportunity after counting stage. There are also a lot of works in which perspective methods of conducting electronic voting are considered, based on such principles as homomorphic encryption, including threshold schemes, mix-net, secret sharing schemes and others [3–16]. However, in most cases, the authors of such works show theoretical calculations, from which the basic structural unit of interaction between parties does not follow, namely, cryptographic protocol. Any

© Springer Nature Switzerland AG 2019
A. Abraham et al. (Eds.): IITI 2018, AISC 875, pp. 49–57, 2019.
https://doi.org/10.1007/978-3-030-01821-4_6

method on which electronic voting is based, no matter how good it is, loses its security if there are any flaws in the structure of cryptographic protocol that lead to various attacks by the intruder. Thus, the goal of this paper is to test the cryptographic protocol in the voting stage from various attacks, such as: attack on parties' authentication, data privacy and replay-attacks using the Avispa tool [17].

2 E-Voting System Based on Blinded Intermediaries

The system architecture [18] is based on the use of the following components: client application for voter - V, AS (authentication server), VS (voting server), encryption application for the passport database and ballots DBE (database encryptor). The general scheme of the interaction of components is shown in Fig. 1.

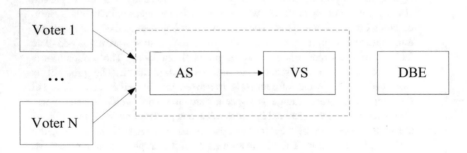

Fig. 1. System architecture.

The basic principle on which the system protocols are based - blinded intermediaries (see Fig. 2). There are 3 interacting sides A, B, C. Using the protocol for generating a common secret key, the session key AB, BC, AC are generated. A encrypts some information info on the AC key, appends an id to it, encrypts it on the AB key and sends this message to B. B in this case is a blinded intermediary, because it can decrypt only the first part of the message with id, and the remainder with info can not. It accepts the message, decrypts and checks if id is in the database and, then redirects the remainder of the message encrypted again on the BC key to the C side.

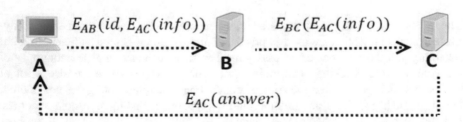

Fig. 2. Scheme based on blinded intermediaries.

C receives the message, decrypts info, encrypts the answer response on the AC key and sends it to A. This principle ensures that: info will be accepted only if id is in the database and that it is impossible to correlate id with info.

Stages of electronic voting in the context of the system:

1. Preparation. At this stage, a database of voters and a ballot are created. This data is encrypted, and officials deliver this data to the appropriate server components of the system.
2. Registration. At this stage, users log in to the system using their identification data, at the moment - using passport data, and they get their anonymous identifier. It should be noted that by using the previously described principle of blind inter-mediaries, it is impossible to correlate open passport data with an anonymous identifier, which ensures the requirement of anonymity.
3. Voting. Users receive a ballot, make their choice and send filled ballot with their anonymous identifier to the server. If such an identifier is present, the vote is accepted, and the verification identifier is sent to user, with which he or she can check vote after counting stage. It is worth noting that it is very important that the user can check his vote after the counting.
4. Counting results and votes checking. At the last stage, the votes are counted, the results are published in the public domain, and any voted user can check his or her vote with a verification identifier.

3 Voting Stage

E-voting system based on blinded intermediaries [20], developed by authors, uses voting cryptographic protocol, simplified scheme of which is shown in Fig. 3.

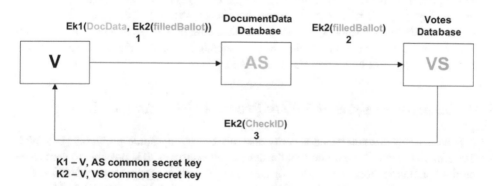

Fig. 3. Simplified scheme of voting cryptographic protocol.

Secret common keys vas, vvs, asvs are generated using the protocol for generating common session key (1)–(3). The server parties generate random numbers and send messages (4), (5), (6) to their recipients. These are used for parties' authentication. Each message is checked for integrity by checking HMAC. The voter forms the message (7), encrypts *filledBallot* with some random numbers on the vvs key, applies *docData* and encrypts everything on the vas key and then sends the message to the AS side. AS decrypts the first part of the message with documentData, checks if it in the database then, sends a message (8) to the VS side. VS receives vote, generates the verification identifier *checkID* and sends it to the voter.

(1) ECDHE (V, AS) = vas;

(2) ECDHE (V, VS) – vvs;

(3) ECDHE (AS, VS) = asvs;

(4) AS -> V: E_{vas} (N_{vs}), HMAC1

(5) VS -> V: E_{vvs} (N_{vs}), HMAC2

(6) VS -> AS: E_{asvs} (N_{asvs}), HMAC3

(7) V -> AS: E_{vas} (N_{vs}, docData, E_{vvs} (N_{vs}, N_v, filledBallot), HMAC4), HMAC5

(8) AS -> VS: E_{asvs} (N_{asvs}, E_{vvs} (N_{vs}, N_v, filledBallot), HMAC4)), HMAC6

(10) VS -> V: E_{vvs} (N_v, N_{vs}, checkID), HMAC7

ECDHE is a Diffie-Hellman protocol on elliptical curves using ephemeral keys. In our case, we use a modified version of ECDHE-RSA, where authentication is done using a signature RSA and a server certificate which help to prevent MIMT (man in the middle) attacks. The protocol description is as follows.

ECDHE:

(1) V -> S: "Hello"

(2) S -> V: DHs, $Sign_{SKs}$ (DHs),Certificate

(3) S: checks Certificate and sign $Sign_{SKs}$ (DHs)

(4) V -> S: DHv

(5) Both sides generate a common session key K for further interaction with a symmetric cipher.

Here V is the client, S is the trusted server that has the certificate, *DHs* is the server secret part, DHv is the client secret part, $Sign_{SKs}$ (DHs) is the signature with the server's private key *SKs*, Certificate is the server certificate. When servers generate common secret key, the same protocol is used, except that both parties exchange certificates and if they are valid, a common session key is generated. The security verification of the voting protocol will be carried out after this stage.

4   Security Analysis of Voting Protocol Using Avispa Tool

Avispa is a tool for automated security analysis of cryptographic protocols [17]. With the help of Avispa, in the context of the developed protocols, it is possible to verify: the parties' authentication, the secrecy of data and protection against replay-attacks. It is impossible to perform integrity checks, in particular, used in protocol CMAC mode (Cipher-based message authentication code) using the Avispa tool. The protocol does not imply the use of timestamps in their classic implementation as a part of message. Instead, the developed system uses a temporary session control by server, in which long live sessions are broke down. In the paper voting stage is analyzed. Three sides are modeled: user, server-intermediary and main server. The protocol will be analyzed after the phase of common session key distribution between the parties. The protocol will be described in CAS+ [19] language, then translated using the Avispa translator into HLPSL [20]. The check will be carried out using the On-the-Fly Model Checking (OFMC) module, where the verification goals are the transmitted data confidentiality and parties' authentication. In the previous work [21] the analysis was made for voting protocol of our old system and now the new protocol, which used in our new system, will be used as a target of verification. Consider the description of the protocol in CAS + at the voting stage.

```
1    protocol EVoting;
2    identifiers
3    V,AS,VS: user;
4    Nas,Nvs,Nasvs,Nv,DocumentData,CheckID,filledBallot  : number;
5    Kvas,Kvvs,Kasvs: symmetric_key;
6
7    messages
8    1. VS -> V  : {Nvs}Kvvs
9    2. VS -> AS : {Nasvs}Kasvs
10   3. AS -> V  : {Nas}Kvas
11   4. V -> AS  : {Nas,DocumentData,{Nvs,Nv,filledBallot}Kvvs}Kvas
12   5. AS -> VS : {Nasvs,{Nvs,Nv,filledBallot}Kvvs}Kasvs
13   6. VS -> V  : {Nvs,Nv,CheckID}Kvvs
14
15   knowledge
16   V:V,AS,VS,Nas,Nvs,Nv,DocumentData,CheckID,Kvas,Kvvs,filledBallot
17   AS:V,AS,VS,Nvs,Nasvs,DocumentData,Kvas,Kasvs
18   VS:V,AS,VS,Nasvs,Nvs,Nv,CheckID,Kvvs,Kasvs,filledBallot
19
20   session_instances
21   [V:v,AS:as,VS:vs,Kvas:kvas,Kvvs:kvvs,Kasvs:kasvs]
22   [V:v,AS:as,VS:vs,Kvas:kvas,Kvvs:kvvs,Kasvs:kasvs];
23
24   intruder_knowledge
25     v,as,vs;
26
27   goal
28     secrecy_of Nvs [V,VS];
29     secrecy_of Nasvs [AS,VS];
30     secrecy_of Nas [V,AS];
31     secrecy_of Nv [V,VS];
32     secrecy_of DocumentData [V,AS];
33     secrecy_of filledBallot [V,VS];
34     secrecy_of CheckID [V,VS];
35     AS authenticates V on Nas;
36     VS authenticates AS on Nasvs;
37     VS authenticates V on Nvs;
38     V authenticates VS on Nv;
```

Three interacting parties are described as roles: V, AS, VS (lines 2–3). The identifiers section describes the objects participating in the protocol: interacting parties (line 3), random numbers for authentication, identifiers (line 4). Symmetric keys are specified that will be used for message encryption (line 5). The messages section (lines 7–13) describes the transfer of messages between roles, which data is transmitted, and on which key it encrypted. The knowledge section (lines 15–18) describes roles' data knowledge during the execution of the protocol. In the session_instances section (lines 20–22), sessions are described. Among the simulated sessions, 2 are allocated, which allow simulating interaction of two clients with the system. This will detect possible attacks on the parties' authentication and replay-attacks. The intruder_knowledge section (lines 24–25) specifies the original knowledge of the intruder. In the goal section (lines 27–38) the secrecy of important values is indicated and the authentication according to the request-response scheme with the transfer of random numbers between the participants. For secrecy of the value, it is necessary that this variable is encrypted and that the encryption key does not come to intruder. In order for one party to authenticate another using the request-response mechanism, it is required that the party wanting to authenticate send a random number to the other party, and that other party in the response message returns this random number. In this protocol there are 4 such actions:

1. AS authenticates V by Nas
2. VS authenticates AS by Nasvs
3. VS authenticates V by Nvs
4. V authenticates the VS to Nv

As for replay-attacks, protection against them is possible due to the presence of a random number at the beginning of each message, which each side checks when message is received. The results of the check using the OFMC module are shown in Fig. 4.

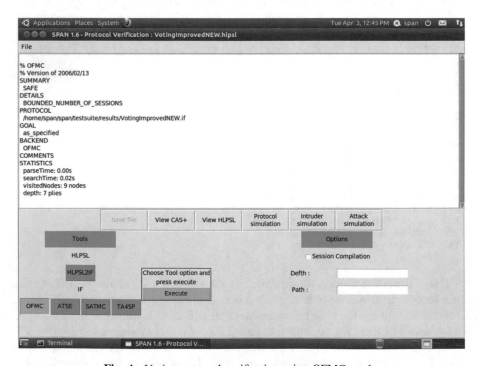

Fig. 4. Voting protocol verification using OFMC mode.

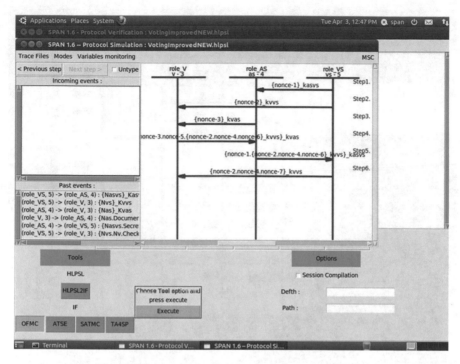

Fig. 5. Voting protocol in "Protocol simulation" mode.

Figure 5 shows the scheme of interaction between the parties at the stage of voting by steps. Fig. 6 shows the interaction scheme in the presence of an intruder (*Intruder_* side, highlighted in red). This scheme is a visual implementation of the attack man in the middle. When transmitting messages during execution, a transition is made from the "Incoming events" area to "Past events", and the format is the direction of message transfer (from whom and to whom) and the message itself. We can see from the simulation results in the field of intercepted data "Intruder knowledge", all transmitted messages are encrypted on keys which intruder doesn't know, and it excludes the possibility in any way to get important information, such as the user's passport data or unique identifier. The record "nonce-N" means some data that is not readable. As a result of the analysis, it was revealed that the voting protocol is safe, ensures the fulfillment of the security objectives (properties) set in the protocol analysis: securing data, authentication of the parties, protection against replay-attacks.

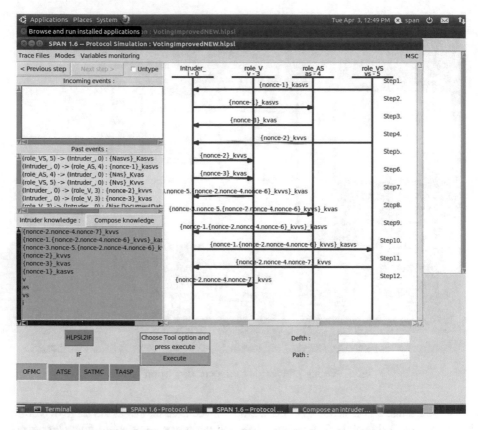

Fig. 6. Voting protocol in "Intruder simulation" mode.

5 Conclusion

The automated security verification tool Avispa was used for security verification of the voting protocol in electronic voting system based on blind intermediaries, in this paper. The protocol was described in the formal languages CAS+ and HLPSL. The secrecy properties of the transmitted data between the interacting parties were analyzed. It was shown that set security objectives: parties' authentication, verification of data privacy and protection from replay attacks were achieved. The scheme of parties' interaction with the help of tools' graphical functional was considered. An analysis of messages that an intruder can intercept was carried out. Based on the graphical representation it was revealed that all transmitted data is secure, because all messages are encrypted on unknown for intruder keys.

Acknowledgments. The work was supported by the Ministry of Education and Science of the Russian Federation grant № 2.6264.2017/8.9.

References

1. Overview of e-voting systems, NICK Estonia. Estonian National Electoral Commission, Tallinn (2005)
2. Dossogne, J., Lafitte, F.: Blinded additively homomorphic encryption schemes for self-tallying voting. J. Inf. Secur. Appl. (2015)
3. Izabachene, M.A.: Homomorphic LWE based e-voting scheme. In: Post-quantum Cryptography: 7th International Workshop, PQCrypto 2016, Fukuoka, Japan, 24–26 February 2016
4. Hirt, M., Sako, K.: Efficient receipt-free voting based on homomorphic encryption. In: International Conference on the Theory and Applications of Cryptographic Techniques, 539–556. Springer, Berlin (2000)
5. Rivest, L.R., et al.: Lecture notes 15: Voting, homomorphic encryption (2002)
6. Adida, B.: Mixnets in Electronic Voting. Cambridge University, Cambridge (2005)
7. Electronic elections: fear of falsification of the results. Kazakhstan today 2004
8. Lipen, V.Y., Voronetsky, M.A.: Lipen DV technology and results of testing electronic voting systems. United Institute of Informatics Problems NASB (2002)
9. Chaum, D.L.: Untraceable electronic mail, return addresses, and digital pseudonyms. Commun. ACM **24**(2), 84–90 (1981)
10. Ali, S.T., Murray J.: An Overview of End-to-End Verifiable Voting Systems (2016). arXiv preprint arXiv:1605.08554
11. Smart, M., Ritter, E.: True trustworthy elections: remote electronic voting using trusted computing. In: International Conference on Autonomic and Trusted Computing, 187–202. Springer, Berlin (2011)
12. Bruck, S., Jefferson, D., Rivest, R.L.: A modular voting architecture ("frog voting"). In: Toward Strustworthy Elections. Springer, Berlin (2010)
13. Jonker, H., Mauw, S., Pang, J.: Privacy and verifiability in voting systems: methods, developments and trends. Comput. Sci. Rev. **10**, 1–30 (2013)
14. Shinde, S.S., Shukla, S., Chitre, D.K.: Secure E-voting using homomorphic technology. Int. J. Emerg. Technol. Adv. Eng. **3**(8), 203–206 (2013)
15. Neumann, S., Volkamer, M.: Civitas and the real world: problems and solutions from a practical point of view. In: Availability, Reliability and Security (ARES), 2012 Seventh International Conference on IEEES, 180–185 (2012)
16. Yi, X., Okamoto, E.: Practical remote end-to-end voting scheme. In: International Conference on Electronic Government and the Information Systems Perspective, 386–400. Springer, Berlin (2011)
17. The AVISPA team, the high level protocol specification language (2006). http://www.avispa-project.org/
18. Babenko, L.K., Pisarev, I.A., Makarevich, O.B.: Secure electronic voting using blinded intermediaries. Journal "Isvestiya SFedU". Technical Sciences, pp. 6–15. Publishing House of ITA SFedU, No. 5, Taganrog (2017)
19. Saillard, R., Genet, T.: CAS+, 21 March 2011
20. Basin, D., Mödersheim, S., Viganò, L.: OFMC: a symbolic model-checker for security protocols. Int. J. Inf. Secur. **4**(3), 181–208 (2004)
21. Babenko, L.K., Pisarev, I.A.: Protocol security analysis of electronic voting system based on blind intermediaries with the Avispa tool. Journal "Isvestiya SFedU". Technical sciences, pp. 227–238. Publishing house of ITA SFedU, No. 7 (192), Taganrog (2017)

Learning Bayesian Network Structure
for Risky Behavior Modelling

Alena Suvorova[1,2(✉)] and Alexander Tulupyev[1]

[1] SPIIRAS, St. Petersburg, Russia
suvalv@gmail.com

[2] National Research University Higher School of Economics, St. Petersburg, Russia

Abstract. Bayesian Belief Networks (BBN) provide a comprehensible framework for representing complex systems that allows including expert knowledge and statistical data simultaneously. We explored BBN models for estimating risky behavior rate and compared several network structures, both expert-based and data-based. To learn and evaluate models we used generated behavior data with 9393 observations. We applied both score-based and constraint-based structure learning algorithms. The score-based structures represented better quality scores according to BIC and log-likelihood, prediction quality was almost the same for data-based models and lower but sufficient for expert-based models. Hence, in case of limited data we can reduce computations and apply expert-based structure for solving practical issues.

Keywords: Bayesian belief network · Structure learning
Machine learning · Behavior models · Risky behavior

1 Introduction

Bayesian Belief Network (BBN) is a type of probabilistic graphical models that represents a set of random variables and their conditional dependencies [7]. It consists of two components: structure and parameters. A network structure is presented in the form of a directed acyclic graph where nodes correspond to the random variables and directed edges represent dependencies among variables. Parameters are represented as a set of conditional probability distributions, one for each variable, characterizing the dependencies represented by the edges [4].

Bayesian Belief Networks provide a comprehensible framework for representing complex systems [3,12]. A major benefit of BBNs is the opportunity to complement empirical data with inputs from other models and expert knowledge [7]. Due to its features BBNs are widely used in decision making in many areas [1,2,12].

One of research areas that deals with the problem of complex relations between variables is individual's risky behavior modelling [5,15]. To make predictions we have to combine expert knowledge and data from different sources (usually incomplete). Earlier studies proposed an approach for risky behavior

© Springer Nature Switzerland AG 2019
A. Abraham et al. (Eds.): IITI 2018, AISC 875, pp. 58–65, 2019.
https://doi.org/10.1007/978-3-030-01821-4_7

modelling based on Bayesian belief network with data about several last behavior episodes [13]. However, the proposed model was heavily based on experts' assumptions about relations between elements of the model. The next step is to explore the importance of these assumptions and to construct a structure automatically.

The purpose of the paper is to compare several network structures for risky behavior modelling, both expert-based and data-based.

2 Initial Model

The model [13] is based on data about the length of intervals between three last episodes of risky behavior and the length of minimum and maximum intervals between episodes during period of interest. The data about episodes in most applications is obtained from respondents' self-reports [15]. We assume that for each respondent occurrence of episodes follows Poisson random process: the occurrence of the next episode is independent from the previous ones, length of interval between concurrent episodes follows exponential distribution. This assumption corresponds to the features of risky behavior [10] and, at the same time, allows less complicated calculations. Adding data about minimum and maximum intervals decreases the influence of recent behavior represented by the last episodes. However, combining all the data about episodes leads to very complicated joint distribution [11] even in case of Poisson random process and requires much more calculation for behavior rate estimate. Any change or revision of the model, again, will require re-calculation of joint distribution.

Figure 1 represented the structure of BBN model proposed in [13]. The model structure is directed acyclic graph $G(V, L)$ with vertices V and edges $L = \{(u, v) : u, v \in V\}$, where $V = \{t_{01}, t_{12}, t_{23}, t_{min}, t_{max}, \lambda, n\}$ and λ (*Rate* on figures) is random variable for the behavior rate; t_i are random variables for the lengths of the i-th last interval between episodes; t_{min} and t_{max} are random variables for the length of minimum and maximum intervals; n is random variable for the number of episodes during the period of interest.

For all further examples we used the following discretization: for the rate variable $\lambda^{(1)} = [0; 0.05)$, $\lambda^{(2)} = [0.05; 0.1)$, $\lambda^{(3)} = [0.1; 0.15)$, $\lambda^{(4)} = [0.15; 0.2)$, $\lambda^{(5)} = [0.2; 0.3)$, $\lambda^{(6)} = [0.3; 0.5)$, $\lambda^{(7)} = [0.5; 1)$, $\lambda^{(8)} = [1; +\infty)$; for the intervals between episodes t_i, $i = 1, 2, 3$, t_{min} t_{max} $t^{(1)} = [0; 0.5)$, $t^{(2)} = [0.5; 1)$, $t^{(3)} = [1; 3)$, $t^{(4)} = [3; 7)$, $t^{(5)} = [7; 14)$, $t^{(6)} = [14; 30)$, $t^{(7)} = [30; 180)$, $t^{(8)} = [180; +\infty)$.

The proposed model showed good prediction quality on the data about behavior on social networking site [14], the discretized data about the last episodes are able to describe the behavior.

3 Experiment Design

3.1 Structure Learning Algorithms

In addition to expert-based structure (Fig. 1) we explored data-based models, that did not involve assumptions about relations between variables. We used

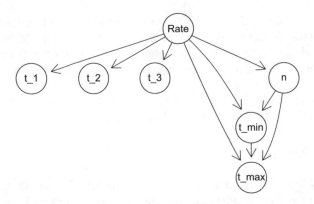

Fig. 1. Expert structure for risky behavior model

R statistical language [8] for all calculations and analysis, particularly, we used *bnlearn* package [9] for all BN-related analysis including learning, predicting, scoring and plotting.

Learning a BN can refer to data-based inference of either the conditional probability parameters for a given structure or the underlying graphical structure itself. In the paper we focused on the second part – structure learning, or, in other words, on finding relations between variables that describe the risky behavior. The parameter estimates were obtained for each structure as part of learning process.

The algorithms for learning the structure of BNs from data can be classified into two types, constraint-based and score-based.

Score-Based Learning Algorithms. Score-based methods consider a number of possible BN structures and assign a score to each that measures how well it explains the observed set of data. The algorithms usually return the single structure that maximizes the score. As the number of potential structures is super-exponential in the number of nodes, most score-based algorithms have employed heuristic search techniques.

In the comparison we used HC Tabu Search algorithm [6] with BIC (Bayesian Information Criteria) score. It is a modified hill-climbing able to escape local optima by selecting a network that minimally decreases the score function.

Constraint-Based Structure Learning Algorithm. Constraint-based methods focus on identifying conditional independence relationships (i.e., Markov conditions) between variables using observed data. These conditional independencies can then be used to constrain the underlying network structure. The conditional independence tests used in constraint-based algorithms in practice are statistical tests on the data set.

As example of constraint-based method we used Grow-Shrink algorithm [6]. It is based on the Grow-Shrink Markov Blanket, the simplest Markov blanket detection algorithm used in a structure learning algorithm [9]. We considered two kinds of conditional independence tests: mutual information test and Jonckheere-Terpstra test. The last one is a trend test for ordinal variables, so it takes into account the order in data. The mutual information measure is proportional to the log-likelihood ratio and is related to the deviance of the tested models [9].

3.2 Data Description

To learn the models we used automatically generated dataset. First, we generate 500 values for behavior rate that followed Gamma distribution with the shape $k = 1.3$ and the scale $\theta = 0.25$. The parameters were chosen to produce more "real-behavior"-like dataset with behavior rate in most cases less than 1 and concentrated around 0.25 episodes per day. Next, for each rate value according to assumptions of the behavior process we generated 15 "respondents" or 15 sequences of behavior episodes each of those for period of 365 days in total. In other words, we had 7500 sequences of episodes. Than we calculated lengths of minimum, maximum intervals between episodes and lengths of intervals between the last three episodes. After deletion of incomplete cases (e.g. cases with only one episode during 365 days) the final dataset included 7412 cases (or "respondents").

Then we repeated these steps and generated the smaller dataset with 1981 cases for testing the models.

4   Results

4.1   Structure Comparison

We compared four network structures: expert-based structure (Fig. 1), score-based structure according to Tabu Search algorithm (Fig. 2) and two constraint-based structures (Figs. 3 and 4). Figure 3 represented the result of Grow-Shrink algorithm with Jonckheere-Terpstra test as the basis for the conditional independence tests for ordered discrete data. Figure 4 represented the result of the same Grow-Shrink algorithm but with mutual information test that did not count for order in data.

The score-based algorithm (Tabu Search) produced a clear structure, close to naive Bayes classifier: behavior rate was related to number of episodes and all other variables depended on the number of episodes only. This structure has simple interpretation because the number of episodes can be directly calculated on the base of the rate.

Constraint-based structures did not have such clear interpretation. Note, the algorithm, that took into account the order, produced more complex structure with more dependencies (for example, the rate and the number of episodes are connected with all other variables) while the second one reduced the number of connections, the number of episodes, the length of minimum interval and the length of the third interval were not important for the rate estimate.

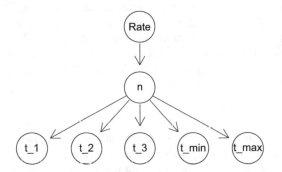

Fig. 2. Model structure (Tabu Search algorithm)

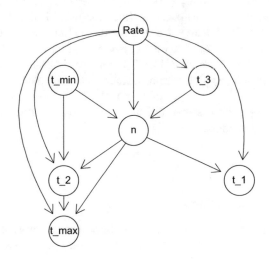

Fig. 3. Model structure (Grow-Shrink algorithm for ordered data)

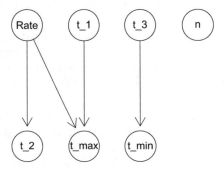

Fig. 4. Model structure (Grow-Shrink algorithm)

Table 1. Structure scores

Algorithm	Train dataset		Test dataset	
	BIC	loglik	BIC	loglik
Expert structure	−75712	−59998	−29113	−15726
Tabu Search	−61845	−60068	−17491	−15713
Grow-Shrink for ordered data	−205471	−58981	−140059	−15260
Grow-Shrink	−77513	−75490	−21456	−19733

Table 1 represented the comparison of structures by the structure scores. As expected, Tabu Search algorithm produced the structure with the highest BIC scores both on train and test datasets (score-based algorithms choose structures with the best scores). Constraint-based algorithm for ordered data showed the best result according to log-likelihood score. Expert-based structure had the middle-range scores, relatively close to the best ones.

4.2   Prediction Quality

The next step in structure comparison was to explore the prediction quality of the models. We learned the parameters of all for models on the train dataset and then made predictions of the rate value on the test dataset. Since the rate was discretized and had eight distinct levels, the prediction become a 8-class classification problem. So we estimated prediction quality in terms of accuracy, precision and recall measures. Prediction metrics of the models are presented in Table 2.

All models predicted more than 83% cases on average correctly. The model with expert-based structure had the lowest accuracy (83.7%), score-based algorithm and constraint-based algorithm that counted for ordered data produced models with almost the same results (87.7% and 87.4% correspondingly). The constraint-based algorithm with mutual information independence test showed the lowest prediction quality among models with data-based structure.

Table 2. Prediction quality

Model	Accuracy	Precision	Recall
Expert structure	0.837	0.591	0.313
Tabu Search	0.877	0.547	0.505
Grow-Shrink for ordered data	0.874	0.511	0.541
Grow-Shrink	0.856	0.443	0.481

5   Conclusion

We proposed the development of the approach for risky behavior modelling in terms of Bayesian belief network based on data about several last behavior episodes. The main question of this paper was to estimate how detailed expert knowledge we needed. In particular, we considered the possibility not to require model structure from experts. To test that we compared the expert-based and the data-based structures.

The score-based structures represented better quality scores according to BIC and log-likelihood, prediction quality was lower for constraint-based models and expert-based models compared to score-based one.

The model with the expert-based structure was not the optimal one but the expert-based structure is pre-defined, we do not need additional data to learn the structure. This property can be useful when we do not have sufficient data source or simple procedure for data collection. We can consider naive Bayes classifier resulted from score-based learning algorithm and simplify the expert structure. Hence, we can reduce computations and apply expert-based (possibly modified) structure for solving practical issues. At the same time, if we have appropriate dataset it is better to learn structure and use model that better fits problem domain.

Acknowledgements. This work was partially supported by the by RFBR according to the research project No. 16-31-60063 and No. 18-01-00626.

References

1. Barton, D.N., Benjamin, T., Cerdan, C.R., DeClerck, F., Madsen, A.L., Rusch, G.M., Villanueva, C.: Assessing ecosystem services from multifunctional trees in pastures using Bayesian belief networks. Ecosyst. Serv. **18**, 165–174 (2016)
2. Boets, P., Landuyt, D., Everaert, G., Broekx, S., Goethals, P.L.: Evaluation and comparison of data-driven and knowledge-supported Bayesian Belief Networks to assess the habitat suitability for alien macroinvertebrates. Environ. Model. Softw. **74**, 92–103 (2015)
3. Chu, Z., Wang, W., Wang, B., Zhuang, J.: Research on factors influencing municipal household solid waste separate collection: Bayesian belief networks. Sustainability **8**(2), 152 (2016)
4. Darwiche, A.: Modelling and Reasoning with Bayesian Networks. Cambridge University Press, Cambridge (2009)
5. Kabir, G., Tesfamariam, S., Francisque, A., Sadiq, R.: Evaluating risk of water mains failure using a Bayesian belief network model. Eur. J. Oper. Res. **240**(1), 220–234 (2015)
6. Margaritis, D.: Learning Bayesian network model structure from data, Ph.D. thesis. Carnegie-Mellon University, School of Computer Science, Pittsburgh (2003)
7. Pearl, J.: Causality: Models, Reasoning, and Inference. Cambridge University Press, Cambridge (2000)
8. R Core Team. R: A language and environment for statistical computing. R Foundation for Statistical Computing, Vienna, Austria (2017). http://www.R-project.org/

9. Scutari, M.: Learning Bayesian networks with the bnlearn R package. arXiv preprint arXiv:0908.3817 (2009)
10. Spiegelman, D., Hertzmark, E.: Easy SAS calculations for risk or prevalence ratios and differences. Am. J. Epidemiol. **162**(3), 199–200 (2005)
11. Stepanov, D.V., Musina, V.F., Suvorova, A.V., Tulupyev, A.L., Sirotkin, A.V., Tulupyeva, T.V.: Risky behavior Poisson model identification: heterogeneous arguments in likelihood. Trudy SPIIRAN [SPIIRAS Proceedings], vol. 23, pp. 157–184 (2012)
12. Su, C., Andrew, A., Karagas, M.R., Borsuk, M.E.: Using Bayesian networks to discover relations between genes, environment, and disease. BioData Mining **6**(1), 6 (2013)
13. Suvorova, A.V., Tulupyev, A.L., Sirotkin, A.V.: Bayesian belief networks for risky behavior rate estimates. Nechetkie sistemy i myagkie vychisleniya [Fuzzy Systems and Soft Computing] **9**(2), 115–129 (2014)
14. Suvorova, A., Tulupyev, A.L.: Evaluation of the model for individual behavior rate estimate: social network data. In: 2016 XIX IEEE International Conference on Soft Computing and Measurements (SCM), pp. 18–20. IEEE (2016)
15. Suvorova, A., Tulupyeva, T.: Bayesian belief networks in risky behavior modelling. In: Proceedings of the First International Scientific Conference Intelligent Information Technologies for Industry (IITI 2016), pp. 95–102. Springer (2016)

On Continuous User Authentication via Hidden Free-Text Based Monitoring

Elena Kochegurova$^{(\boxtimes)}$, Elena Luneva, and Ekaterina Gorokhova

National Research Tomsk Polytechnic University,
Lenina Avenue, 30, 634050 Tomsk, Russia
kocheg@mail.ru

Abstract. This paper investigates the stages and specific features of continuous user authentication by hidden monitoring of keystroke dynamics when creating a free text. The stages include extraction of informative characteristics of keyboard rhythm, creation and update of user profiles and identification of efficiency criteria. A software application was developed for the project. The authors further analyzed existing algorithms for user identification based in metric distances. Previously proved features of keystroke dynamics were scaled with regard to frequency of use of Russian and English letters in free texts.

Keywords: Keystroke dynamics · Continuous authentication
Behavioral biometrics · Feature selection · Classification

1 Introduction

The number of information leaks has been increasing in the world; leak dynamics has grown by a third in 2017 as compared to 2016 according to InfoWatch. That said, company employees and corporate management are responsible for information leakage in 55% of the cases. Therefore, the need to authenticate users who have access to public information resources becomes imminent.

Biometric authentication methods are widely used along with technical and organizational measures to protect information, i.e. physiological and behavioral biometrics are employed in addition to password and property methods. However, physiological characteristics (DNA, fingerprints, vein patterns, etc.) are unique and remain the same life long, while behavioral characteristics (gait, voice, handwriting) can change over time.

It is known that keystroke dynamics (or keyboard rhythm, KR) is determined by the individual rhythm of typing and can be used for biometric authentication [1]. As a behavioral characteristic, KR has a dynamic nature. Besides, behavioral characteristics include semi-constant and random components that are conditioned by physiology and emotions. Compared to physiological characteristics, behavioral features of KR are more resistant to forgery; however, they can be difficult to identify with high accuracy [2].

When a standard keyboard is utilized, recognition of users depends on the type of created texts and can be static or dynamic/continuous. In static authentication, a user is recognized via structured/predefined texts; in dynamic authentication, verification is based on a free text typed by a user in any software application and the user identity is continuously verified during the session in a hidden monitoring mode. Static

© Springer Nature Switzerland AG 2019
A. Abraham et al. (Eds.): IITI 2018, AISC 875, pp. 66–75, 2019.
https://doi.org/10.1007/978-3-030-01821-4_8

identification is mainly used for primary authentication, while the main goal of hidden monitoring is to establish the authenticity of users and their emotional profiles in the course of working in corporate systems.

This paper aims to study continuous authentication and expand known approaches to static authentication for dynamic user recognition based on free and long texts. The study further aims to increase reliability and efficiency of hidden continuous identification in the space of extracted keystroke features.

The main organizational patterns of obtaining and recognizing input data in any KR-based automatic user identification are:

- collection of KR samples;
- selection of keyboard timing characteristics;
- creation of user templates (profiles);
- selection of recognition efficiency indicators;
- recognition algorithms and methods.

2 State of the Art and Relevance of Keyboard Authentication

Recently, keyboard authentication has attracted research interest and used in security systems because of its low cost and ease of integration with existing systems. This type of authentication requires no other equipment than a keyboard, which is available with all computers and high-performance software.

Since the first studies of keyboard rhythm, a sufficient number of reviews of keystroke dynamics are now available. Topical classifiers based on statistical methods and neural networks were analyzed in 2004–2010 [3, 4]. It was found that probabilistic methods were more computationally attractive, but had lower identification accuracy. Most of the papers of that time described user static identification, i.e. authentication based on a password or a fixed text.

Besides researching standard PC keyboards, user authentication via touch keyboards, mobile devices and web services were analyzed in 2011–2018 [5–12]. The number of publications on keyboard authentication is growing quite rapidly as shown in Fig. 1, where the data before 2012 are obtained from [7, 8] and the data for 2012-present time are collected from the main scientometric databases (DB).

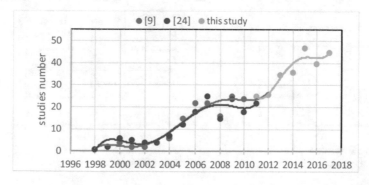

Fig. 1. Growth in the number of publications on keystroke dynamics

Recently, the majority of studies are focused on the applied nature of keyboard biometrics. Some papers are devoted to gender and age identification [13], others are associated with mixed identification, e.g. authentication based on keyboard rhythm and the use of a mouse or facial features [14]. This approach allows to recognize psychoemotional states of people to a great extent.

It should be noted that keyboard biometrics cannot completely replace password authentication. However, keyboard identification has advantages over static authentication including hidden monitoring, low implementation cost, high identification degree and the ease of integration with other security systems, which makes it a beneficial additional security tool.

Three subsystems are employed in keyboard identification: (a) data collection and retrieval, (b) formation of a keystroke profile, and (c) user identification.

3 Data Collection and Keystroke Profiles

Data can be collected by a special software without the use of additional hardware. When a text is typed, an operational system records an ANSI code that is associated with a particular key, i.e. the time when the key is pressed and released. Keystrokes can be measured with a millisecond accuracy, which allows collecting primary data on user keystroke dynamics.

Up to 28 muscles are used during a keystroke. A KR pattern obtained from a multidimensional control problem proves that KR is unique and potentially important for user identification.

A keystroke pattern is based on a sufficiently large set of indicators that characterize a typing rhythm, duration and delay of keystrokes [7, 8, 15, 16, 21]: the hold time of a key; the inter-key time, the presence of overlaps, the typing error rate, the degree of arrhythmia when typing, the typing speed and the specific use of service keys.

Most studies of keystroke dynamics analyze the time between two successive keystrokes, the so-called digraphs. There are two main digraph characteristics: dwelt time (DT) and flight time (FT). DT and FT can be illustrated by Figs. 2a and b, which show two basic notations to graphically represent the timing characteristics.

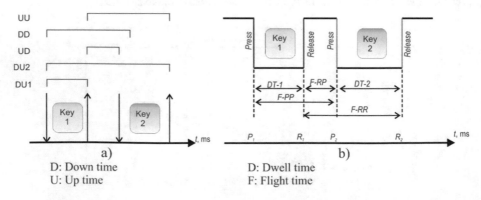

Fig. 2. Features of keyboard rhythm

Hold time (DT/DU − 41%) and latency (UD + DD + UU − 49%) are considered the most important characteristics of keystroke patterns.

The selected features of keystroke dynamics are collected and processed to generate a user profile. Keystroke features are commonly extracted from structured and pre-defined text.

A number of additional issues arise when data are collected continuously [2]:

- the amount of data required to generate a pattern. According to [7], short texts of less than 1000 characters are examined in 57% and long texts in 24% of studies; the remaining 19% analyze the texts containing only figures or unknown content.
- generation and use of public databases of user profiles. Most public databases contain data collected from passwords and fixed texts and, therefore, are of little use for continuous monitoring.
- complete absence of such databases with user profiles for the Russian language. Therefore, comparison of data in similar conditions is complicated.

4   Efficiency Evaluation of Authentication

Efficiency of authentication refers to the accuracy of the method to identify a genuine domain user and prevent unauthorized access.

The following scenarios are possible when KR samples are compared [15, 17]:

True Accept (TA): two samples belong to the same user and the system defines the samples as similar − an expected scenario;
True Reject (TR): two samples belong to different users and the system defines the samples as different − another expected scenario;
False Reject (FR): two samples belong to the same user but the system defines the samples as different − a true hypothesis is rejected;
False Accept (FA): two samples belong to different users but the system defines the samples as similar − a false hypothesis is accepted.

Efficiency criteria used in different keystroke dynamics studies can be classified into four main error frequency indices.

False Rejection Rate (FRR) − type I error, which defines the percentage of cases when an authorized user is erroneously rejected

$$FRR = \frac{Number\ of\ false\ rejection}{Total\ number\ of\ genuine\ match\ attemps} \tag{1}$$

False Acceptance Rate (FAR) − type II error, which defines the percentage of cases when unauthorized users are accepted.

$$FAR = \frac{Number\ of\ false\ matches}{Total\ number\ of\ imposter\ attemps} \tag{2}$$

As shown in Fig. 3a, FRR and FAR demonstrate opposite behavior and depend on the sensitivity of the algorithm (its thresholds).

The trade-off between FAR and FRR is determined by objectives of an applied task. For complex penetration of unauthorized users (low FAR values), a high threshold of sensitivity is required. However, this can lead to a high false rejection rate.

The trade-off between FAR and FRR can be depicted by a parametrically given curve of DET (Detection Error Trade-off) [12, 18, 19], Fig. 3b. A particular case of DET curve is an Equal Error Rate (EER). EER is independent on the sensitivity of the identification algorithm. The lower EER is, the more efficient the authentication system is for one threshold value.

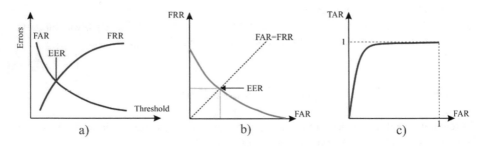

Fig. 3. Efficiency criteria in keyboard authentication

Receiver Operating Characteristic (ROC) is used less common [12, 18, 19] and demonstrates the trade-off between TAR and FAR at different threshold values, Fig. 3c. In the ideal case, TAR = 1, FAR = 0.

5   Classification Algorithms and Methods

A user is recognized by a selected classification algorithm at the last stage of keyboard authentication. Research of recent reviews [2, 7–12, 15, 16] showed that

- algorithms for dynamic and static authentication are based on similar principles;
- there are still few works on dynamic authentication [15, 20];
- most researchers use short or predefined texts.

Algorithms of static and dynamic authentication can be divided into three major groups in the viewpoint of pattern recognition: (a) estimation of metric distances, (b) statistical methods, and (c) machine learning. Relative frequency of use of different KR classification methods is shown in Fig. 4 in descending order [5, 6].

The largest group of recognition methods (23%) is based on an assessment of proximity between the user's current and reference profiles by calculating some known metric distance and comparing it with a threshold value. Historically, statistical and probabilistic methods were the first methods used in KR recognition. Nowadays, they are still relevant and applied in 29% of research.

The next most frequent method (16%) is artificial neural networks (INS). They evolved into a separate category among machine learning methods of pattern recognition; however, this group of KR recognition also includes a number of known algorithms, such as fuzzy logic, decision trees and evolutionary computing.

Cluster analysis in the recognition of KR can be used to combine the features of the analyzed template with a similar keyboard profile.

The division of the KR recognition methods by categories is rather intuitive. The same classifiers can be based on different teaching methods and metric distances.

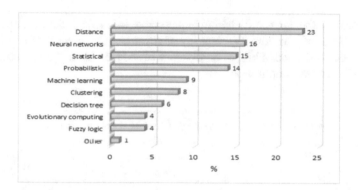

Fig. 4. Relative frequency of use of keyboard recognition methods

6   Experiment, Results and Discussion

The experiment was aimed to implement a continuous authentication of standard keyboard users and included all stages from collecting information about keystrokes in the Russian and English languages to recognizing authorized and unauthorized users. Approximately thirty university students and professors with medium and good typing speed volunteered to participate in the experiment.

The software architecture included two components: the client and the server [17]. The client part was responsible for the collection and transmission of data on keystrokes. The server component generated keyboard profiles, updated the domain database and authenticated domain users.

A free text was typed in the client interface of the experiment application or any other program (test editor, browser line, etc.). The users were usually not informed about hidden monitoring of their use of keyboards.

Data with KR samples were transmitted to the server in packs of at least 200 characters. Then the volume of received keystrokes reached 1000 characters and the last 1000 characters were constantly updated during the user's session. Data was transmitted via TCP sockets.

The server calculated the mean values of the selected keyboard timing characteristics in each current session for each key. It further updated the user's keyboard profile in the domain database. Profile fragments for two users for several letters of the Russian alphabet are given in Table 1 on the example of dwell time (DT) that was calculated in milliseconds.

Table 1. Structure and features of keyboard profile

Key	a	б	о	п	я	a	б	о	п	я
	User 1 (nvb16)					User 2 (ksk12)				
Mean	76	85.5	77.5	74.0	81.4	83.9	81.7	67.5	88.8	95.1
SD	7.8	70.1	9.6	30.7	38.1	16.8	31.9	15.1	62.7	88.2
Max	82.3	106	83.2	89.5	95.5	91.8	135	74.6	107	117.3
Min	71.9	60	70.3	65.8	71	77.2	64.4	62.3	73.6	82.2

The data on the keyboard rhythm represented in Table 1 demonstrate a certain discrepancy in user behaviors. Discrepancy of keystroke dynamics between six users is graphically demonstrated in Fig. 5 for all letters of the Russian alphabet except for "ф", "ъ" and "ё" that are seldom used.

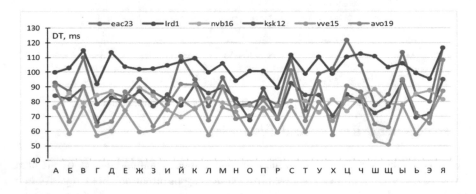

Fig. 5. Mean dwell time, ms

Figure 5 shows that the curves reflecting keystroke patterns of different users differ from each other. This confirms the possibility to authenticate users by input rhythm as it determines keyboard rhythm.

The algorithm of KR-based user identification includes the following steps:

- generate a vector for the keyboard rhythm,
- select a measure of proximity of the user's keyboard rhythm to the reference pattern,
- select efficiency indicators and develop a decision rule for user identification.

Studies cited above proved that KR timing characteristics are independent from each other and from the frequency of using particular letters of the alphabet, although the frequency range of repeating letters in texts was relatively wide for the experiment group and made [0.013%–10.98%] for Russian letters and [2.03%–12.02%] for English letters.

The experiment showed the dependence of a keyboard rhythm on how often letters appear in texts. Seldom used letters are not associated with muscle memory and require additional attention and time to be pressed. Therefore, it is advisable to use a letter frequency distribution when generating a KR vector. Therefore, the Russian and English letters were divided into groups with frequency ranges of the ratio 0.5/0.3/0.2 as shown in Table 2.

Similar normalizing factors of 0.5/0.3/0.2 were introduced for the timing characteristics of each group.

Table 2. Distribution of letters by frequency

Russian alphabet		English alphabet	
Frequency range %	Letters	Frequency range %	Letters
[10.98–5.47]	о/е/а/и/н/т/с	[12.02–6.02]	e/t/a/o/i/n/s/r
[4.75–1.59]	р/в/л/к/м/д/п/у/я/ы/ь/г/з/б	[5.92–2.09]	h/d/l/u/c/m/f/y/w
[1.45–0.013]	я/й/х/ж/ш/ю/ц/щ/э/ф/ъ/ё	[2.03–0.07]	g/p/b/v/k/x/q/j/z

Key efficiency indicators of keystroke dynamics were calculated to evaluate the quality of user recognition. The obtained indicators FRR, FAR, ROC and DET are given in Fig. 6a, b and c, respectively. The threshold of pattern similarity varied in the range from 0.1 to 20 ms; Fig. 6a demonstrates only its informative part. Figure 6a also depicts a half-sum of indicators $(FAR + FRR)/2$, which is an attractive identification in a number of problems.

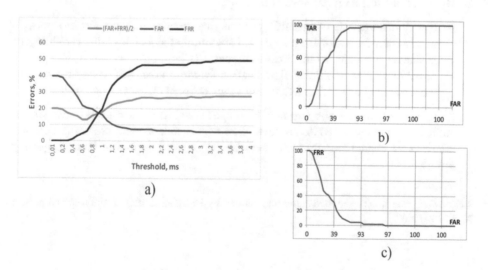

Fig. 6. Mean dwell time in different sessions, ms

The most popular method of recognizing users from their KR is still to estimate the distance (metric) between the current and reference KR samples. This paper analyzed the most common proximity measures: Euclidean Distance and Manhattan Distance. Metric distances were calculated with and without due regard to frequency coefficients.

Table 3 demonstrates the minimum efficiency criteria obtained for different metric distances and a vector criterion for letter frequency.

Table 3. Efficiency criteria

	Euclidean distance		Manhattan distance	
	Without regard to frequency	With regard to frequency	Without regard to frequency	With regard to frequency
FRR	2.8%	2.7%	2.6%	2.4%
FAR	0%	0%	0%	0%
FRR + FAR	13.6%	12.7%	13.3%	12.3%
Algorithm accuracy	86.4%	87.3%	86.9%	87.7%

7   Conclusion

The analysis of recent research of keystroke identification has shown that:

1. Behavioral biometrics-based user identification can be organized with the use of free texts generated in any software application.
2. Dynamic identification has several benefits:

   - continuous upgrade of dynamic patterns that describe user emotional status;
   - generation of a vector for a keyboard rhythm that associates with variability of keystroke dynamics to the greatest extent;
   - possibility to use additional classification features in recognition algorithms, e.g. the frequency of specific letters and other national characteristics of texts.

3. The study of identification results for experienced computer users have shown sufficient accuracy of 87%. At this, accuracy is independent from the selected metric distance and slightly increases when the scaling factors of letter frequency are used.

Acknowledgment. The reported study was funded by RFBR according to the research project № 18-07-01007.

References

1. Yampolskiy, R.V., Govindaraju, V.: Behavioural biometrics: a survey and classification. Int. J. Biom. **1**(1), 81–113 (2008)
2. Pisani, P.H., Lorena, A.C.: Emphasizing typing signature in keystroke dynamics using immune algorithms. Appl. Soft Comput. **34**, 178–193 (2015)
3. Peacock, A., Ke, X., Wilkerson, M.: Typing patterns: a key to user identification. IEEE Secur. Priv. **2**(5), 40–47 (2004)
4. Shanmugapriya, D., Padmavathi, G.: A survey of biometric keystroke dynamics: approaches, security and challenges. Int. J. Comput. Sci. Inf. Secur. **5**(1), 115–119 (2009)
5. Banerjee, S.P., Woodard, D.L.: Biometric authentication and identification using keystroke dynamics: a survey. J. Pattern Recognit. Res. **7**(1), 116–139 (2012)
6. Karnan, M., Akila, M., Krishnaraj, N.: Biometric personal authentication using keystroke dynamics: a review. Appl. Soft Comput. **11**(2), 1565–1573 (2011)
7. Teh, P.S., Teoh, A.B.J., Yue, S.: A survey of keystroke dynamics biometrics. Sci. World J. **2013**, 1–24 (2013)
8. Pisani, P.H., Lorena, A.C.: A systematic review on keystroke dynamics. J. Braz. Comput. Soc. **19**(4), 573–587 (2013)
9. Mondal, S., Bours, P.: A study on continuous authentication using a combination of keystroke and mouse biometrics. Neurocomputing **230**, 1–22 (2016)
10. Vasiliev, V.I., Lozhnikov, P.S., Sulavko, A.E., Eremenko, A.V.: Hidden biometric identification technologies of users of computer systems (review). Inf. Secur. Issues **3** (110), 37–47 (2015)
11. Teh, P.S., Zhang, N., Teoh, A.B., Chen, K.: A survey on touch dynamics authentication in mobile devices review article. Comput. Secur. **59**, 210–235 (2016)
12. Mahfouz, A., Mahmoud, T.M., Eldin, A.S.: A survey on behavioral biometric authentication on smartphones. Res. Artic. J. Inf. Secur. Appl. **37**, 28–37 (2017)
13. Pentel, A.: Predicting age and gender by keystroke dynamics and mouse patterns. In: Proceedings of UMAP 2017 Adjunct, Publication of the 25th Conference on User Modeling, Adaptation and Personalization, pp. 381–385 (2017)
14. Lozhnikov, P., Sulavko, A., Buraya, E., Pisarenko, V.: Authentication of computer users in real-time by generating bit sequences based on keyboard handwriting and face features. Cybersecur. Issues **3**(21), 24–34 (2017)
15. Kim, J., Kim, H., Kang, P.: Keystroke dynamics-based user authentication using freely typed text based on user-adaptive feature extraction and novelty detection. Appl. Soft Comput. **62**, 1077–1087 (2018)
16. Morales, A., Fierrez, J., Tolosana, R., Ortega-Garcia, J., Galbally, J., Gomez-Barrero, M., et al.: KBOC: keystroke biometrics ongoing competition. In: Proceedings 8th IEEE International Conference on Biometrics: Theory, Applications, and Systems, pp. 1–6 (2016)
17. Kochegurova, E.A., Gorokhova, E.S., Mozgaleva, A.I.: Development of the keystroke dynamics recognition system. J. Phys.: Conf. Ser. **803**(1), 1–7 (2017)
18. Alpar, O.: Frequency spectrograms for biometric keystroke authentication using neural network based classifier. Knowl.-Based Syst. **116**, 163–171 (2017)
19. Goodkind, A., Brizan, D.G., Rosenberg, A.: Utilizing overt and latent linguistic structure to improve keystroke-based authentication. Image Vis. Comput. **58**, 230–238 (2017)
20. Alsultan, A., Warwick, K.: Keystroke dynamics authentication: a survey of free-text methods. Int. J. Comput. Sci. Issues **10**(4), 1–10 (2013)
21. Ali, M.L., Monaco, J.V., Tappert, C.C., Qiu, M.: Keystroke biometric systems for user authentication. J. Signal Process. Syst. **86**, 175–190 (2017)

Synthesis and Learning of Socially Significant Behavior Model with Hidden Variables

Aleksandra V. Toropova[1,2(✉)] and Tatiana V. Tulupyeva[1,2]

[1] St. Petersburg Institute for Informatics and Automation of RAS,
St. Petersburg, Russia
alexandra.toropova@gmail.com
[2] St. Petersburg State University, St. Petersburg, Russia

Abstract. Socially significant behavior model with hidden variables is suggested as means for processing unreliable data. Two models with hidden variables are considered: the one with synthesized structure and the other with structure defined expertly. Both models are learned using automatically generated set of data. The models are compared with each other and with the original socially significant behavior model.

Keywords: Bayesian belief network · BBN
Bayesian belief network with hidden variables
Socially significant behavior model

1 Introduction

In many problems of psychology, sociology and other sciences that study human behavior, it becomes necessary to estimate the intensity of socially significant behavior of a person (i.e. behavior that has any effect on other people).

The term behavior rate is determined as the mean number of behavior episodes happened during a particular period. Respondents' self-reports about their behavior are used as the initial data for such an estimation [7].

There are several procedures to obtain data about different type behavior episodes. The "diary method" [2, 4] includes records of all respondents' actions during the day. Then an expert calculates the number of certain type behavior episodes for a given period using obtained data. However, such research is expensive; its organization is difficult and takes a lot of time.

Another way for the number of episodes estimate is behavior rate use. It based on a condition that this behavior is considered as a random process of a particular class. In this case, a new problem arises: we need to estimate behavior rate on the base of the respondent's "one-stage" self-reporting, i.e. on the base of the responses to the questionnaire or the results of the interview. Note that these sets of questions are based on information stored in the respondent's memory, and, of course, the deeper the retrospective, the more difficult to the respondents answer questions and the more they make errors.

A. Abraham et al. (Eds.): IITI 2018, AISC 875, pp. 76–84, 2019.
https://doi.org/10.1007/978-3-030-01821-4_9

Two methods are being used for a self-report nowadays have some disadvantages [3]. The first way is asking a direct question: "How many times have you been doing that for the last month (three months, year)?" the respondents' answers to these questions usually are not correlated with the real values. For example, if we ask ourselves the question: "How many times have I been drinking coffee during the last year?", we hardly give a correct answer. Thus, this method has a low confidence.

The second method—Likert scales—use qualitative, not quantitative values [3, 13]: "never", "rarely", "some-times", "often", "always" and other. It is easy to form the questions and to get the responses, but these answers do not have useful information about the number of episodes: "often" for one person may be "rarely" for another one. In practice, these scales are often quantified, but there is no reliable hypothesis for this quantification, the resulting calculations do not lead to the behavior rate at all. Other likert scale weakness is central tendency bias, because participants may avoid extreme response categories [1].

The interview about last episodes of respondent's behavior solves Likert scales' problems described above. Questionnaire produces data about several behavior episodes only and about extreme intervals between them. As a result, we have to analyze incomplete data. A Poisson random process [12] describes the behavior type.

In [8, 9] an approach to the socially significant behavior rate estimate was proposed based on Bayesian belief networks and data obtained from interviews about last episodes of respondent's behavior. Note that Bayesian belief networks [11] has special advantages: when we consider more complex relations and as a result include more nodes to the model we just have to rearrange our model and the software does most part of calculations. Suggested model is based on data about the last three, minimum and maximum intervals between episodes of the behavior. To portray themselves in more socially favorable light respondents can be not honest and give a priori false data. In addition, because of the work of human memory, respondents can unconsciously make mistakes in their answers. Thus, the intensity estimate calculated using the model of socially significant behavior might turn out to be inaccurate.

In [10] we used evidence coherence diagnosis to evaluate a respondent's reliability to deal with this problem. In this paper we use another approach to obtain more accurate results.

We suggest extending socially significant behavior model with hidden variables. These hidden variables characterize the real data about respondents' last episodes of risky behavior.

We consider two socially significant behavior models with hidden variables (one with defined structure and one learned from the generated set of data) and compare them with the original model.

2   Original Socially Significant Behavior Model

Figure 1 shows a generalized socially significant behavior model $M = (G(V, L), \mathbf{P})$ as a Bayesian belief network [11]. The model structure is represented by the directed graph $G(V, L)$, where $V = \{t_{01}, t_{12}, t_{23}, t_{min}, t_{max}, \lambda, n\}$ is corresponded to the set of nodes, $L = \{(u, v) : u, v \in V\}$ is corresponded to the set of directed links between nodes. In other words, Fig. 1 shows random elements included in the model and relations between them.

Rate is a random variable representing the behavior rate, t_{ij} are random variables characterizing the lengths of the interval between the i-th and j-th to the end episodes. With an assumption that behavior was a Poisson random process random variables were exponentially distributed. The additional information was obtained by including minimum and maximum intervals between episodes (t_{min} and t_{max} respectively). n is a random variable representing a number of episodes during the study period.

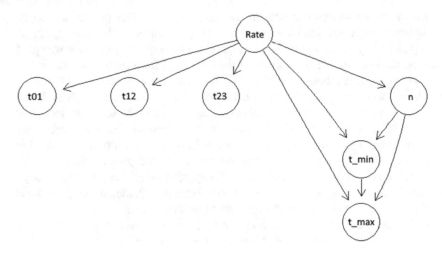

Fig. 1. Socially significant behavior model

3   Socially Significant Behavior Model with Hidden Variables

Figure 2 shows extended risky behavior model. Now we consider t_{01}, t_{12}, t_{23}, t_{min} and t_{max} as hidden variables. The thing is that we actually do not know the real last risky behavior episodes, we only know respondents' answers which can be not correct or even intentionally changed by respondents because of social unapprovement of such behavior. Thus the nodes t_{01}^0, t_{12}^0, t_{23}^0, t_{min}^0 and t_{max}^0 express respondents' answers.

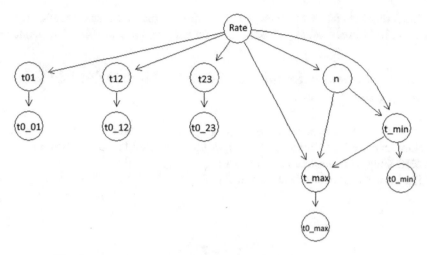

Fig. 2. Socially significant behavior model with hidden variables

4 Data Description

To synthesize and learn the model, a program was developed that generates "episodes of behavior" in accordance with the model theoretical assumptions, i.e. in accordance with the Poisson model of behavior.

Firstly 300 intensity values were generated, corresponding to the values of a random variable distributed by gamma distribution with parameters $k = 1.1$, $\theta = 0.3$ [6]. On the one hand, most of the values are less than 0.5, which is correlated with many examples of real behavior, on the other there are values for all the intervals to which the value λ is divided by sampling.

Then 20 "respondents" are generated for each intensity value. "Respondents" are sequences of points, distances between these points are exponentially distributed with a corresponding intensity value. From each such sequence we select an initial data: the length of the intervals between the last three points, the minimum and maximum interval for a length of 365 "days"; sequences that do not have at least two points in this interval are deleted.

To generate "noisy" respondents' data, at intervals that depend on each of the episodes, we chose random variables in such a way that the deviation between the moment of the interview and the last episode is no more than a quarter, between the last and the penultimate deviation is no more than half, and between the penultimate and the pre-previous one differs less than twice.

To the minimum and maximum intervals we added a normal noise.

This approach to the generation of noisy data makes sense, because it reflects a real life: the earlier an event occurred (in this case the episode of socially significant behavior) the more difficult it is to recall when exactly it happened.

The final set includes 5870 "respondents", also for each of them additional intensity value is known, which allows to compare it with the final estimation.

Similarly, test data was generated for further comparison and evaluation of the models (100 intensity values, 15 sequences for each value, total 1484 "respondents").

5 Learned Socially Significant Behavior Model with Hidden Variables

To build the Bayesian belief network [11] from the generated data the score-based structure learning Hill-Climbing (HC) algorithm was used [5]; the measure of quality was BIC (Bayesian Information Criterion).

The resulting structure is shown in Fig. 3.

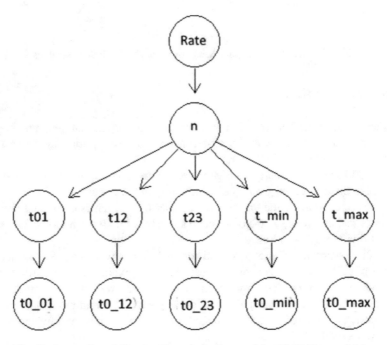

Fig. 3. Learned socially significant behavior model with hidden variables

It should be noted that this structure has a simple interpretation: the intensity of behavior is determined through the number of episodes of behavior that occurred during the considered period (in the current example, 365 days). However, it is impossible to obtain an explicit number of episodes for most examples of behavior. Thus, n is a hidden variable, defined through the initial data, and which determines, in turn, the desired value λ.

6 Comparison of Models

To fully define the models (Figs. 1, 2 and 3), further automatic learning of the parameters of the Bayesian networks was performed.

Let us compare the obtained models. According to the construction algorithm, on the training sample the measure of the quality of the structure shown in Fig. 3 is -55264 for BIC and -53099 for the maximum likelihood measure, the structure shown in Fig. 2 has -70194 for BIC and -53277 for the maximum likelihood measure and the structure shown in Fig. 1 has -53293 for BIC and -37027 for the maximum likelihood measure.

On the testing set the structure in Fig. 3 has -15435 and -13270, the structure in Fig. 2 has -27504 and -13268, the structure in Fig. 1 has -24068 and -10379 for BIC and for the maximum likelihood measure respectively.

However, since the main purpose of the proposed models is to estimate the intensity, the next stage of model comparison is to evaluate the quality of the predictions according to each of these models.

Data showing correspondence between the predicted and the initial intensity values for the proposed models are presented in Table 1 (socially significant behavior model with hidden variables), Table 2 (learned socially significant behavior model with hidden variables) and Table 3 (the original model of socially significant behavior).

Note that, with the given discretization of the variable, the problem of estimating the intensity of individual behavior is a classification problem for 10 disjoint classes.

In Table 4 we can see comparison of all the models using the most common prediction quality metrics, i.e. accuracy, precision and recall. Socially significant behavior model with hidden variables defined expertly has better quality scores than the other models, the original model has the lowest scores, but they differ slightly.

Table 1. Socially significant behavior model with hidden variables prediction.

		Intensity rate									
		$\lambda^{(1)}$	$\lambda^{(2)}$	$\lambda^{(3)}$	$\lambda^{(4)}$	$\lambda^{(5)}$	$\lambda^{(6)}$	$\lambda^{(7)}$	$\lambda^{(8)}$	$\lambda^{(9)}$	$\lambda^{(10)}$
Initial value	$\lambda^{(1)}$	2	6	3	3	1	0	0	0	0	0
	$\lambda^{(2)}$	0	5	4	3	0	2	0	0	0	0
	$\lambda^{(3)}$	0	7	23	18	4	7	1	0	0	0
	$\lambda^{(4)}$	0	2	12	22	5	13	6	0	0	0
	$\lambda^{(5)}$	0	0	9	14	8	19	24	1	0	0
	$\lambda^{(6)}$	1	0	0	8	2	7	22	5	0	0
	$\lambda^{(7)}$	0	0	2	6	5	30	166	130	6	0
	$\lambda^{(8)}$	0	0	1	0	0	3	48	464	69	0
	$\lambda^{(9)}$	0	0	0	0	0	0	1	141	92	21
	$\lambda^{(10)}$	0	0	0	0	0	0	0	4	12	14

Table 2. Learned socially significant behavior model with hidden variables prediction.

		Intensity rate									
		$\lambda^{(1)}$	$\lambda^{(2)}$	$\lambda^{(3)}$	$\lambda^{(4)}$	$\lambda^{(5)}$	$\lambda^{(6)}$	$\lambda^{(7)}$	$\lambda^{(8)}$	$\lambda^{(9)}$	$\lambda^{(10)}$
Initial value	$\lambda^{(1)}$	2	6	5	2	0	0	0	0	0	0
	$\lambda^{(2)}$	0	3	4	6	0	1	0	0	0	0
	$\lambda^{(3)}$	3	2	22	17	7	8	1	0	0	0
	$\lambda^{(4)}$	0	5	11	20	4	10	9	1	0	0
	$\lambda^{(5)}$	0	2	2	20	7	17	27	0	0	0
	$\lambda^{(6)}$	0	0	2	4	1	11	23	4	0	0
	$\lambda^{(7)}$	0	0	1	7	3	27	154	149	3	1
	$\lambda^{(8)}$	0	0	1	0	0	2	45	468	68	1
	$\lambda^{(9)}$	0	0	0	0	0	0	1	148	90	16
	$\lambda^{(10)}$	0	0	0	0	0	0	0	3	14	13

Table 3. Original socially significant behavior model prediction.

		Intensity rate									
		$\lambda^{(1)}$	$\lambda^{(2)}$	$\lambda^{(3)}$	$\lambda^{(4)}$	$\lambda^{(5)}$	$\lambda^{(6)}$	$\lambda^{(7)}$	$\lambda^{(8)}$	$\lambda^{(9)}$	$\lambda^{(10)}$
Initial value	$\lambda^{(1)}$	4	5	5	1	0	0	0	0	0	0
	$\lambda^{(2)}$	1	3	5	3	0	2	0	0	0	0
	$\lambda^{(3)}$	0	3	17	29	2	7	2	0	0	0
	$\lambda^{(4)}$	0	2	2	32	2	15	7	0	0	0
	$\lambda^{(5)}$	0	0	2	22	2	26	23	0	0	0
	$\lambda^{(6)}$	0	0	2	8	0	7	24	4	0	0
	$\lambda^{(7)}$	0	0	0	5	1	33	165	139	2	0
	$\lambda^{(8)}$	0	0	1	0	0	0	50	456	78	0
	$\lambda^{(9)}$	0	0	0	0	0	0	0	152	78	25
	$\lambda^{(10)}$	0	0	0	0	0	0	0	2	15	13

Table 4. Comparison of common prediction quality metrics.

	Accuracy	Average accuracy	Precision	Recall
Original model	0.52	0.904	0.52	0.32
Model with hidden vars	0.54	0.908	0.54	0.35
Learned model with hidden vars	0.53	0.906	0.53	0.33

7 Conclusion

Socially significant behavior model with hidden variables was suggested as means for processing unreliable data. Two models with hidden variables were considered: the one with synthesized structure and the other with structure defined expertly. Both models were compared with each other and with the original model.

It was shown that the model with hidden variables gives more accurate predictions in comparison with the original model and learned socially significant behavior model with hidden variables.

Note, however, that in general the quality of all the models is quite high. Thus, of the proposed models can be used to solve practical problems. The choice can be conditioned by a specific task.

The results obtained can be used in various scientific fields, the subject of which is a human behavior, for example, in sociology and epidemiology.

Acknowledgments. The research was carried out in the framework of the project on state assignment SPIIRAS No. № 0073-2018-0001, with the financial support of the RFBR (project №18-01-00626).

References

1. Bertram, D.: Likert scale are the meaning of life. CPSC 681-Topic report. http://poincare. matf.bg.ac.rs/~kristina/topic-dane-likert.pdf. Accessed 21 Apr 2018
2. Bolger, N., Davis, A., Rafaeli, E.: Diary methods: capturing life as it is lived. Annu. Rev. Psychol. **54**, 579–616 (2003)
3. Frowler, F.J.: Improving survey questions: design and evaluation. In: Applied Social Research Methods Series, vol. 38, p. 200. Sage Publications, Thousand Oaks (1995)
4. Graham, C., Catania, J., Brand, R., Duong, T., Canchola, J.: Recalling sexual behavior: a methodological analysis of memory recall bias via interview using the diary as the gold standard. J. Sex Res. **40**(4), 325–332 (2003)
5. Scutari, M.: Learning Bayesian networks with the bnlearn R package. arXiv preprint arXiv: 0908.3817 (2009)
6. Suvorova, A.V., Tulupyev, A.L.: Bayesian belief network structure synthesis for risky behavior rate estimation. In: Informatsionno-upravliaiushchie sistemy [Information and Control Systems] (1) 116–122 (2018) (In Russian) https://doi.org/10.15217/issn1684-8853. 2018.1.116
7. Suvorova, A.V., Tulupyev, A.L., Sirotkin, A.V.: Bayesian belief networks in problems of estimating the intensity of risk behavior. J. Russ. Assoc. Fuzzy Syst. Soft Comput. **9**(2), 115–129 (2014)
8. Suvorova, A.V., Tulupyeva, T.V., Tulupyev, A.L., Sirotkin, A.V., Pashchenko, A.E.: Probabilistic graphical models socially significant behavior of the individual, taking into account incomplete information. Proc. SPIIRAS **3**(22), 101–112 (2012)
9. Suvorova, A.V.: Socially significant behavior modeling on the base of super-short incomplete set of observations. Inf.-Meas. Control. Syst. **9**(11), 34–38 (2013)
10. Toropova, A.: Data coherence diagnosis in BBN risky behavior model. In: *Proceedings of the First International Scientific Conference Intelligent Information Technologies for Industry* (IITI'16), pp. 95–102. Springer International Publishing (2016)

11. Tulupyev, A.L., Sirotkin, A.V., Nikolenko, S.I. Bayesian Belief Networks: Logical-probabilistic Inference in the Acyclic Directed Graph, 400 p. Publishing House, St. Petersburg University Press, St. Petersburg (2009)
12. Van Vliet, C., Van der Ploeg, C., Kidula, N., Malonza, I., Tyndall, M., Nagelkerke, N.: Estimating sexual behavior parameters from routine sexual behavior data. J. Sex Res. **35**(3), 298–305 (1998)
13. Wardle, J.: Eating style. A validation study of the dutch eating behaviour questionnaire in normal subjects and women with eating disorders. J. Psychosom. Res. **31**(2), 161–169 (1987)

Adaptation of the Nonlinear Stochastic Filter on the Basis of Irregular Exact Measurements

Marianna V. Polyakova[1]([✉]), Sergey V. Sokolov[1],
and Anna E. Kolodenkova[2]

[1] Rostov State Transport University (RSTU), Rostov-on-Don, Russia
poliakova.marianna@yandex.ru
[2] Samara State Technical University (SamSTU), Samara, Russia
anna82_42@mail.ru

Abstract. The problem of adaptive discrete nonlinear filtration on the basis of irregular exact measurements is solved analytically. The possibility of irregular exact measurements allows to define precisely in a step of their emergence the coefficient of strengthening and coefficient of adaptation of the filter providing a zero error of estimation. Similar reorganization of parameters of an algorithm of estimation increases its convergence sharply that is illustrated by the corresponding example.

Keywords: Irregular exact measurements · Complexed measuring systems
The adaptive nonlinear (generalized) Kallman's filter · Strengthening coefficient
Adaptation coefficient

1 Introduction

The purpose of the integration of various information and measuring systems (for example, navigation) is the use of information on the same or functionally connected parameters received from various measuring instruments, for increase of accuracy and reliability of the processed measurements (for example, when determining the location of an object) [1]. So, practically in all modern navigation complexes inertial measurements when the main information arrives from accelerometers and gyroscopes are demanded. The Inertial Navigation Systems (INS) are self-sufficient, additional measurements aren't required for them, but at an autonomous operating mode of INS is unlimited and with a high speed errors of positioning increase and collect [2]. Applying INS this factor forces to use navigation correction, one of main types of which for INS are satellite measurements which main lack is their irregular loss leading to a possibility only of incidental correction [3–5].

Today the mathematical apparatus of Kalman's filtration which allows to estimate a vector of a condition of an object for noisy measurements with the minimum mean

This work is performed with assistance of a grant of the Russian Federal Property Fund № 18-07-00126.

A. Abraham et al. (Eds.): IITI 2018, AISC 875, pp. 85–91, 2019.
https://doi.org/10.1007/978-3-030-01821-4_10

square mistake is the most widely used for the solution of problems of complex processing of stochastic information. When applying the filter of Kallman his parameters, as a rule, differ from calculated ones. It often leads to the inadmissible growth of errors of estimation. Due to this circumstance there is a need of constant correction of parameters of the filter – that is to his adaptation. One of the most effective approaches to creation of adaptive filters is the approach on the basis of a possibility of use exact, generally irregular, measurements of a vector of a condition of an object (for example, for INS it can be satellite measurements, reperny tags of road infrastructure, etc.) [6–8].

Now adaptive correction of the filter of Kallman is carried out by direct replacement of the current estimates of a vector of a condition with the exact measurements corresponding to them with invariable parameters of an algorithm of estimation. However according to this approach the growth of errors of estimation on the time interval is determined by the moment of the following exact measurement, it doesn't decrease [6–8].

In this regard there is a problem of adaptation parameters directly of the algorithm of estimation on the received irregular exact measurements that will allow on time intervals between them increase significantly the accuracy of estimation of a vector of a condition of an object.

2   Problem Definition

Exact measurements are performed in irregular discrete timepoints, and to solve our problem we will consider further discrete option of the most often used estimation algorithm – nonlinear (generalized or expanded) the filter of Kallman. Assessment of a vector of a condition of the X_k system in k timepoint is determined by a formula [9]:

$$\hat{X}_k = F_k(\hat{X}_{k-1}) + K_k(Z_k - h_k[F_k(\hat{X}_{k-1})]), \tag{1}$$

where $\hat{X}_k$ – assessment of the vector of a condition of the system in k timepoint;

$F_k(X_{k-1})$ – vector nonlinear function of system;
$F_k(\hat{X}_{k-1})$ – the extrapolated assessment of a vector of a condition;
K_k – coefficient of strengthening of the filter;
Z_k – a vector of measurements;
$h_k(X_k)$ – vector function of observation of parameters of a state;
$h_k[F_k(\hat{X}_{k-1})]$ – the extrapolated assessment of a vector of observations.

The vector of measurements is determined by a formula:

$$Z_k = H_k \cdot X_k + V_k,$$

where V_k – the casual aligned Gaussian sequence with a dispersive matrix of R_k.

Due to the fact that the formulated task is a problem of adaptation of an algorithm of assessment, we use further adaptive version of the filter of Kallman in which the coefficient of strengthening is defined as [10, 11]:

$$K_k = P_{k/k-1} \cdot H_k^T \cdot (H_k \cdot P_{k/k-1} \cdot H_k^T + R_k)^{-1}, \tag{2}$$

$$P_{k/k-1} = \mu_k \cdot \Phi_k \cdot P_{k-1} \cdot \Phi_k^T + Q_k, \tag{3}$$

$$P_k = (E - K_k \cdot H_k) \cdot P_{k/k-1},$$

where $H_k = \left. \frac{\partial h_k(X)}{\partial X} \right|_{X=F_k(\hat{X}_{k-1})}$; $\Phi_k = \left. \frac{\partial F_k(X)}{\partial X} \right|_{X=\hat{X}_{k-1}}$;

$P_{k/k-1}$ – the extrapolated covariation matrix;
μ_k – diagonal matrix of coefficients of adaptation of the filter;
P_k – covariation matrix in k timepoint;
Q_k – a dispersive matrix of noise of system;
E – single matrix.

Based on the presented filter intensification coefficient form, we will formulate a problem of its adaptation on precise measurements as a problem of finding of a matrix μ_k from a coincidence condition at the appropriate time of estimates $\hat{X}_k$ (1) with precise state vector of the system X_k (precise measurements).

3 Solution of a Task

For the solution of an objective we will use further the complete expression of the Kalman's intensification coefficient received by substitution (3) in (2):

$$K_k = (\mu_k \cdot \Phi_k \cdot P_{k-1} \cdot \Phi_k^T + Q_k) \cdot H_k^T$$
$$\times \left[H_k \cdot (\mu_k \cdot \Phi_k \cdot P_{k-1} \cdot \Phi_k^T + Q_k) \cdot H_k^T + R_k \right]^{-1},$$

where we will enter designation $(\mu_k \cdot \Phi_k \cdot P_{k-1} \cdot \Phi_k^T + Q_k) \cdot H_k^T = \gamma_k$ for the convenience of the subsequent decision and we will write down expression of coefficient K_k as follows

$$K_k = \gamma_k \cdot (H_k \cdot \gamma_k + R_k)^{-1}. \tag{4}$$

In this case we receive the Eq. (1):

$$\hat{X}_k - F_k(\hat{X}_{k-1}) = \gamma_k (H_k \gamma_k + R_k)^{-1} (Z_k - h_k[F_k(\hat{X}_{k-1})]) \tag{5}$$

and relative to the matrix μ_k, entering the matrix γ_k, it represents the nonlinear vector equation which solution by traditional numerical methods demands multiple application of very expensive procedure of an inversion of matrix. For a possibility of its analytical decision we will carry out the following constructions.

As under the terms of a problem of $X_k - F_k(\hat{X}_{k-1}) = X_*$, $Z_k - h_k[F_k(\hat{X}_{k-1})] = Z_*$, entering designations $\hat{X}_k = X_k$ we will present the Eq. (5) as:

$$X_* = \gamma_k \cdot (H_k \cdot \gamma_k + R_k)^{-1} \cdot Z_*. \tag{6}$$

Multiplying both members of Eq. (6) by an inverse matrix $[\gamma_k(H_k\gamma_k + R_k)]^{-1}$, we have:

$$(H_k \cdot \gamma_k + R_k) \cdot \gamma_k^{-1} \cdot X_* = Z_*,$$

or

$$R_k \cdot \gamma_k^{-1} \cdot X_* = Z_* - H_k \cdot X_*. \tag{7}$$

Continuing further multiplication of both members of Eq. (7) to matrix $\gamma_k \cdot R_k^{-1}$, we will lead it to the form, the linear relatively γ_k, and, therefore, and concerning a μ_k:

$$X_* = \gamma_k \cdot R_k^{-1} \cdot (Z_* - H_k \cdot X_*). \tag{8}$$

For the final decision of this equation relatively μ_k we will open expression $\gamma_k = (\mu_k \cdot \Phi_k \cdot P_{k-1} \cdot \Phi_k^T + Q_k) \cdot H_k^T$ and we will receive:

$$X_* = (\mu_k \cdot \Phi_k \cdot P_{k-1} \cdot \Phi_k^T \cdot H_k^T \cdot R_k^{-1} + Q_k \cdot H_k^T \cdot R_k^{-1}) \times (Z_* - H_k \cdot X_*). \tag{9}$$

Entering for simplification of the subsequent conclusion of designation $\Phi_k \cdot P_{k-1} \cdot \Phi_k^T \cdot H_k^T \cdot R_k^{-1} = \Delta_{k,k-1}$ и $Q_k \cdot H_k^T \cdot R_k^{-1} = U_k$, let's present the Eq. (9) in the form:

$$X_* = \mu_k \cdot \Delta_{k,k-1} \cdot (Z_* - H_k \cdot X_*) + U_k \cdot (Z_* - H_k \cdot X_*). \tag{10}$$

Denoting vectors $\Delta_{k,k-1} \cdot (Z_* - H_k \cdot X_*) = \Delta_*$, $U_k \cdot (Z_* - H_k \cdot X_*) = U_*$, let's lead the Eq. (10) to the following form:

$$X_* - U_* = \mu_k \cdot \Delta_*. \tag{11}$$

The Eq. (11) easily allows the analytical decision concerning all elements of a scalar matrix μ_k if consider a possibility of submission of the work $\mu_k \cdot \Delta_*$ in the form $\Delta_{* \, diag} \mu_{K \, vect}$, where

$$\Delta_{* \, diag} = \begin{vmatrix} \Delta_{*1} & 0 & 0 & ....0 \\ 0 & \Delta_{*2} & 0 & ....0 \\ & & .... & \\ 0 & 0 & 0 & ....\Delta_{*n} \end{vmatrix} ; \mu_{k \, vect} = \begin{vmatrix} \mu_{k1} \\ \mu_{k2} \\ \\ \mu_{kn} \end{vmatrix},$$

$\Delta_{*i}, \mu_{ki}, i = 1, \ldots, n$ – elements, respectively, of a vector Δ_* and a matrix μ_k.

We have the required equation of a vector of elements of coefficient of adaptation in a form in this case:

$$\Delta_{*diag}^{-1}(X_* - U_*) = \mu_{k\,vect},\tag{12}$$

where Δ_{*diag}^{-1} – the return matrix which is easily calculated analytically due to its diagonal form:

$$\Delta^{-1}_{*diag} = \begin{vmatrix} \dfrac{1}{\Delta_{*1}} & 0 & 0 & .... & 0 \\[2mm] 0 & \dfrac{1}{\Delta_{*2}} & 0 & .... & 0 \\[2mm] & & .... & & \\[2mm] 0 & 0 & 0 & .... & \dfrac{1}{\Delta_{*n}} \end{vmatrix}.$$

Thus, the found solution of the nonlinear vector Eq. (5) in the form of (12) allows to solve analytically an objective of adaptation of a nonlinear Kalman's algorithm of assessment on precise measurements.

There is a possibility of the alternate use of the received μ_k after determination of the current value of a matrix: or to keep its value in expression (3) invariable before receiving the following vector of precise measurements, or to leave a single matrix μ_k also before receiving a new vector of precise measurements, considering that the value of a matrix P_k is already corrected. It is obvious that this issue has to be resolved for each concrete object taking into account its features [12].

Let's consider the first option of adaptation as the most common in the example given below illustrating overall effectiveness of the offered approach.

4 Example

The equations of the movement of the center of mass of an object along the sphere of Earth under the measurement of the current parameters of the movement have such form:

$$\dot{\varphi} = \frac{V_y}{(r+h)},\tag{13}$$

$$\dot{\lambda} = \frac{V_x}{\cos\varphi(r+h)},\tag{14}$$

where φ, λ – the geographical latitude and longitude of an object; r – Earth radius; V_y, V_x – object speed projections to the corresponding axes of geographical system of coordinates; h – object height.

Further we believe in an example that an object moves from a point with coordinates $\varphi_0 = 0,85$ radian, $\lambda_0 = 0,35$ radian, during time interval [0; 1500 c] with a constant speed of $V = 20$ м/c on a loxodromic track with an azimuthal corner $A = 0,2$ radian on the Earth's surface.

The navigation system of an object is complexed, on the basis of weak integration of inertial and satellite NS. With an interval 20, 15 and 30 s. in NS the satellite measurements which are considered exact come at other timepoints with a step $\tau = 0,1$ s. measurements of inertial NS on channels λ and φ are used. The matrix of noise of measurements has a value: $R_k = \begin{vmatrix} 3,8 \cdot 10^{-10} & 0 \\ 0 & 8,6 \cdot 10^{-10} \end{vmatrix}$.

Numerical model of operation of process of estimation of state vector (13), (14) with use of a traditional nonlinear Kalman's filter and the adaptive filter offered in this version was carried out. In the first case in the result of process of model operation the errors of estimation are the following ones: on the latitude – 72 m and on longitude – 64 m, and in the second case, respectively 1,35 m – on latitude and 0,9 m – on longitude.

5 Conclusion

Low values of errors of estimation under the organization of the adaptive filter according to the developed technique (1,35 m – on latitude and 0,9 m – on longitude) allow to draw a conclusion about expediency of application of this approach. This algorithm can be applied in information measuring systems of a wide class owing to the simplicity and accuracy.

References

1. Surkov, V.O.: The directions of upgrading of functioning of navigation systems for the relative frame land objects at the solution of navigation tasks. Young Sci. **13**, 209–211 (2015). https://moluch.ru/archive/93/20939/. Accessed 26 Mar 2018
2. Sokolov, S.V., Pogorelov, V.A.: Bases of Synthesis of Multistructural Strapdown Navigation Systems: The Manual. Fizmatlit (2009). 184 p
3. Poddubny, V.V.: Restrektivnaya filtration in navigation systems. Bull. Tomsk. State Univ. **275**, 202–215 (2002)
4. Solovyov, Y.A.: Systems of Satellite Navigation. Jeko-Trendz, Moscow (2000). 270 p
5. Litvin, A.M., Malyugina, A.A., Miller, A.B., Stepanov, A.N., Chirkin, D.E.: Types of mistakes in inertial navigation systems and methods of their approximation. Inf. Al Process. **14**(4), 326–339 (2014)
6. Reznichenko, V.I., Maleev, P.I., Smirnov, M.Yu.: Satellite correction of stretching conditions of sea objects. Navig. Hydrogr. **27**, 25–32 (2008)
7. Tsyplakov, A.A.: Introduction to model operation in the state space. Quantile **9**, 1–24 (2011)
8. Sjolin, V.A.: System of inertial sensors for navigation in rooms. The youth scientific and technical messenger. MSTU of N.E. Bauman. Online Magazine **4** (2015). http://sntbul.bmstu.ru/doc/778220.html. Accessed 04 July 2017
9. Polyakova, M.V., Sokolova, O.I.: Increase of accuracy of the adaptive filtering on the basis of use of non-cycle exact observations. The collection of reports of VII International Scientific and Technical Conference "Technology of Development of Information Systems (TDIS)", pp. 61–64. SFU publ., Taganrog (2016)

10. TDoA based UGV Localization using Adaptive Kalman Filter Algorithm [An electronic resource]. http://www.sersc.org/journals/UCA/vol2_no1/Lpdf. Accessed 10 June 2017
11. Adaptive Kalman filter for navigation sensor fusion [An electronic resource]. http://ebookbrowse.com/gdoc.php?id326155547&url=0761dcbcc180093566925b023b9f88e7. Accessed 15 June 2017
12. Polyakova, M.V., Bayandurova, A.A., Sokolov, S.V.: Use of irregular exact measurements in a problem of an adaptive filtration. Adv. Intell. Comput. **679**, 379–387 (2018)

Pattern Recognition and Emotion Modeling

Visual Analysis of Information Dissemination Channels in Social Network for Protection Against Inappropriate Content

Anton Pronoza, Lidia Vitkova, Andrey Chechulin$^{(\boxtimes)}$,
and Igor Kotenko

St. Petersburg Institute for Informatics and Automation of the Russian Academy
of Sciences, 14-th Liniya, 39, St. Petersburg 199178, Russia
pronoza@gmail.com, iskinlidia@gmail.com,
{chechulin, ivkote}@comsec.spb.ru

Abstract. The paper presents an approach to a visual analysis of links in social networks for obtaining information on channels of inappropriate information dissemination. The offered approach is based on the formation of the knowledge base about communication between users and groups, interactive display of propagation paths of inappropriate information and the visual analysis of the received results to detect sources and repeaters of inappropriate information. The interconnections of users and groups in social networks allow the construction of connectivity graphs, and the facts of the transmission of inappropriate information through these channels provide an opportunity to identify the ways of malicious content dissemination. The results of experiments confirming the applicability of the proposed approach are outlined.

Keywords: Social network · Visualization · Protection against information
Inappropriate information · Communication graph · Visual analytics

1 Introduction

Modern problems in the field of information and psychological security of society and the state require experts to develop new approaches for monitoring and countering threats emanating from the information space of social networks. Modern social networks are not only a means of communication but also a tool for extracting knowledge about subjects, objects, and connections between users and groups. It is also a platform for information dissemination analysis. Analysis of dissemination of inappropriate information in social networks is currently an urgent task. Modern investigations, such as "Social networks in Russia - figures and trends," suggest that the number of "writing" authors only in May 2017 in the social network "VKontakte" was 25.7 million, while they generated 310 million messages [1].

In fact, one of the areas of information security today is protection from unwanted information. It is the emergence of a variety of sources containing unwanted content, messages with information that violate the law that poses new challenges for information security specialists. Therefore, in social networks, one of the tasks of the analysis of information flows is to detect repeaters, distribution channels and sources of information. Visual analysis of information distribution channels in social networks allows to identify the main information flows containing undesirable content. In this paper, the authors propose an approach for analysis of information objects in social networks (users, groups, public pages, events) and channels for information dissemination on the basis of visual analytics.

Based on information about reposts in social networks, the presented approach allows us to identify the sources of inappropriate information distribution. Using data of reposts and views of information objects, the approach allows you to display the ways of disseminating information in social networks and their characteristics. Applicability of the proposed approach is confirmed by the results of the conducted experiments. The novelty of the approach consists in the method of collecting data on information objects and the method of displaying the obtained results, which allows us to conduct visual analysis efficiently. It is assumed that the proposed approach can be used to increase the level of protection of users in social networks from inappropriate information. The contribution of the paper is lies in the approach for intelligent analysis of information distribution channels.

It should be noted that the proposed approach to visualization allows us to characterize the channels of distribution of undesirable information at the same time as a medium of transmission, and as objects with their own characteristics.

So, the interrelationships of objects in social networks allow constructing graphs, and visualizing ways of information dissemination through the same repeaters to identify the ways of inappropriate content. This research contribution lies in enhancing the protection against inappropriate information in the social network space by the intelligent analysis of information distribution channels.

The paper is organized as follows. The second section provides a review of related works. The third and fourth sections describe the proposed methods for data collection and visualization. In the fifth section, the results of the conducted experiments and an example of a visual analysis of information distribution are presented. The sixth section presents an analysis of the results obtained. The seventh section completes the paper by summing up the results and determines the future research.

2   Related Work

This paper is a continuation of the previous investigation of the authors. A general approach to the classification of web pages was presented in [2]. It is based on the analysis of various aspects of web pages. The main aspect that was used to determine the category was the text of the web page. However, as a result of the experiments, the authors concluded that the use of text is not suitable for the analysis of such categories of websites as "news", "blogs", "social networks," etc. This is due to the fact that the web pages of these categories can contain simultaneously texts on various topics. To identify such categories, it was suggested to use structural features of web pages [3]. A general approach combining analysis of text content, structural features, URL addresses is presented in [4]. It allows us to determine the category of a web page with high enough accuracy, which was proved by the results of experiments. However, when working with the category "social networks" it was found that an important information is not only the category of a particular undesirable information object (messages, groups, etc.), but also ways to disseminate this information.

Identifying sources and repeaters of inappropriate information can significantly improve the effectiveness of counteraction methods. The main area of the related works is the analysis of social networks based on streaming methods and on concepts SNA (Social Network Analysis). Zaden et al. [5] found that analyzing the missing nodes in the connection that arise or may arise in social networks can improve the efficiency of predictions and the modeling of events. This paper relies on the analysis of fuzzy structured dynamical systems. Although graphs are also used, the authors' goal is not to analyze the flows or improve the quality of information representation.

Among the studies aimed at analyzing the interrelationships of objects, it is worth mentioning the papers [6, 7]. The authors developed a stack of technologies for the analysis of social networking sites. The main components of the stack are searching for implicit user groups, identifying users of various social networks, defining demographic attributes, measuring information impact between users in social networks, and generating graphs that visualize the structures of user groups.

Que et al. [8] investigate methods of countering the agitation of extremist organizations in Twitter. The goal of the authors is to find an approach that would cover the maximum number of objects already carried away by the ideas of extremism.

Such research direction as SNA is actively developing. SNA is a way of studying social networks as a set of entities between which there are certain relationships. In SNA, the magnitude of the positive correlation of a social network node is characterized by such indicators as a degree, an eigenvector, a measure of closeness and centrality. In the process of analysis, the most influential nodes and their connections with other objects are singled out. Opsahl et al. [9] describe social connectivity using three different indices: the degree centrality, the proximity to the center of the network, the relationship with the center of other nodes.

Hickethier et al. [10] hypothesize that in a social network it is possible to divide the users into clusters of an informal organization and visualize these clusters. Visualization based on clustering raises the level of information perception when you receive a graph of the social network, allows you to separate the graph into color groups and highlight the main connections of objects. The authors introduce in the network interaction such key terms as "owner", "designer" and "contractor".

Sudhahar et al. [11] offer an approach to the automated analysis of presidential elections in the US using big data and network analysis. As a result of the work of the parser developed by them, a network with positive and negative facets between the entities and a visual map of the dissemination of information on the main candidates for the pre-election race and their division in the social network are built.

Thus, as the scale of social media and the number of users grow, the analysis of social networks becomes an important tool in the field of social computing. Inappropriate information often spreads on the user's pages and in their posts. Visualization of distribution paths can help to automate the analysis of information flows and enhance the level of protection against inappropriate information in social networks. Analysis of related works shows that the main focuses are qualitative or quantitative characteristics of the links of nodes in social networks, the clustering of the data obtained, system-atization, and storage. At the same time, the investigation of information dissemination channels to enhance the protection of social network users from inappropriate infor-mation are not mature research area.

3   Data Gathering

One of the major steps in social network analysis is a development of the data gathering algorithm and the architecture of the storage subsystem. Storing (caching) the infor-mation collected allows to reduce load on social network when making extra experi-ments and avoid a great number of locks. The possibility to store heterogencous data with different levels of hierarchy is an important requirement to the storage subsystem that reduces the dependency from the data gathering algorithm. The data gathering algorithm searches for required data in storage subsystem and interconnects with social network servers only if there is not any or the information is outdated.

In terms of social network, the algorithm can collect the information about its objects (a user, a group, a post, a photo, etc.) or about relationships between them (the objects are "friends", the objects are in the same group, the objects are authors, etc.). It needs to be mentioned, that it is possible to collect data about the object of the social network in general (a user profile, a group description, etc.) and about particular information objects (e.g. a user's post has such attributes as text, number of views, list of attached documents, etc.) that represent the information space of the object in social network (e.g. it's "wall", discussions, etc.).

The relationship called "repost" exists in many social networks. It is a direct copy of an information object from parent object to a new one, saving references both on the new and the parent object. This relationship seems to be one of the most effective ways of infor-mation dissemination channels detection in social networks, mostly because it contains the information about its source and receiver and the content of the information object remains unchanged. In most cases, the sources and receivers of information objects are users and groups of social network. Depending on the information object location in repost chain, the source or the object can be missing.

It should be mentioned that there are other techniques to detect information dis-semination channels that are not based on its functional capabilities but on the attributes of information objects themselves. For example, if an information object represents the same unique text or image, which appeared on different "walls" at different time, then it is possible that there exists an information dissemination channel between those "walls" and the order of information spreading is defined by the chronology of appearance information objects. The searching for such channels requires a lot of resources, so it remains relevant to search for possible information dissemination channels using basic functional capabilities of social network first.

Thus, the first step for data gathering to analyze information spreading channels is to collect data about "repost" relationships between two objects of a social network.

The data gathering algorithm proposed in this paper has three steps: (1) Detection of all information objects I_r, that are in "repost" relationship with other objects for the specified time period P. (2) Gathering information about all the sources $S = \{s_i\}$ of

information objects $i \in I_r$ (looking for information objects in "repost" chain "down").
(3) Iterative information gathering about all receivers of information objects: $T(s_i) = \cup\, T(s_k)$ (looking for information objects in "repost" chain "up").

At the end of data gathering all information objects union inside the object of social network they belong to. Thus, a directed graph is constructed. Its nodes are such objects of social network as a user or a group, and its edges represent the "repost" relationship between them. The direction of an edge represents the movement of an information object from its source to its receiver.

The constructed graph allows for a number of additional definitions. The *source* of information is the first source in the spreading chain of information object. The *repeater* of information is any link in the chain except the source and its extreme links.

While transferring information from its source to each link in the chain one can observe the fact of "attenuation" or "distortion" of original sense when the "reposted" information appears in the information space of sequential object of social network.

To evaluate the fact of "attenuation" or "distortion" of original sense in new information space the keywords method [12] is proposed.

The evaluation algorithm consists of three steps: (1) Text data from all initial information objects is combined into one text array for which a set of keywords K_I is calculated. (2) For each receiver j all text data, found in information objects that were created in information space of the receiver for the specified time period P, is combined into one text array for which a set of keywords K_j is calculated. (3) For each receiver j a number of keywords that exists both in sets K_I and K_j is calculated: $K_j' = |K_j \cap K_I|$. Calculated value K_j' indicates the degree of similarity between information spaces of sources and receivers. For convenience, the value K_j' can be represented as a percentage.

Collected data is suitable for permanent storage and machine calculations, however, in raw form, they are extremely difficult to understand and analyze by a person.

4 Visualization

To study the results obtained at the end of the data collection phase, it is advisable to develop a graphical scheme of their presentation, providing the possibility of visual analysis. There are many ways of related heterogeneous objects graphical representation [13], however, to represent the objects of the social network and their relationships it is advisable to use a graph. In particular, to represent nodes one can use such attributes as form, color and size. The representation on relationships can be provided using such edges attributes as direction, weight and color. Additional information can be represented in node's label if needed.

Collected data can also be used to calculate additional values that are relevant to information saturation of the visualization. It should be noted that all values, calculated for the object of a social network $q \in Q$ or the "repost" relationship $r \in R$, will depend on the specified time period P. For example, using "a number of views" indicator of the information object one can obtain the information about the audience that saw this object (audience coverage). The bigger the average value of views $V_q(t)$ of the information object is, the bigger audience is influenced by the object q.

Another important indicator of the information object q is the value of generated content uniqueness $U_q(t)$. This value is calculated as the ration of the number of information objects that are in "repost" relationships with other information objects, to the number of all information objects. This indicator helps to determine whether this object of a social network is an information consumer, aggregator or repeater.

An indicator $F_r(t)$ evaluates information flow between two objects in a social network. It is equal to the difference between the number of "reposts" between those objects. The direction of information flow depends on the sign of the indicator.

In this paper the method of calculating the indicator $C_q(t)$, which means the "attenuation" or "distortion" of original sense in the information space of object q of a social network, is proposed. Using this indicator one can find how the context of original information changes while moving from one link in the "repost" chain to another.

Using all given indicators one can point out the following types of objects of a social network involved in information interaction: *the information source* – the object that is a starting point for information content of other objects and has high level of unique content; *the information repeater* – the object that is repeated by other objects involved in information interaction with low level of unique content; *the information aggregator* – the object with low level of unique content but big audience; *the information consumer* – the object with low level of unique content and poor audience with no further distribution of the content.

When visualizing data, it is necessary to balance the informativeness and simplicity of perception of the output data [14]. That is why it is important to develop the data visualization scheme and filtering methods.

A set of proposed indicators and the ways of their visualization on a graph are as follows: Vertex (size) – Average number of views $V_q(t)$; Vertex (shape) – Type of the object q of a social network; Vertex (color) – Generated Content uniqueness $U_q(t)$; Vertex (label) – "Attenuation" or "distortion" of original sense $C_q(t)$; Edge (weight) – Information flow $F_r(t)$; Edge (direction) – Direction of information flow; Edge (color) – Generated content uniqueness $U_r(t)$.

An edge's direction can be considered as either the information about the source and the receiver of information objects or the information about the influence of one object on another. Further in this paper it is assumed that an edge's direction is an information influence of the source on the receiver.

The technique described allows one to show all the elements of the information spreading scheme in a clear form. The approach to visualize information spreading scheme based on graphs gives the possibility to visually determine the main sources and repeaters of information and to find certain paths of information flow.

The novelty of the approach is the way of matching the graph's properties and data obtained, which can improve the efficiency of visual analysis. It also opens up the possibility to develop an algorithm for further countering the spread of dangerous information.

5   Experiments

To evaluate the efficiency of the algorithms proposed and informativeness of the visualization technique presented earlier in the paper it is reasonable to show the results of the experiments. The aim of the experiments was to define the information sources, receivers, aggregators and repeaters, to reveal the information dissemination channels and to find the objects of social network where one can see the "attenuation" or "distortion" of original sense using visual analytics.

The data for the experiments was collected from the most popular social network in Russia "VKontakte". As an input data the group "Nationalists and ..." with 10 additional groups were selected. A time period P was selected as 7 days before the experiments took place. Later on, for the whole information space of the input data (475 058 signs) the set of keywords was formed. All data was collected "as is" by the proposed algorithm, with no impact on groups, users or information flows.

The general view of the graph for the "repost" relationships is in Fig. 1. The vertexes, which are not sources of information for other objects of the social network, are filtered to simplify the figure. Also there neither labels or shapes of vertexes. The graph gives one a whole picture of the information flows between the objects in the social network. Let us consider a subgraph that contains the information about the most active participants in the social interaction (Fig. 2).

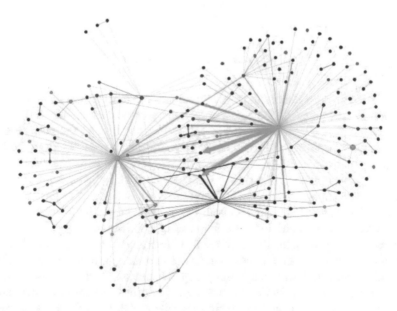

Fig. 1. The graph of the "repost" relations

To indicate the type of object q of the social network the vertex form is used. In particular, the "diamond" form is used to represent a group ("Nationalists and …", "The Committee Nation …", etc.) and the "circle" form is used to represent a user ("Alexey R.", "Elena K.", etc.) in the subgraph shown below.

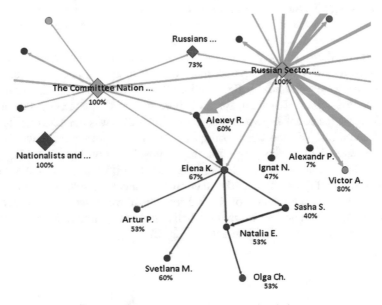

Fig. 2. The subgraph of the "repost" relations

The size of a vertex defines the average number of views of posts published on the wall of a user or a group. In the figure above, the groups "Nationalists and …", "The Committee Nation …", "Russian Sector …", "Russian …" and the user "Elena K." have the biggest coverage of the audience. The color of a vertex describes the content uniqueness generated by a user or a group. In the presented color scheme a range of colors from green (unique content) to red (reposts) is used. In particular, the content of the groups "The Committee Nation …" and "Russian Sector …" is the most unique. The group "Russian …" disseminate both the unique and reposed content. The same colors are used for links and indicates a spreading content uniqueness.

One can also see from the figure above that the groups "The Committee Nation …" and "Russian Sector …" are the main sources of information (in this case, the first one influences on the latter), meanwhile the group "Nationalists and …" is the aggregator with a greatest readers amount and it has a great information impact on the audience.

The user "Elena K." is an information repeater which takes the information from the groups "The Committee Nation …", "Russian Sector …" and the user "Alexey R." and transmits it to users "Artur P." and others. Despite of a large number of reposts this user also generates unique content. It should be noted that during this experiment the user "Elena K." was renamed to "Alena K.".

While studying the information spreading channels one can track the path from the main sources of the information to the information consumers, that passes through user "Elena K.". The information "attenuation" can be noted. For example, the keywords for the group "Russian Sector ..." are "elections, child, Kemerovo, Putin, Kremlin, Moscow, Russia, nation, Ukraine, Vic, media", meanwhile the keywords for user "Olga Ch." are "elections, child, Kemerovo, Putin, Kremlin, Moscow, education, essence, power, region, administration". One can assume that the context of reposted information changes in the user's information space and the original one "attenuates".

Thus, the visual analysis allows one to find the main information spreading channels and define the key nodes in the information transmitting network (the sources, the consumers, the aggregators and the repeaters).

6 Discussion

The starting point for collecting information and visualizing its distribution channels is the ratio of "repost" between two information objects, which is fixed by means of the social network itself. At the same time, with this approach, the clone messages prepared by users manually and appearing on the social network for their authorship disappear from the field of view. This way of distributing information is much less convenient, but it gives an advantage to users who want to stay "in the shadows" and at the same time distribute inappropriate content. The open question remains, connected with the input data for analyzing the information dissemination. Having only a general idea of potentially dangerous sources of information, it becomes necessary to repeatedly apply the proposed algorithm using the information obtained at the previous iteration to identify the true centers and ways of disseminating information.

Detection of sources, repeaters, aggregators, and users of information does not allow calculating information attacks, carried out, as a rule, decentralized with the help of bots. This, in turn, limits the scope and target audience of the proposed approach.

The presentation of information using graphs can be uninformative in the case of large volumes of information. To do this, it is necessary to develop a universal set of measures to reduce the connectivity of the graph, to allocate its key nodes and key channels, to receive details on demand. However, despite the above disadvantages, this approach seems to be a promising area in the study of the target audience on which the information flow is directed. For example, the approach makes it possible to conduct behavioral factors investigation during a given period in the information space of an object, the dynamics of which will allow to see the audience growth and evaluate the involvement of information impact objects.

7　Conclusion

In this paper, we presented an approach to a visual analysis of relationships in social networks. It is assumed that this approach will provide information on channels for the dissemination of inappropriate information, with the recognition of sources, repeaters, aggregators, and users. The immediate area of application of the proposed approach is the information security of the state: increasing the level of protection against inappropriate information in social networks, analyzing the attacked audience, identifying ways to disseminate information, finding the source of information attacks. The obtained results allow to carry out investigation in the field of visual analysis of social networks. Also, the developed approach can be applied not only for user protection tasks. For example, in marketing, it will allow to predict the effectiveness of advertising companies, assess the speed of market reaction to a product, involvement. In further it is planned to conduct additional experiments aimed at analyzing information flows around groups with forbidden legislator content. It is also promising to use heat maps of involvement. Currently, the authors explore the possibilities of combining the current model of visualization of information flows with a map of the involvement of impact objects. In addition, it seems important to classify the types of nodes of information dissemination, to separate agitation sources from simple repeaters. Identifying the signs of such nodes will allow to fulfill the segmentation of social network objects and increase the effectiveness of counteracting inappropriate effects.

Acknowledgements. The work is performed by the grant of RSF #18-11-00302 in SPIIRAS.

References

1. Social networks in Russia, summer 2017: figures & trends. https://www.cossa.ru/289/166387/. Accessed 11 Apr 2018
2. Kotenko, I., Chechulin, A., Shorov, A., Komashinsky, D.: Analysis and evaluation of web pages classification techniques for inappropriate content blocking. In: 14th Industrial Conference on Data Mining, LNAI, vol. 8557, pp. 39–54. Springer (2014)
3. Novozhilov, D., Kotenko, I., Chechulin, A.: Improving the categorization of web sites by analysis of html-tags statistics to block inappropriate content. In: 9th International Symposium on Intelligent Distributed Computing, pp. 257–263. Springer (2016)
4. Kotenko, I., Chechulin, A., Komashinsky, D.: Categorisation of web pages for protection against inappropriate content in the internet. Int. J. Internet Protoc. Technol. **10**(1), 61–71 (2017)
5. Zadeh, L., Abbasov, A., Shahbazova, S.: Fuzzy-based techniques in human-like processing of social network data. Int. J. Uncertain. Fuzziness Knowl. Based Syst. **23**(Suppl. 1), 1–14 (2015)
6. Gomzin, A., Kuznetsov, S.: Methods of construction of socio-demographic profiles of Internet users. Program. Comput. Softw. J. **27**(4), 129–144 (2015)
7. Drobyshevskiy, M., Korshunov, A., Turdakov, D.: Parallel modularity computation for directed weighted graphs with overlapping communities. Program. Comput. Softw. J. **28**(6), 153–170 (2016)

8. Que, F., Rajagopalan, K., Zaman, T.: Penetrating a Social Network: The Follow-back Problem (2018). https://arxiv.org/pdf/1804.02608.pdf. Accessed 15 Apr 2018
9. Opsahl, T., Agneessens, F., Skvoretz, J.: Node centrality in weighted networks: Generalizing degree and shortest paths. Soc. Netw. J. **32**(3), 245–251 (2010)
10. Hickethier, G., Tommelein, I.D., Lostuvali, B.: Social network analysis of information flow in an IPD-project design organization. In: 21th Annual International Group for Lean Construction (2013)
11. Sudhahar, S., Veltri, G., Cristianini, N.: Automated analysis of the US presidential elections using Big Data and network analysis. Big Data Soc. J. **2**(1), 1–28 (2015)
12. Paraphrasing and synonymy of text. www.paraphraser.ru. Accessed 15 Apr 2018
13. Kolomeec, M., Chechulin, A., Kotenko, I.: Review of methodological primitives for the phased construction of data visualization model. SPIIRAS Proc. **5**, 232–257 (2015)
14. Kolomeec, M., Chechulin, A., Pronoza, A., Kotenko, I.: Technique of data visualization: Example of network topology display for Security monitoring. J. Wirel. Mob. Netw. Ubiquitous Comput. Dependable Appl. **7**(1), 41–57 (2016)

The Problem of the Anomaly Detection in Time Series Collections for Dynamic Objects

S. G. Antipov, V. N. Vagin$^{(\boxtimes)}$, O. L. Morosin, and M. V. Fomina

Department of Application Mathematics, National Research University "MPEI",
Moscow, Russian Federation
antisergey@gmail.com, vagin@appmat.ru,
omorsik@gmail.com, m_fomina2000@mail.ru

Abstract. In this paper the approaches to processing temporal data is considered. The problem of the anomaly detection among sets of time series is setting up. The algorithms TS-ADEEP and TS-ADEEP-Multi for anomaly detection in time series sets for the case when the learning set contains examples of several classes are proposed. The method for improving the accuracy of anomaly detection, due to "compression" of these time series is used.

Keywords: Time series · Inductive notion formation · Exclusion detection
Classification

1 Introduction

One of the most problem class, the decision of which demands intelligent support of computer systems, is manage problems of complex technical objects. The main feature of such manage objects is that they are dynamic ones have ability for developing. The states of such objects and systems may change over time. Therefore one needs to develop methods and algorithms allowing to take into consideration a time factor under analysis of such object behavior.

The structure of the work: there are viewed the problems arising under time series processing; task setting up an anomaly detection in time series sets; the anomaly detection algorithm in series collections; experimental results and conclusions.

2 The Task of an Anomaly Detection

2.1 Problems Arising Under Time Series Processing

Modern Data mining systems solve the problem of extracting new knowledge, hidden dependencies, from large data sets [1]. The most common case of temporal data analysis is a knowledge extraction from time series [2–4].

At the same time, an intelligent analysis of time series on a certain interval can consist in the search for some trends in the state of the process being studied.

Time series are used in many different areas (technology, economics, medicine, banking, etc.) and they describe different processes that occur over time.

© Springer Nature Switzerland AG 2019
A. Abraham et al. (Eds.): IITI 2018, AISC 875, pp. 106–117, 2019.
https://doi.org/10.1007/978-3-030-01821-4_12

The problem of time series mining is important for solving the following tasks of process analysis:

- to give a process state prediction depending on the qualitative assessment of a current or previous state;
- to bring out the presence of typical and abnormal event sorts;
- to bring about qualitative changes of a researched process on the basis of time series analysis and to find time series trends or other change identification [5].

The problems of detecting trends, their qualitative assessment and the forecast based on analysis of time series are of particular relevance due to the continuous growth of real-time data from the specific and complex technical objects, for example, sensors whose values change over time.

Consider the case where the object's behavior is assessed on the basis of particular parameter values observations. In general, the time series TS is an ordered sequence of values $TS = <ts_1, ts_2, ..., ts_i, .., ts_m>$ describing the flow of a long process, where the index i corresponds to a time label, and ts_i values can be sensor indications, product prices, exchange rates and so on. The example of the data forming the time series is shown in Table 1, where the i-th point corresponds to the ts_i value obtained at time i (the time t is assumed to be discrete in the range of from 0 to 9).

Table 1. The example of time series

Time i	0	1	2	3	4	5	6	7	8	9
Parameter values ts_i,	−1.07	0.13	0.85	0.96	0.81	0.84	−0.07	−1.01	−0.90	−1.14

2.2 Setting up the Anomaly Detection Task in Time Series Sets

The anomaly detection problem [6] is set up as the task of searching for patterns in data sets that do not satisfy some typical behavior. The ability to find abnormalities in a data set is important in a variety of subject areas: in the complex technical system analysis (e.g. satellite telemetry), in network traffic analysis, in medicine (analysis of MRT images), in the banking industry (fraud detection) and etc.

The anomaly, or "blowout" is defined as an element that is explicitly selected from the data set which it belongs to and differs significantly from other elements of the sample. Informally, the task of anomaly detection in time series sets is formulated as follows. There is a collection of time series describing some processes. This collection is used to describe normal processes. It is required to construct a model on the basis of the available data that is a generalized description of normal processes and allows distinguishing between normal and abnormal processes.

For the approach of anomaly detection, there is usually a description of the normal operation of a system, for example, a set of system states with no problems. But, as a rule, we have no possibilities to receive the description of a situation with failures on an object in the full volume.

When learning on such data it is necessary to build a model of the normal system operation, that in the future could predict whether the current situation is "normal" or "abnormal", i.e. whether there is any failures or not currently.

The task is complicated by the fact that a set of input data is limited and does not contain any examples of abnormal processes. It does not specify a criteria by which it would be possible to distinguish the "normal" and "abnormal" time series. In this regard, it is difficult to accurately assess the quality of the algorithm (the percentage of correctly detected anomalies, the number of false positives and the number of missed abnormalities). In addition, many algorithms working well for some data sets will not fit well for other subject areas. It may also vary a criterion for determination of the correct time series.

Let there be a set of objects, where each object has a time series:

$TSStudy = <tsstudy_1, tsstudy_2, ..., tsstudy_m>$ is a learning set. Each time series in a learning set is an example of normal process flow. Based on the time series analysis of $TSStudy$, one needs to build a model to refer a testing set of time series $TSTest = <tstest_1, tstest_2, ..., tstest_p>$ to "normal" or "abnormal" by some criterion.

This problem should be divided into two cases [6]: the first case, when a learning set contains examples of a single class; the second case, when a learning set contains examples of several classes. In the first case, the fact of membership of these objects to a class of the learning set is important, it requires to define a "border" in some way, whereby time series are parts of the class represented by the learning set (not an anomaly) or does not belong to it (an anomaly). In the second case, one needs to further define an object belongs to a particular class. The task of anomaly searching for these two cases are considered here.

2.3   The Anomaly Detection Algorithm in Time Series Collections

There is proposed the method for the anomaly detection in time series sets, that is a modification of the method based on an exact exclusions description. The original formulation of this problem is given in [7]: for the given set of objects I, one needs to get an exclusion-set I_x. To do this, there are introduced:

- the function of dissimilarity $D(I_j)$, $I_j \subseteq I$: defined on $P(I)$, where $P(I)$ is a set of all subsets I_j for I receiving positive real values;
- the cardinality function $C(I_j)$: $I_j \subseteq I$, defined on $P(I)$ and receiving positive real values such that for any two subsets $I_k \subseteq I$, $I_l \subseteq I$, $(k \neq l)$ $I_k \subset I_l \Rightarrow C(I_k) < C(I_l)$;
- the smoothing factor $SF(I_j) = C(I \setminus I_j) \cdot (D(I) - D(I \setminus I_j))$, that is calculated for each $I_j \subseteq I$.

Then $I_X \subset I$ will be considered as an exclusion-set for I with respect to $D(I)$, and $C(I)$, if its smoothing factor $SF(I_x)$ is maximal [7].

Informally, an exclusion-set is a smallest subset of I, that makes the largest contribution to its dissimilarity. A smoothing factor shows how much the dissimilarity of a set I can be reduced, if from it to exclude a subset I_j. A dissimilarity function can be any function that returns low values if elements of a set are similar to each other and higher values if elements are dissimilar.

This method was adapted for the task of the anomaly detection in collections of time series. Let us introduce the following changes.

As a set of objects I we use $TSStudy \cup \{tstest_j\}$ for each $tstest_j \in TSTest$. The function of dissimilarity of I_j for time series is defined as follows. Let $I_j \subseteq I$ be a subset of time series. Each time series is considered as a vector of its values. Further in short, any individual time series is designated by $a \in I_j$. Calculate an average value on vector coordinates $\bar{I}_j$. Then the dissimilarity function for time series is computed as a sum of squared distances between $\bar{I}_j$ and vectors $a \in I_j$:

$$D(I_j) = \tfrac{1}{N} \cdot \sum\nolimits_{a \in I_j} \left| a - \bar{I}_j \right|^2 \text{ where } \bar{I}_j = \sum\nolimits_{a \in I_j} \tfrac{a}{|I_j|} \text{ and } N - \text{the number of elements } I_j$$

.

The cardinality function is given by the formula $C(I \setminus I_j) = 1 / |I_j| + 1$. The formula for calculating the smoothing factor is $SF(I_j) = C(I \setminus I_j) \cdot (D(I) - D(I_j))$.

If an exclusion-set for $I = TSStudy \cup \{tstest_j\}$ contains $tstest_j$, then $tstest_j$ is an anomaly.

To determine anomalies in sets of time series based on the method described above the algorithm TS-ADEEP was developed. The algorithm TS-ADEEP for the case with a learning set with only one class is given in details in [8].

To deal with the situation when a study set contains objects of several classes, the algorithm TS-ADEEP-Multi was developed, that is a generalization of the algorithm TS-ADEEP. The generalization is quite obvious: dividing learning set into subsets containing examples of only one class and consistently applying to them and to each time series from a test set of the algorithm TS-ADEEP, one can determine whether the considered time series is anomaly. For cases when time series is an anomaly for each subset, it is an anomaly for the whole learning sample. The algorithm TS-ADEEP-Multi for the case with a learning set with several classes is given in details in [9].

3 Experimental Results

3.1 The Description of Data Sets in Experiments

Simulation of the anomaly detection process was carried out on both artificial and real data. As artificial data, the classical time series descriptions used in the scientific literature were taken: "cylinder-bell-funnel" [10] and "control chart" [11]. As real data, we used time series sets from the repository [12], such as "analysis of the daily physical activity of a person ACDL", "the recognition of suitable and defective wafer semiconductors", spectrographic analysis of various food products (Coffee, Olive oil, Beef) [12]. For the task of detecting unauthorized access to data, a set of "Traffic" was generated - data collected using special traffic analysis systems when transferring files through various protocols was used [13].

All the data used in experiments were preliminarily processed, consisting of two stages: normalization and subsequent discretization of the normalized time series with the transition to the symbolic data representation, where the alphabet size (the number of used symbols) depending on the task could vary. The process of preliminary data processing is based on the ideas of the SAX algorithm [14].

3.2   Experimental Results for Familiar Data Sets

The simplest way to evaluate how well a possible decision function (such as a system of decision rules or a decision tree) works is to test it on a test set. A well-known technique is the partitioning of the learning set, i.e. using two-thirds of this set for learning, and the remaining part - to assess the generalization quality. But such partitioning reduces the size of a learning set, and, consequently, the possibility of an excess of suitable decision functions increases. Next, some verification techniques will be described to avoid these problems.

When cross-validating, a learning set is divided into k disjoint subsets of equal sizes: $U_1,..., U_k$. For each subset U_i, training is conducted on the union of all the remaining subsets and then the error coefficient ε_i for U_i subset is determined by an experimental method. (The error ratio is a relation of the number of classification errors made in U_i, to the number of examples in U_i). Then for the classification model learned on all examples of U, an error coefficient value that can be expected on the new examples will be equal to the average of all ε_i.

In order to assess the efficiency of the algorithm TS-ADEEP, one can start from the following assumption: the anomaly detection is essentially a classification of the objects under consideration to one of the classes - "normal" or "abnormal", and it is not necessary to define exactly to what class this object belongs. Unfortunately, the task is complicated because in the presence of several "normal" or " abnormal " classes, one cannot use this algorithm since it is intended for detecting anomalies in sets with a single class. Thus, comparing the accuracy of the anomaly detection with the accuracy of classification on the same data sets may in some approximation allow us to assess the efficiency of the algorithm TS-ADEEP. The data sets, on which the experiments were conducted, are taken from data collections [10–12].

For comparison, we will use the following well-known algorithms [15]:

- K nearest neighbors method (Knn);
- algorithm for building a decision tree C4.5;
- Bayesian networks (NB);
- the multi-layer perceptron, logistic regression (MLP);
- Random Forest (RF) algorithm;
- logistic regression + decision trees (LMT);
- support vector method (SVM).

Table 2 shows the results obtained by different classification algorithms in solving the problem of finding anomalies on test data.

The bold digits designate best results.

To assess the efficiency of the TS-ADEEP-Multi algorithm, one can proceed from the same assumptions as with the TS-ADEEP algorithm. The anomaly detection by means of the algorithm, in effect, is a reference of the objects under consideration to one of normal or abnormal classes, and not all of them are known in advance. It should be taken into account that in the search for anomalies, the objects division must be carried out for the smaller number of classes than there is in data sets (anomalies and all other classes). Unfortunately, the same fact complicates the task by that it is impossible to use information concerning all classes of the subject domain in question. The efficiency assess of the algorithm TS-ADEEP-Multi is given in Table 3.

Table 2. The comparison of the accuracy of anomaly detection by the algorithm TS-ADEEP with known algorithms (sets of time series with one class are considered)

Data set	Knn	NB	C4.5	MLP	RF	LMT	SVM	TS-ADEEP (average)
Coffee	75.00	67.86	57.14	96.43	75.00	**100.00**	96.43	82.14
CBF	85.00	**89.67**	67.33	85.33	83.56	77.00	87.67	78.89
Olive oil	76.67	76.67	73.33	**86.67**	**86.67**	83.33	**86.67**	81.67
CC	88.00	96.00	81.00	91.33	86.00	92.00	92.33	**98.03**
Beef	60.00	50.00	56.67	73.33	50.00	80.00	66.67	**81.33**
Average	76.93 (5)	76.04 (7)	67.09 (8)	**86.61** **(1)**	76.25 (6)	86.47 (2)	85.95 (3)	84.41 (4)

Table 3. The comparison of the accuracy of anomaly detection by the algorithm TS-ADEEP-Multi with known algorithms (we consider time series sets with several classes)

Data set	Knn	NB	C4.5	MLP	RF	LMT	SVM	TS-ADEEP-Multi (average)
CBF	85.00	**89.67**	67.33	85.33	83.56	77.00	87.67	77.89
CC	88.00	**96.00**	81.00	91.33	86.00	92.00	92.33	91.49
Face (four)	87.50	84.09	71.59	87.50	78.41	77.27	**88.64**	80.40
Average	86.83 (4)	**89.92** **(1)**	73.31 (8)	88.05 (3)	82.66 (6)	82.09 (7)	89.55 (2)	83.26 (5)

The bold digits designate best results.

The results presented in Tables 2 and 3 lead to the conclusion that on some data sets (both artificial and real) the proposed algorithms show results that are better than the algorithms listed above. For tasks such as traffic analysis, control cards, and a number of others, the algorithms TS- ADEEP and TS-ADEEP-Multi have the explicit advantage [8, 9].

Let's consider several real problems, that are rather complicated, since assigning an object to one of the given classes requires analysis of several time dependencies. Such tasks include the tasks of recognizing the physical activity of a person and analyzing the quality of semiconductor wafers.

3.3   Results of Recognizing the Physical Activity of a Person

The data set "(«Activities of Daily Living Recognition with Wrist-worn Accelerometer Data Set») («ADL») [16] from the repository of the University of California at Irvine [12] contains accelerometer indications. They correspond to various actions that a person can perform. These actions, that are designated as "Human Motion Primitives", include the following: brush your teeth, climb stairs, comb your hair, go down the stairs, drink water from a glass, eat meat (with a fork and knife), eat soup (by a spoon), get out of bed, go to bed, pour water, sit on a chair, get up from a chair, call, go. When recording accelerometer readings, the acceleration projections are stored on three perpendicular axes (they are denoted below by the X axis, the Y axis, the Z axis) and, therefore, there is a set of three time series for each action that describes this action.

In this case, the task is to recognize or classify objects (actions, situations) represented by a set of time series. According to the presented time series, it is necessary to define what type of activity a person performed.

Table 4. The classification accuracy by the TS-ADEEP algorithm

Parameter	The classification accuracy, %.the TS-ADEEP algorithm
Classes «Get up from a chair», «Sit on a chair»	
1 (X axis)	97.09
1 (Y axis)	65.96
1 (Z axis)	98.84
Classes «Get out of bed», «Go to bed»	
1 (X axis)	99.01
1 (Y axis)	58.42
1 (Z axis)	96.04
Classes «Climb stairs», «Go on the stairs»	
1 (X axis)	70.63
1 (Y axis)	71.43
1 (Z axis)	65.87

Using all three parameters for the classification on given data sets on average can give an accuracy of object classification higher than using only one parameter (see Table 4). Nevertheless, the use of data on one axis allows to obtain a classification accuracy close to 100% (X, Z axis).

3.4   Results of Recognizing the Defects of Wafers

The data set "wafer" from the repository of the University of California at Riverside [15], contains time series corresponding to the sensor indications in the production of semiconductor wafers. The semiconductor plate is a semi-finished product in the technological process of semiconductor devices and microcircuits production. It is a thin (250–1000 μm) plate made of semiconductor material with a diameter of up to 450 mm, on the surface of which, using the operations of planar technology, an array of discrete semiconductor devices or integrated circuits is formed. After creating the necessary semiconductor structure, the plate is cut into individual crystals (chips).

The production of such plates (etching) is a complex technological process involving more than 250 processing steps, each of which can deteriorate characteristics or reliability, reduce the yield of the product or even reject, if the parameters have exceeded the required limits. The most critical in monitoring the production of semi-conductor wafers are 6 parameters [17]: the radio frequency power of the direct wave, the radio frequency power of the reflected wave, the pressure in the chamber, the radiation intensity of the plasma with a wavelength of 405 nm, the radiation intensity of the plasma with a wavelength of 520 nm. Among these parameters, experts singled out two ones that, according to the results of experiments, showed the most accurate results for determining qualitative and defective products: this is the radiation intensity of a plasma with a wavelength of 405 nm and 520. Analysis of the six parameters described above makes it possible to distinguish between classes of quality and defective plates.

Table 5. The classification accuracy of (%) of objects for the "wafer" data set depending on the selected parameter by the TS-ADEEP algorithm

Parameter	The classification accuracy (%)
1	91.06
2	91.49
3	87.79
4	90.71
5	90.63
6	90.89

A software simulation of the anomaly search process was performed to classify objects from the "wafer" data set. Practically, using the TS-ADEEP algorithm, the problem of dividing the objects presented into classes of qualitative and defective plates was solved (see Table 5). Defective plates were considered as anomalies. It was established that each parameter from the considered, representing the time series, allows to distinguish between these two classes of products with an accuracy close to 90%. It has been established that none of the six parameters has a decisive advantage over the others.

3.5   Results of Traffic Analysis in a Network Information Exchange

At present, the use of foreign-made components lies at the heart of the production of hardware and software for most computer systems. However, at the same time there is a threat of information leakage through the use of functionality that negatively affects the safety of the information processed (hereinafter referred to as malicious code or malware).

Such malicious functionality of the hidden channel organization of information leakage bypassing known security means can be used to organize information hidden channels when transferring protected information over computer networks.

Transmission of coded data via communication channels is carried out by converting bit sequences into electromagnetic signals. Data represented by bits or bytes is transmitted at a rate determined by the number of bits per unit of time. These physical layer parameters of network protocols, such as a clock frequency, an encoding method, a transmission scheme and a signal spectrum are defined by standards that are developed by competent organizations. The process of physical transmission of data at a certain interval can be viewed as a time series.

The basis for the detection of malicious software functions in the information exchange system will be traffic analysis: if the parameters of a typical information exchange for a certain protocol are known, then the abnormal pattern of behavior in the computer network detected during traffic analysis in the exchange over this protocol can indicate that in the analyzed system, there is a bookmark of the above type, and the anomalies in the traffic are caused by the actions of such programs.

The possibility to apply algorithms for searching anomalies in collections of time series is investigated for the task of detecting cases of atypical information exchange in the network, that presupposes the presence of malicious programs. The task is complicated by the fact that it is extremely difficult to obtain a representative sample that is sufficiently accurate and at the same time describes all possible variants of the information system behavior. It should be noted that it is easier to obtain a sample for the normal behavior of the information system than for the abnormal behavior, since the normal behavior can be modeled in the laboratory, while abnormal behavior is extremely rare. Moreover, the anomalous behavior is dynamic in nature, and hence new types of anomalies that could not be represented in the initial learning sample can appear.

There is proposed the following solution of the problem. There are reference models represented by time series, that reflect the changes in protocol parameters depending on the types of information exchange. For comparison, time series are used that represent the actual behavior of the information system during data exchange. The comparison of these two models of information exchange is aimed at finding patterns of behavior in the information transmission other than typical, that is, they are anomalous.

As an illustration of the method, FTP file transfer protocol was chosen. The method can be extended to other standard protocols of information interaction, having specifications in the form of standards or widely distributed de facto and having a description in open sources.

Based on the analysis of network traffic when transferring files via FTP in various conditions (including the simultaneous transmission of several files), a data set was obtained, that is a learning sample for creating a model of "reference" data transmission.

A special stand was assembled to obtain the data. At this stand, data were transmitted over the network between two computers, both via the FTP protocol and a mixture of protocols. For example, simultaneously with the transfer of the file, the network was scanned using the ICMP protocol (PING command), arbitrary UDP communication - so-called "information noise" for the reference traffic. Only the length of the transmitted data packet was fixed. The reverse transfer that accompanies the exchange (the need for reverse transfer is determined by the lower layer protocol (TCP)) was not fixed [13].

The following data transfer options were investigated:

- FTP transfer (standard).
- Simultaneous transfer via FTP and ping protocols (FTP traffic was analyzed).
- Simultaneous transmission over FTP and UDP protocols (analyzed FTP traffic).

Having information of this kind about data transmission over the network, it is necessary to define whether data transmission is "suspicious", what may indicate a possible compromise of the network infrastructure, the availability of software and / or hardware backdoors.

As test data, among others, there were used specially generated time series simulating the transfer of unauthorized data.

By changing the alphabet size parameters and the time series dimensionality, an optimal representation of the time series can be obtained for their use by the algorithms TS-ADEEP and TS-ADEEP-Multi.

Table 6 shows the results of anomaly recognition in the transmission of data on the above protocols. As can be seen from the results, for the problem under consideration, it was possible to achieve an classification accuracy of the anomalies up to 100%.

Table 6. The accuracy of the anomaly detection in data sets "Traffic" with one class for the TS-ADEEP algorithm

		Dimensionality of time series						
		210	150	100	50	30	20	10
The number of alphabet symbols	5	71,43	82,14	60,71	64,29	60,71	46,43	67,86
	10	92,86	96,43	100	96,43	82,14	85,71	64,29
	15	92,86	100	100	96,43	92,86	82,14	85,71
	20	92,86	100	100	96,43	92,86	82,14	82,14
	25	92,86	100	100	96,43	92,86	82,14	92,86
	30	92,86	100	100	96,43	92,86	92,86	82,14
	40	92,86	100	100	96,43	92,86	92,86	92,86
	50	92,86	100	100	96,43	92,86	92,86	92,86

When analyzing simultaneously all the traffic types considered in the experiments, the TS-ADEEP-Multi algorithm was used. Here the problem is complicated by the fact that normal behavior can correspond to one of several classes. The results presented in Table 7 show that even for this case (several classes), it is possible to achieve the accuracy of anomaly classification up to 100% in the selection of the time series normalization parameters.

Despite the fact that using TS-ADEEP and TS-ADEEP-Multi algorithms aids to achieve the high classification accuracy in the search for anomalies, these algorithms require a large amount of computation, so their use in analyzing a large amount of incoming data may require the introduction of additional heuristics that reduce the search [13].

Table 7. The accuracy of anomaly detection in data sets "Traffic" with several classes for the TS-ADEEP-Multi algorithm

		Dimensionality of time series						
		210	150	100	50	30	20	10
The number of alphabet symbols	5	85,71	89,29	57,14	64,29	60,71	46,43	67,86
	10	96,43	96,43	100	96,43	96,43	85,71	67,86
	15	92,86	100	100	96,43	92,85	82,14	67,86
	20	96,43	100	100	96,43	96,43	96,43	75,00
	25	96,43	100	100	96,43	96,43	82,14	82,14
	30	96,43	100	100	96,43	96,43	96,43	92,86
	40	96,43	100	100	96,43	96,43	96,43	92,86
	50	96,43	100	100	96,43	96,43	96,43	96,43

4   Conclusion

The problem of finding anomalies among sets of time series is considered. In the developed algorithms TS-ADEEP and TS-ADEEP-Multi for the anomaly detection in time series sets, the cases were studied when a learning set contained examples of one and several classes of objects. The programmed modeling of the proposed algorithms was carried out. The comparison of the results shown by these algorithms with the results of a number of other algorithms capable of solving similar problems is given. Practical results obtained using TS-ADEEP and TS-ADEEP-Multi algorithms showed that these algorithms are among the five most successful ones, and on separate data sets, they showed the best results in the classification accuracy. With the help of these algorithms, a number of practical problems have been successfully solved.

Acknowledgment. This work was supported by grants from the Russian Foundation for Basic Research № 15-01-05567, 17-07-00442.

References

1. Vagin, V., Golovina, E., Zagoryanskaya, A., Fomina, M.: Exact and plausible inference in intelligent systems. In: Vagin, V., Pospelov, D. (eds.) 712 p. FizMatLit, Moscow (2008). (in Russian)
2. Roddick, J.F., Spiliopoulou, M.: A bibliography of temporal, spatial and spatio-temporal data mining research. SIGKDD Explor. Newsl. 1(1), 34–38 (1999). http://doi.acm.org/10.1145/846170.846173
3. Lin, W., Orgun, M.A., Williams, G.J.: An overview of temporal data mining. In: Proceedings of the 1st Australasian Data Mining Workshop, Sydney, Australia, pp. 83–90 (2002)
4. Antunes, C.M., Oliveira, A.L.: Temporal data mining: an overview. In: Eleventh International Workshop on the Principles of Diagnosis (2001)
5. Perfilieva, L., Yarushkina, N., Afanasieva, T., Romanov, A.: Time series analysis using soft computing methods. Int. J. Gen. Syst. 42(6), 687–705 (2013)

6. Chandola, V., Banerjee, A., Kumar, V.: Anomaly detection - a surey. ACM Comput. Surv. **41**(3), 1–72 (2009)
7. Arning, A., Agrawal, R., Raghavan P.: A linear method for deviation detection in large databases. In: Proceedings of KDD 1996, pp. 164–169 (1996)
8. Antipov, S., Fomina, M.: Problem of anomalies detection in time series sets. Prog. Prod. Syst. (2), 78–82 (2012). (in Russian)
9. Fomina, M., Antipov, S., Vagin, V.: Methods and algorithms of anomaly searching in collections of time series. In: Proceedings of the first International Scientific Conference Intelligent Information Technologies for Industry (IITI 2016), vol. 1, pp. 63–73. In Series Advances in Intelligent Systems and Computing, vol. 450. Springer (2016)
10. Saito, N.: Local feature extraction and its application using a library of bases. Ph.D. thesis, Yale University, December 1994. 244 p
11. Pham, D.T., Chan, A.B.: Control chart pattern recognition using a new type of self organizing neural network. Proc. Instn. Mech. Engrs. **212**(1), 115–127 (1998)
12. UCI Repository of Machine Learning Datasets. http://archive.ics.uci.edu/ml/
13. Antipov, S.G., Vagin, V.N., Fomina, M.V.: Detection of data anomalies at network traffic analysis. In: Open Semantic Technologies for Intelligent Systems - Conference Proceedings, Minsk, Belarus, pp. 195–198 (2017)
14. Lin, J., Keogh, E., Lonardi, S., Chiu, B.: A symbolic representation of time series, with implications for streaming algorithms. In: Proceedings of the 8th ACM SIGMOD Workshop on Research Issues in Data Mining and Knowledge Discovery, pp. 2–11 (2003)
15. Bruno, B., Mastrogiovanni, F., Sgorbissa, A., Vernazza, T., Zaccaria, R.: Analysis of human behavior recognition algorithms based on acceleration data. In: IEEE International Conference on Robotics and Automation (ICRA), pp. 1602–1607 (2013)
16. Yanping, C., Eamonn, K., Bing, H., et al.: The UCR Time Series Classification Archive–2015, July 2015. www.cs.ucr.edu/~eamonn/time_series_data
17. Olszhewski, R.: Generalized Feature Extraction for Structural Pattern Recognition in Time-Series Data. Ph.D thesis. School of Computer Science, Carnegie Mellon University, Pittsburgh (2001). 125 p

Prediction and Detection of User Emotions Based on Neuro-Fuzzy Neural Networks in Social Networks

Giovanni Pilato[1][✉], Sergey A. Yarushev[2], and Alexey N. Averkin[3]

[1] ICAR-CNR, Arcavacata, Italy
giovanni.pilato@cnr.it
[2] Plekhanov Russian University of Economics, Moscow, Russia
sergey.yarushev@icloud.com
[3] Dorodnicyn Computing Centre, FRC CSC RAS, Moscow, Russia
averkin2003@inbox.ru

Abstract. In this paper we propose a neuro-fuzzy method for emotions prediction. On one hand we suggest a taxonomy-based detection of user joyful interests through the use of semantic spaces and, on the other hand, we propose a neuro-fuzzy method for prediction of emotions used in Twitter posts. Catching the attention of a new acquaintance and empathize with her can improve the social skills of a robot. For this reason, we illustrate here the first step towards a system which can be used by a social robot in order to "break the ice" with a new acquaintance.

1 Introduction

One of the most relevant steps in making new acquaintances in the "engagement" phase, which is a very complicated phenomenon involving both cognitive and affective components, including attention and enjoyment [1,6]. For this reason, there has been a growing interest about this specific phase in the human-machine-interaction (HMI) field [2].

With the term "engagement" we refer to "starting or intention to start an interaction". In particular, we believe that in making new acquaintances "first impressions are everything". For this reason, finding common interests to "talk about", can make it possible to start an empathetic interaction between an human and a robot, improving the human-machine interaction effectiveness. In order to trigger both attention and enjoyment, given these premises, it could be useful to design a social robotic system which tries to discover topics that can be interesting for the just met interlocutor, attempting to understand what might raise a sentiment of joy in order to catch an empathetic attention of the user. As a matter of fact, the knowledge of the topics of interest and the "joyful" subjects for the user can lead the first stages of a conversational interaction that allows the robot to ease the engagement phase, instead of a standard and overly prepared interaction between a robot and an human user.

© Springer Nature Switzerland AG 2019
A. Abraham et al. (Eds.): IITI 2018, AISC 875, pp. 118–125, 2019.
https://doi.org/10.1007/978-3-030-01821-4_13

To achieve this objective, the robot can be able to access the social network posts of the new acquaintance trying somehow to detect her/his interests, which let arise a joyful feeling in her/him to start an, hopefully, interesting conversation for the user [3].

Social networks represent maybe the best place to gather information about people's opinions, as a matter of fact, social media users generally express personal thoughts and to discuss with others about specific subjects [4,19]. These opinions are actually valuable to understand and classify the emotion of an event, a product, a person, etc. and analyze his trend [5,13,14].

During the last decade the use of emoji has increasingly pervaded Social Media platforms by providing users with a rich set of pictograms useful to visually complement and enrich the expressiveness of short text messages. Nowadays this novel, visual way of communication represents a de-facto standard in a wide range of Social Media platforms including fully-fledged portals for user-generated contents like Twitter, Facebook and Instagram as well as instant-messaging services like WhatsApp. As a consequence, the possibility to effectively interpret and model the semantics of emojis has become an essential task to deal with when we analyze Social Media contents.

Even if over the last few years the study of this new form of language has been focusing a growing attention, at present, the body of investigations that deal with emojis is still scarce, especially when we consider their characterization from a Natural Language Processing (NLP) standpoint. In general, exciting and highly relevant avenues for research are still to explore with respect to emoji understanding, since emojis represent often an essential of Social Media texts and thus ignoring or misinterpreting them may lead to misunderstandings in comprehending the intended meaning of a message. The ambiguity of emojis raises an interesting question in human-computer interaction: how can we teach an artificial agent to correctly interpret and recognise emojis' use in spontaneous conversation? The main motivation behind this question is that an AI system able to predict emojis could contribute notably to better natural language understanding and thus to different natural language processing tasks such as generating emoji-enriched social media content, enhancing emotion/sentiment analysis systems, and improving retrieval of social network material, and ultimately improving user profiling.

In this paper we illustrate the design of a system which can be used for detection emotions from social media content and prediction user emotions based on neuro-fuzzy network from tweets. The proposed system will be used for prediction of emoji based on tweets on one hand and for detection user emotion on other.

2   The System

The proposed system is composed of a set of modules interacting in order to catch the attention of the user. The system exploits a training phase, where a semantic space S is induced from Twitter data and a joyful-topic-detection process, which exploits the Twitter ID of the user in order to retrieve her posts

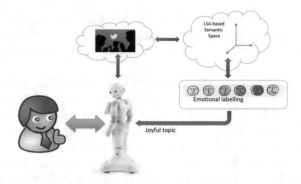

Fig. 1. Joyful-topic-detection process

and trying to catch the interests of the user that somehow let arise a "joy" emotion. The topic detection is obtained by mapping the user tweets to the IAB taxonomy [8,15] (Fig. 1).

2.1 The IAB Taxonomy

The Interactive Advertising Bureau (IAB) Tech Lab Content Taxonomy is a taxonomy which is also an international standard to map contextual business categories [8,9]. This taxonomy is particularly suited for being used by companies in the market, it is standardized and industry-neutral. These characteristics can be effectively exploited for profiling an user interests. We have used this solution just for convenience, being the approach applicable on different targets.

2.2 Emotion Detection Module

This module is responsible for the detection of emotions in tweets. To perform this task, we have taken into consideration the six Ekman fundamental emotions [7]: *anger, disgust, fear, joy, sadness* and *surprise*, exploiting an emotions lexicon obtained from the Word-Net Affect Lexicon, as it has been illustrated in [17,18]. The module exploits a methodology that has been described in [3,15]. The technique is based on the Latent Semantic Analysis paradigm (LSA), a methodology that is capable of giving a coarse sub-symbolic encoding of word semantics [12] and of simulating several human cognitive phenomena [11]. The LSA procedure is based on a term-document occurrence matrix $\mathbf{A}$, whose generic element a_{ij} represents a function of the occurrences of a term in a document corpus. Let K be the rank of $\mathbf{A}$. The factorization named Singular Value Decomposition (SVD) holds for the matrix $\mathbf{A}$:

$$\mathbf{A} = \mathbf{U}\Sigma\mathbf{V}^T \tag{1}$$

Let R be an integer > 0 with $R < N$, and let $\mathbf{U}_R$ be the $M \times R$ matrix obtained from $\mathbf{U}$ by suppressing the last $N - R$ columns, let Σ_R be the matrix obtained from Σ by suppressing the last $N - R$ rows and the last $N - R$ columns; let $\mathbf{V}_R$

be the $N \times R$ matrix obtained from $\mathbf{V}$ by suppressing the last $N - R$ columns. Then:

$$\mathbf{A}_R = \mathbf{U}_R \Sigma_R \mathbf{V}_R^T \tag{2}$$

$\mathbf{A}_R$ is a $M \times N$ matrix of rank R, and it is the best rank R approximation of $\mathbf{A}$ (among the $M \times N$ matrices) with respect to the Euclidean metric. The i-th row of the matrix $\mathbf{U}_R$ may be considered as representative of the i-th word. The columns of the $\mathbf{U}_R$ matrix represent the R independent dimensions of the $\Re^{\Re}$ space S. Each j-th dimension is weighted by the corresponding value σ_j of Σ_R. Any document d can be mapped into a Data Driven semantic space, by computing a vector $\mathbf{d}$ whose i-th component is the number of times the i-th word of the vocabulary, corresponding to the i-th row of $\mathbf{U}_R$, appears in d. The mapping is computed by using the following formula:

$$\mathbf{d}_R = \mathbf{d}^T \mathbf{U}_R \Sigma_R^{-1} \tag{3}$$

The emotional lexicon has been split into six lists, each one associated to one of the basic Ekman emotions {*anger, disgust, fear, joy, sadness, surprise*}. Fixed an emotion e, a set of sentences has been properly constructed following the procedure that has been illustrated in [15].

The methodology exploits six sets $E_{anger}, E_{disgust}, \cdots, E_{surprise}$ of vectors constituting the sub-symbolic coding of words or sentences associated to a particular emotion. The generic vector belonging to one of the sets is denoted as $\mathbf{b}_i^{(e)}$ where $e \in \{$ *"anger"* , *"disgust"* , *"fear"*, *"joy"* , *"sadness"*, *"surprise"*$\}$ and i is the index that identifies the i-th $\mathbf{b}_i^{(e)}$ in the e set. Specifically, $\mathbf{b}_i^{(e)}$ is computed according to formula (3). Analogously, any textual content t of a tweet can be mapped into the same space by computing a vector $\mathbf{t}$ whose i-th component in the number of times the i-th word of the vocabulary, corresponding to the i-th row of $\mathbf{U}_R$, appears in t. Also this mapping is done by using formula (3). Once the tweet t is mapped into the "conceptual" space as a vector $\mathbf{t}_R$, its emotional fingerprint is obtained by calculating for each set E_e, fixed $\mathbf{t}_R$, the weight:

$$w_e = \max cos(\mathbf{t}_R, \mathbf{b}_i^{(e)}) \tag{4}$$

Once all the six w_e weights are computed, the vector $\mathbf{f}_t$, associated to the vector $\mathbf{t}_R$, and by consequence to the tweet t, is calculated as:

$$\mathbf{f}_t = \left[\frac{w_{(anger)}}{\sqrt{\sum_e w_e^2}}, \frac{w_{(fear)}}{\sqrt{\sum_e w_e^2}}, \cdots, \frac{w_{(surprise)}}{\sqrt{\sum_e w_e^2}} \right] \tag{5}$$

The vector $\mathbf{f}_t$ constitutes the *emotional fingerprint* of the tweet t in the six-dimensional emotional hypersphere where all tweets can be mapped and grouped.

2.3 Adaptive Neuro-Fuzzy Inference System

The Adaptive Neuro-Fuzzy Inference System (ANFIS) ANFIS is the abbreviation Adaptive Neuro-Fuzzy Inference System - an adaptive network of fuzzy output. Proposed in the early nineties [10], ANFIS is one of the first variants of hybrid neural-fuzzy networks - a neural network of direct signal propagation of a special type. The architecture of the neural-fuzzy network is isomorphic to the fuzzy knowledge base. Neuro-fuzzy networks use differentiated implementations of triangular norms (multiplication and probabilistic OR), as well as smooth functions. This allows the use of cross-fuzzy neural networks, rapid algorithms for learning neural networks, based on the method of back propagation of errors. The architecture and rules for each layer of the ANFIS network are described below. ANFIS implements the Sugeno fuzzy inference system in the form of a five-layer neural network of direct signal propagation. The system works as follows:

- the first layer is the terms of the input variables;
- the second layer is antecedents (parcels) of fuzzy rules;
- the third layer is the normalization of the degree of implementation of the rules;
- the fourth layer is the conclusion of the rules;
- the fifth layer is the aggregation of the result, du according to different rules.

The network inputs in a separate layer are not allocated. Figure 2 shows an example of an ANFIS network with two input variables (x_1 and x_2) and four fuzzy rules.

We introduce the following notation, which is necessary for the following presentation:

$x_1, ..., x_n$- network inputs;
y - network output;
R_r: If $x_1 = a_{1,r}$ and ... and $x_n = a_{n,r}$, then $y = b_{0,r} + b_{1,r}x_1 + ... + b_{n,r}x_n$- fuzzy rule with serial number r;
m is the number of rules, $r = \overline{1, m}$;
$a_{i,r}$ is a fuzzy term with the membership function $\mu_r(x_i)$, used for the linguistic evaluation of the variable x_i in the r-th rule ($r = \overline{1, m}, i = \overline{1, n}$);
$b_{q,r}$ are real numbers in the conclusion of the r -th rule ($r = \overline{1, m}, \tau_i^* = \overline{\frac{\tau_r}{(\sum_{j=\overline{1,m}\tau_i})}}$).

The ANFIS network functions as follows [16]:

Layer 1. Each node of the first layer represents one term with a bell-like membership function. The inputs of the network $x1, x2, ..., x_n$ are connected only with their terms. The number of nodes of the first layer is equal to the sum of the powers of the term-sets of the input variables. The output of the node is the degree of belonging of the value of the input variable to the corresponding fuzzy term:

$$\mu_r(x_i) = \frac{1}{1 + |\frac{x_i-c}{a}|^2 b}, \qquad (6)$$

where a, b and c are the configurable parameters of the membership function.

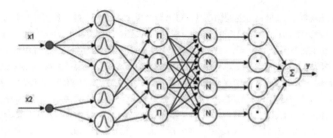

Fig. 2. Example of ANFIS architecture.

Layer 2. The number of nodes of the second layer is m. Each node of this layer corresponds to one fuzzy rule. The node of the second layer is connected to those nodes of the first layer, which form the antecedents of the corresponding rule. Therefore, each node of the second layer can receive from 1 to n input signals. The output of the node is the degree of execution of the rule, which is calculated as the product of the input signals. Denote the outputs of nodes of this layer by τ_r, $r = (1, m)$.

Layer 3. The number of nodes of the third layer is also equal to m. Each node in this layer calculates the relative degree of fuzzy rule execution:

$$\tau_r^* = \frac{\tau_r}{\sum_{j=\overline{1,m}} \tau_j} \tag{7}$$

Layer 4. The number of nodes of the fourth layer is also equal to m. Each node is connected to one node of the third layer, as well as to all inputs of the network (in Fig. 2 links with inputs are not shown). The fourth layer node calculates the contribution of one fuzzy rule to the network output:

$$y_r = \tau_r^*(b_{0,r} + b_{1,r}x_1 + \ldots + b_{n,r}x_n). \tag{8}$$

Layer 5. The only one node of this layer summarizes the contributions of all rules:

$$y = y_1 + \ldots y_r \ldots + y_m. \tag{9}$$

Typical procedures for learning neural networks can be used to configure an ANFIS network because it uses only differential functions. Usually a gradient descent combination is used in the form of an algorithm for back propagation of an error and a method of least squares. The error back propagation algorithm configures the rules of antecedents of the membership functions. The method of least squares evaluates the coefficients of the conclusions of the rules, since they are linearly related to the output of the network. Each iteration of the setup procedure is performed in two stages. At the first stage, a training sample is fed to the inputs, and the optimal parameters of the nodes of the fourth layer are found by the discrepancy between the desired and actual network behavior by the iterative least squares method. In the second stage, the residual is transferred from the network output to the inputs, and the parameters of the nodes of the first layer are modified by the method of back propagation of the error.

In this section we presented neuro-fuzzy network which can be used for prediction of user emotions from tweets. It is concept for the future research. We suggest to use neuro-fuzzy network for prediction of user emotions because emotions has a fuzzy nature. We put in the inputs of neural network selected keywords from tweets and in output we predict the six emotions: *anger, disgust, fear, joy, sadness* and *surprise*. In neuro-fuzzy network we can develop a fuzzy rules for better emotion prediction. In our future work we will to present working model and first results.

3   Conclusions and Future Works

We have illustrated a preliminary work on a system that makes use of both semantic spaces and neuro-fuzzy networks to enhance the human-robot interaction.

The system uses as an emotion detection module based on semantic spaces induced by the Latent Semantic Analysis paradigm to reach this goal. A conversational engine guides the initial process and continues with$t$ the conversation.

We have presented the ANFIS model which can be used for user emotion prediction. In our future work we will to present working model and first results.

Acknowledgments. This work was supported by the Russian Foundation for Basic Research (Grant No. 17-07-01558).

References

1. Brethes, L., Menezes, P., Lerasle, F., Hayet, J.: Face tracking and hand gesture recognition for human-robot interaction. In: IEEE International Conference on Robotics and Automation, vol. 2, pp 1901–1906. IEEE (2004)
2. Corrigan Lee, J., Peters, C., Küster, D., Castellano, G.: Engagement perception and generation for social robots and virtual agents. In: Toward Robotic Socially Believable Behaving Systems, vol. I, Intelligent Systems Reference Library 105, pp. 29–51. Springer (2016)
3. Cuzzocrea, A., Pilato, G.: Taxonomy-based detection of user emotions for advanced artificial intelligent applications. In: de Cos Juez, F., et al. (eds.) Hybrid Artificial Intelligent Systems. HAIS 2018. Lecture Notes in Computer Science, vol. 10870. Springer, Cham (2018)
4. D'Avanzo, E., Pilato, G.: Mining social network users opinions' to aid buyers' shopping decisions. Comput. Hum. Behav. **51**, 1284–1294 (2014)
5. D'Avanzo, E., Pilato, G., Lytras, M.D.: Using twitter sentiment and emotions analysis of Google trends for decisions making. Program **51**(3) (2017)
6. Delaherche, E., Dumas, G., Nadel, J., Chetouani, M.: Automatic measure of imitation during social interaction: a behavioral and hyperscanning-EEG benchmark. Patt. Recognit. Lett. (2014)

7. Ekman, P., Friesen, W.V.: Constants across cultures in the face and emotion. J. Personal. Soc. Psychol. **17**, 124 (1971)
8. Interactive Advertising Bureau (IAB) Contextual Taxonomy. http://www.iab. net/. Accessed Dec 2017
9. Kanagasabai, R., Veeramani, A., Ngan, L.D., Yap, G.E., Decraene, J., Nash, A.S.: Using semantic technologies to mine customer insights in telecom industry. In: International Semantic Web Conference (Industry Track) (2014)
10. Jang, J.-S.R.: ANFIS: adaptive-network-based fuzzy inference system. In: IEEE Transactions on Systems, Man, and Cybernetics, pp. 665–685. IEEE (1993)
11. Landauer, T.K., Dumais, S.T.: A solution to Plato's problem: the latent semantic analysis theory of acquisition, induction, and representation of knowledge. Psychol. Rev. **104**(2), 211–223 (1990)
12. Landauer, T.K., Foltz, P.W., Laham, D.: An introduction to latent semantic analysis. Discourse Process. **25**, 259–284 (1998)
13. Liu, B.: Sentiment analysis and subjectivity. In: Indurkhya, N., Damerau, F.J. (eds.) Handbook of Natural Language Processing, pp. 627–665. CRC Press (2010)
14. Pang, B., Lee, L., Vaithyanathan, S.: Thumbs up? Sentiment classification using machine learning techniques. In: Proceedings of the ACL-02 Conference on Empirical Methods in Natural Language Processing, vol. 10, pp. 79–86. Association for Computational Linguistics (2002)
15. Pilato, G., D'Avanzo, E.: Data-driven social mood analysis through the conceptualization of emotional fingerprints. Procedia Comput. Sci. (2018, in press)
16. Averkin, A.N., Yarushev, S.: Hybrid approach for time series forecasting based on ANFIS and fuzzy cognitive maps. In: 2017 XX IEEE International Conference Soft Computing and Measurements (SCM), pp. 379–381. IEEE (2017)
17. Strapparava, C., Mihalcea, R.: Semeval-2007 task 14: affective text. In: Proceedings of the 4th International Workshop on Semantic Evaluations, pp. 70–74. Association for Computational Linguistics (2007)
18. Strapparava, C., Mihalcea, R.: Learning to identify emotions in text. In: SAC 2008 Proceedings of the 2008 ACM Symposium on Applied Computing (2008)
19. Terrana, D., Augello, A., Pilato, G.: Facebook users relationships analysis based on sentiment classification. In: Proceedings of 2014 IEEE International Conference on Semantic Computing (ICSC), pp. 290–296 (2014)

Deep Learning in Vehicle Pose Recognition on Two-Dimensional Images

Dmitry Yudin$^{(\boxtimes)}$ and Ekaterina Kapustina

Belgorod State Technological University named after V.G. Shukhov, Kostukova
Str. 46, Belgorod 308012, Russia
ydin.da@bstu.ru

Abstract. The paper describes usage of deep neural network architectures such
as VGG, ResNet and InceptionV3 for the classification of small images. Each
image may contain one of four vehicle pose categories or background. An
iterative procedure for training a neural network is proposed, which allows us to
quickly tune the network using wrongly classified images on test sample.
A dataset of more than 23,000 marked images was prepared, of which 70% of
images were used as a training sample, 30% as a test sample. On the test sample,
the trained deep convolutional neural networks are ensured the recognition
accuracy for all classes of at least 93.9%, the classification precision for different
vehicle poses and background was from 85.29% to 100.0%, the recall was from
81.9% to 100.0%. The computing experiment was carried out on a graphics
processor using NVIDIA CUDA technology. It showed that the average processing time of one image varies from 3.5 ms to 15.9 ms for different architectures. Obtained results can be used in software for image recognition of road
conditions for unmanned vehicles and driver assistance systems.

Keywords: Image recognition · Classification · Vehicle pose · Deep learning
Convolutional neural network

1 Introduction

At the present time, the technologies related to unmanned vehicles, driver assistance
systems, video surveillance systems for violations on the roads are actively developing.
They require the solution of the task of recognizing vehicles on images. At the same
time, such a task should, as a rule, be solved by an onboard visual vision system in real
time. During the vehicle recognition on the image in addition to its presence/absence it
is desirable to determine the car pose.

There are 3D-model based methods of vehicle matching in large variations of pose
and illumination [1–3]. They do not need training and re-rendered vehicle images in
any other viewpoints and illumination conditions can be obtained from just one single
input 2D image. But for reliable operation in real road conditions such methods are
difficult to apply, since they require the presence of 3D models of all possible vehicles.

Methods of vehicle pose estimation based on 3D-reconstruction are described in
detail in [4–6]. Each single vehicle model consists of a cloud of 3D points in a class-
centered coordinate frame. Along with these 3D models authors store a collection of

© Springer Nature Switzerland AG 2019
A. Abraham et al. (Eds.): IITI 2018, AISC 875, pp. 126–137, 2019.
https://doi.org/10.1007/978-3-030-01821-4_14

regions obtained from the set of training images at different scales. Each image region (patch) is associated with a particular 3D position and a particular viewpoint and scale. These patches are the basic elements of nonparametric model. Estimation of car pose was carried out using reconstructed model and support vector machine [4]. The limitation of such methods is also the necessity of having a detailed data set which contains different vehicle 3D models associated with regions of two-dimensional images.

Unsupervised generation of a viewpoint annotated car dataset from videos is described in the paper [7]. It provides explicit 3D reconstruction to automatically generate viewpoint and bounding box annotation for a video. But for all vehicles on the road it is difficult to prepare such videos.

A lot of researches are devoted pose estimation based on database of images acquired at a car show on which cars were rotating on a platform [8–11]. However, this method of preparing a data set for pose recognition is difficult to implement for all vehicle types.

Key point analysis proposed in the paper [12] may be used to produce reliable method of vehicle pose recognition. This method requires the creation of a data set, where 61 key points corresponding to the various vehicle design elements are marked for each image. This approach is extremely time-consuming.

Method of 3D object localization tested on the KITTI and TANGO datasets is presented in [13]. Estimated ellipsoids of detected vehicle may give information about its pose. This approach requires further expansion and testing in various road conditions.

Detail survey of advances on application of deep learning for recovering object pose is presented in the paper [14]. It contains information about main popular datasets and vehicle pose recognition methods based on deep learning approaches which can automatically extract features from images. The authors of the paper note the prospects for further research in this field.

Paper [15] describes Discriminative Components Fully Convolutional Network which allows find special vehicle image segments: wheel, headlight, side rear view mirror, front air-intake lattice. This requires less time-consuming procedure for creating labeled dataset with appropriate annotation masks than method from [12].

Deep learning approach based on 3D boxes as CNN Input is proposed in the paper [16]. It provides new dataset which contains images from fixed surveillance cameras.

Vehicle pose recognition task may be solved as classification task. Deep learning is the most effective tool for classification problems of graphic images. It can use different architectures of convolutional neural networks and different methods of their training. For example, in 2014, over 20.000 image categories were recognized from the ImageNet database using the Very deep convolutional network (VGG) architecture [17], the GoogLeNet architecture were applied for the same task, based on Inception units [18]. In 2015 the architecture of the Residual Deep Neural Network (ResNet) [19] is appeared. These architectures show different quality of the image classification on different datasets. So for each specific task we require make experimental verification of different deep architectures of neural networks.

2 Task Formulation

In this paper we will solve the task of determining one of the four vehicle pose categories ("rear view", "rear view at an angle", "front view", "front view at an angle", see the Fig. 1) or the fact of vehicle absence on the image (background), see Fig. 2.

We developed the data set which contains 23267 images obtained from in-vehicle cameras, manually cropped and labeled [20]. The training sample is 70% of the total number of available images, the test sample 1 contains 20% of data set, the test sample 2 includes 10% of data set (see Table 1). Samples contain images of different sizes from 38×38 to 1000×1000 pixels that are obtained from different cameras in different shooting conditions.

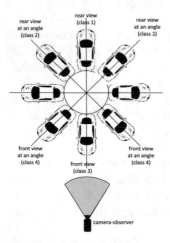

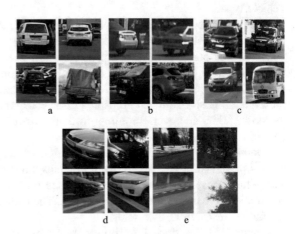

Fig. 1. Vehicle pose categories used in prepared dataset.

Fig. 2. Images corresponding to classes 1–5: a – class 1 ("rear view"), b – class 2 ("rear view at an angle"), c – class 3 ("front view"), d – class 4 ("front view at an angle"), e – class 5 ("background")

Table 1. Training and testing samples of used dataset

Number of class	Training sample	Test sample 1	Test sample 2
Class 1 ("rear view")	2947	843	426
Class 2 ("rear view at an angle")	2767	790	395
Class 3 ("front view")	2394	684	342
Class 4 ("front view at an angle")	3130	894	447
Class 5 ("background")	5046	1441	721
Total:	16284	4652	2331

To solve the task it is necessary to develop various variants of deep neural network architectures and to test them on the available data set. Next, we need to select the best architecture that will provide best performance and the highest quality measures of image classification: accuracy, precision and recall [21].

3 Classification of Vehicle Pose Categories Using Deep Convolutional Neural Networks

In this paper to solve formulated task we investigate the application of a deep convolutional neural networks of three architectures:

- VGGm architecture which is similar to VGG [17, 22]. Its structure is shown in Fig. 3 and contains 6 convolutional layers, 3 max pooling layers and two output dense layers. This approach allows, for example, on the first layer to get the color features of the image, information about the boundary points, corners, sections of straight lines, on the second layer – to get their combinations, on the third layer – to reduce the image dimension twice both for speeding up the calculations and for reducing sensitivity to image shift. Further blocks similarly allow receive features of a higher level of abstraction. Thus, first 9 layers provide automatic feature extraction on the image. The last two fully connected layers allow us to classify founded features and determine one of five image classes;
- ResNetm architecture which is similar to ResNet [19]. Its structure is shown in Fig. 4 and contains 3 convolutional blocks, 5 identity blocks, 2 max pooling layers, 1 average pooling layer and one output dense layer. First 11 layers and blocks provide automatic feature extraction and the last one fully connected layer allows us to find one of five image classes corresponding to input image;
- InceptionV3m architecture which is simplified InceptionV3 structure [18]. Its structure is shown in Fig. 5 and contains 4 "conv2d_bn" blocks (convolutional layer with batch normalization layer), 5 different mixed blocks, 1 max pooling layer, 1 global average pooling layer and one output dense layer. First 10 blocks provide automatic feature extraction and the last one fully connected layer allows us to get one of five image classes.

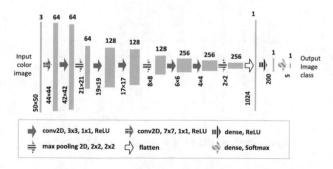

Fig. 3. VGGm architecture.

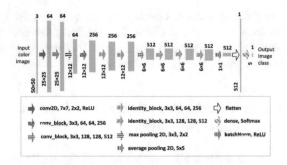

Fig. 4. ResNetm architecture.

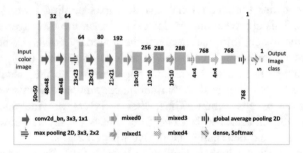

Fig. 5. InceptionV3m architecture.

Output layer in all architecture has 5 neurons with "Softmax" activation function. All input images are pre-scaled to a size of 50 × 50 pixels and subjected to small shift and scaling operations. Neural networks works with color (three-channel) images.

To train the neural network we have used "categorical crossentropy" loss function, Stochastic Gradient Descent (SGD) as training method with 0.0001 learning rate. Accuracy is used as classification quality metric during training. The batch is consisted of 16 images.

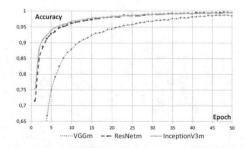

Fig. 6. Deep neural networks training on 50 epochs.

The training process of deep neural networks is shown in Fig. 6. The training experiment was carried out for 50 learning epochs using our developed software tool [23] implemented on Python 3.5 programming language. We can see that InceptionV2m and ResNetm networks have similar speed and accuracy, while VGGm needs more epochs to achieve high accuracy level.

The calculations had performed using the NVidia CUDA technology on the graphics processor of the GeForce GTX 960 graphics card with 2.00 GB of RAM.

Table 2 shows the results of the vehicle pose recognition on training and test samples using VGGm, ResNetm and InceptionV3m architectures.

Analysis of the obtained results shows the highest accuracy on all samples for Inception V3 architecture. It also has the greatest values of precision (P) and recall (R) for almost all categories (classes) of vehicle pose. ResNetm is better only in terms of recognition recall of 3rd class on test sample 2. VGGm is better in terms of recognition precision of 3rd and 4th classes respectively on test samples 1 and 2.

On training sample the accuracy metric for InceptionV3m architecture is 0.9999, recognition precision and recall of each class are above 0.9996. At the same time on the test samples accuracy, precision and recall fall. This is especially evident for classes 2, 3 and 4. Recognition recall of 2nd class is less than 0.33 on test sample 1. Recognition precision of 3rd and 4th classes on test sample 1 respectively equal to 0.8558 and 0.6257. On the test sample 2 the recognition quality is slightly better. This indicates that test samples contain new objects which are significantly different in shape, color and other characteristics from the objects in the training set.

4 Fine-Tuning of Deep Convolutional Neural Networks for Quality Increasing of Vehicle Pose Recognition

In this section we will examine the fine-tuning of trained neural networks with the addition of wrongly classified images from the test sample 1 into the training sample. This should lead to improving the recognition quality of the deep neural networks on all image datasets. To check it we propose an iterative procedure, shown in Fig. 7. Tuning iterations are repeated if the recognition quality on the test sample is not enough.

Using this tuning procedure a computer experiment was performed for all architectures. Deep neural networks fine-tuning process is shown in Fig. 8. The highest accuracy on 10 epochs was achieved by InceptionV3m architecture.

Results of vehicle pose classification by tuned neural networks are shown in the Table 3. It demonstrates that for all architectures classification quality have increased on both test samples and remained about the same level for the training sample. For example for InceptionV3m architecture:

- recall of 2nd class on test sample 1 has increased from 0.3266 to 0.9886;
- precision of 3rd class on test sample 1 has increased from 0.8558 to 0.9740;
- precision of 4th class on test sample 1 has increased from 0.6257 to 0.9806;
- accuracy on test sample 2 has increased from 0.9318 to 0.9391;
- accuracy on training sample has slightly changed from 0.9999 to 0.9996.

Table 2. Quality of vehicle pose recognition on different samples for 50 training epochs

Sample	Acc.	Class 1		Class 2		Class 3		Class 4		Class 5	
		P	R	P	R	P	R	P	R	P	R
VGGm architecture											
Training sample	0.9952	0.9986	**1.0000**	0.9967	0.9917	0.9992	0.9904	0.9864	0.9942	0.9960	0.9972
Test sample 1	0.8304	0.9892	0.9810	0.9227	0.2418	**0.8805**	0.8830	0.5710	0.9351	0.9630	0.9750
Test sample 2	0.8850	0.9508	0.9977	0.9710	0.6785	0.7275	0.8509	**0.8608**	0.8300	0.9112	0.9820
ResNetm architecture											
Training sample	0.9982	0.9997	0.9993	0.9982	0.9993	0.9950	0.9983	0.9978	0.9952	0.9990	0.9986
Test sample 1	0.8368	0.9882	0.9905	0.9696	0.2823	0.7587	0.8918	0.6136	0.8971	0.9707	0.9875
Test sample 2	0.9095	0.9682	**1.0000**	0.9905	0.7949	0.8231	**0.8977**	0.8049	0.8031	0.9457	0.9903
InceptionV3m architecture											
Training sample	**0.9999**	**1.0000**	0.9997	**1.0000**	0.9996	**0.9996**	**1.0000**	**0.9997**	**1.0000**	**1.0000**	**0.9986**
Test sample 1	**0.8624**	**0.9988**	**0.9941**	**0.9923**	0.3266	0.8558	0.9108	**0.6257**	**0.9687**	**0.9903**	**0.9903**
Test sample 2	**0.9318**	**0.9861**	0.9977	**0.9971**	**0.8608**	**0.8449**	0.8918	0.8276	**0.8591**	**0.9782**	**0.9958**

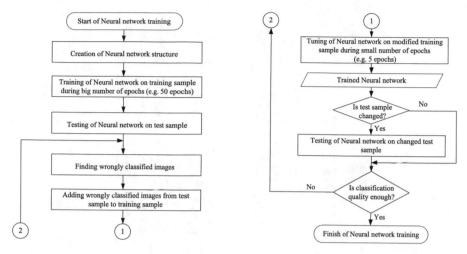

Fig. 7. Flow chart of proposed iterative procedure of deep neural network training using wrongly classified images on testing dataset.

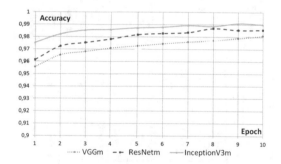

Fig. 8. Deep neural networks fine-tuning on 10 epochs.

Thus the best from proposed architectures – InceptionV3m has the following classification results:

- for "rear view" (class 1) precision varies from 97.71% to 100.00%, recall is 100.00%,
- for "rear view at an angle" (class 2) precision varies from 99.45% to 100.00%, recall varies from 91.14% to 100.00%,
- for "front view" (class 3) precision varies from 85.29% to 99.96%, recall varies from 93.27% to 99.96%,
- for "front view at an angle" (class 4) precision varies from 93.37% to 99.84%, recall varies from 81.88% to 100.00%,
- for "background" (class 5) precision varies from 93.61% to 99.98%, recall varies from 99.58% to 99.93%.

Table 3. Results of neural networks tuning using wrongly classified images on test sample 1

Sample	Acc.	Class 1		Class 2		Class 3		Class 4		Class 5	
		P	R	P	R	P	R	P	R	P	R
Tuning of VGGm net with adding of 828 wrongly classified images to training sample											
Training sample	0.9909	0.9990	0.9993	0.9935	0.9902	0.9807	0.9979	0.9752	0.9936	0.9998	0.9814
Test sample 1	0.9536	0.9953	0.9976	0.9476	0.9165	0.8958	0.9927	0.8999	0.9150	**0.9985**	0.9535
Test sample 2	0.9043	0.9444	0.9977	0.9735	0.7443	0.7512	0.9094	0.8828	**0.8591**	**0.9507**	0.9626
Tuning of ResNetm net with adding of 789 wrongly classified images to training sample											
Training sample	0.9982	0.9993	**1.0000**	**1.0000**	0.9949	0.9958	**0.9996**	0.9955	0.9978	0.9994	0.9986
Test sample 1	0.9755	0.9941	0.9976	0.9693	0.9582	0.9415	0.9883	0.9559	0.9452	0.9972	0.9847
Test sample 2	0.9142	0.9552	**1.0000**	0.9808	0.7772	0.8093	0.9181	0.8726	0.8277	0.9395	0.9903
Tuning of InceptionV3m net with adding of 640 wrongly classified images to training sample											
Training sample	**0.9996**	**1.0000**	**1.0000**	**1.0000**	**1.0000**	**0.9996**	0.9996	**0.9984**	**1.0000**	**0.9998**	**0.9988**
Test sample 1	**0.9882**	**0.9965**	**1.0000**	0.9949	0.9886	0.9740	0.9839	0.9806	0.9620	0.9911	**0.9993**
Test sample 2	**0.9391**	0.9771	**1.0000**	0.9945	0.9114	0.8529	0.9327	0.9337	0.8188	0.9361	0.9958

We have tested the performance of neural networks for vehicle pose recognition task (Table 4). Average classification time of one image for the most accurate architecture InceptionV3m is 5 time higher than for VGGm architecture. InceptionV3m network also has the highest number of trainable parameters (3432453 parameters) and training time (6331.9 s on 50 epochs).

Table 4. Performance of neural networks for vehicle pose recognition task

Neural network architecture	Number of trainable parameters	Training time on 50 epochs, sec	Average classification time of one image, ms
VGGm	1949173	2192.0	3.5
ResNetm	1452549	4311.2	11.2
InceptionV3m	3432453	6331.9	15.9

5   Conclusions

It follows from the Table 3 that the applied architectures of a deep neural network show high quality indicators for the considered task of classifying four vehicle poses categories and the background. The most accurate architecture is InceptionV3m. It gave classification accuracy on training sample higher than 0.999.

We have proposed and implemented network tuning procedure using dataset with merged training sample and wrongly classified images from test sample 1. As a result we have obtained that all quality measures have significantly increased for test sample 1 and slightly improved for test sample 2. So this tuning procedure allows us iterative expand training dataset with only wrongly classified images from new test samples received by machine vision systems. This expanded dataset leads to increase in classification quality of new images.

An important aspect for the further application of the considered approaches is the average classification time of one image. It varies from 3.5 ms for VGGm to 15.9 ms for InceptionV3m. This suggests that the proposed neural networks can be integrated into a software complex operating in real time.

To further studies on the paper topic it is necessary to expand the training and test samples to cover as many various vehicles in different conditions as possible and to add more vehicle poses.

Acknowledgment. This article is written in the course of the grant of the President of the Russian Federation for state support of young Russian scientists № MK-3130.2017.9 (contract № 14.Z56.17.3130-MK) on the theme "Recognition of road conditions on images using deep learning".

References

1. Hou, T., Wang, S., Qin, H.: Vehicle matching and recognition under large variations of pose and illumination. In: International Conference Computer Vision and Pattern Recognition, pp. 24–29 (2009)
2. Prokaj, J., Medioni, G.: 3-D model based vehicle recognition. In: Workshop on Applications of Computer Vision (WACV), pp. 1–7 (2009)
3. Jayawardena, S., Hutter, M., Brewer N.: A novel illumination-invariant loss for monocular 3D pose estimation. In: International Conference on Digital Image Computing: Techniques and Applications, pp. 37–44 (2011)
4. Glasner, D., Galun, M., Alpert, S., Basri, R., Shakhnarovich G.: Viewpoint-aware object detection and pose estimation. In: IEEE International Conference on Computer Vision, pp. 1275–1282 (2011)
5. Penate-Sanchez, A., Moreno-Noguer, F., Andrade-Cetto, J., Fleuret, F.: LETHA: learning from high quality inputs for 3D pose estimation in low quality images. In: IEEE Second International Conference on 3D Vision, pp. 517–524 (2014)
6. Novotny, D., Larlus, D., Vedaldi, A.: Learning 3D object categories by looking around them. In: IEEE International Conference on Computer Vision, pp. 5228–5237 (2017)
7. Sedaghat, N., Brox, T.: Unsupervised generation of a viewpoint annotated car dataset from videos. In: IEEE International Conference on Computer Vision, pp. 1314–1322 (2015)
8. Ozuysal, M., Lepetit, V., Fua P.: Pose estimation for category specific multiview object localization. In: IEEE Conference on Computer Vision and Pattern Recognition, CVPR 2009, pp. 778–785 (2009)
9. Teney, D., Piater, J.: Continuous pose estimation in 2D images at instance and category levels. In: International Conference on Computer and Robot Vision, pp. 121–127 (2013)
10. Pedersoli, M., Tuytelaars, T.: A scalable 3D HOG model for fast object detection and viewpoint estimation. In: IEEE Second International Conference on 3D Vision, pp. 163–170 (2014)
11. Bakry, A., Elgaaly, T., Elhoseiny, M., Elgammal, A.: Joint object recognition and pose estimation using a nonlinear view-invariant latent generative model. In: IEEE Winter Conference on Applications of Computer Vision (WACV), pp. 1–9 (2016)
12. Gu, H.-Z., Lee, S.-Y.: Car model recognition by utilizing symmetric property to overcome severe pose variation. Mach. Vis. Appl. **24**, 255–274 (2013)
13. Rubino, C., Crocco, M., Del Bue, A.: 3D object localisation from multi-view image detections. IEEE Trans. Pattern Anal. Mach. Intell. **40**(6), 1281–1294 (2018)
14. Li, W., Luo, Y., Wang, P., Qin, Z., Zhou, H., Qiao, H.: Recent advances on application of deep learning for recovering object pose. In: Proceedings of the 2016 IEEE International Conference on Robotics and Biomimetics, Qingdao, China, pp. 1273–1280 (2016)
15. Sochor, J., Herout, A., Havel, J.: BoxCars: 3D boxes as CNN input for improved fine-grained vehicle recognition. In: IEEE Conference on Computer Vision and Pattern Recognition, pp. 3006–3015 (2016)
16. Xue, Y., Qian, X.: Vehicle detection and pose estimation by probabilistic representation. In: ICIP 2017, pp. 3355–3359 (2017)
17. Szegedy, C., Vanhoucke, V., Ioffe, S., Shlens, J., Wojna Z.: Rethinking the inception architecture for computer vision. In: ECCV, arXiv:1512.00567 (2016)
18. Kaiming, H., Xiangyu, Z., Shaoqing, R., Jian S.: Deep residual learning for image recognition. In: ECCV, arXiv:1512.03385 (2015)

19. Simonyan, K., Zisserman, A.: Very deep convolutions for large-scale image recognition. In: ICLR, arXiv:1409.1556 (2015)
20. Yudin, D.A., Kapustina, E.O.: Dataset containing four car pose categories and background (2017). https://yadi.sk/d/xjQKIoyU3NNVyt. Last Accessed 11 May 2018
21. Olson, D.L., Delen D.: Advanced Data Mining Techniques, 1st edn. Springer, Heidelberg (2008)
22. Yudin, D., Knysh, A.: Vehicle recognition and its trajectory registration on the image sequence using deep convolutional neural network. In: The International Conference on Information and Digital Technologies, pp. 435–441 (2017)
23. DeepClassificationTool. Deep image classification tool based on Keras. https://github.com/yuddim/deepClassificationTool. Last Accessed 11 May 2018

Feature Extraction of High-Frequency Patterns with the a Priori Unknown Parameters in Noised Electrograms Using Spectral Entropy

Nikolay E. Kirilenko[1](✉), Igor V. Shcherban'[2],
and Andrey A. Kostoglotov[1]

[1] Rostov State Transport University, Rostov-on-Don 344038, Russia
nikolai-kirilenko@mail.ru
[2] Southern Federal University, Rostov-on-Don 344090, Russia

Abstract. **Statement of the problem**: the article looks into a class of problems which require identification of hidden regularities in adjustment of the bioelectrical activity of living organisms registered against various stimuli, through search and temporal localization of patterns in noised electrograms containing useful information. One of the approaches to solving such problems is based on an analysis of the Shannon entropy calculated based on the components of the power spectrum and called the spectral entropy function. It is found that under the conditions providing that the patterns in question pertain to high-frequency rhythms, and the boundaries of their energy spectra are a priori unknown, the criterial functions of spectral entropy are of low sensitivity. **Purpose of the research**: to develop cost functions of entropy analysis which are sufficiently sensitive for searching for high-frequency patterns with a priori unknown parameters in noised electrograms. **Results**: development of a cost function that makes it possible to find the frequency sub-band where the spectral components of the patterns in question maximally contribute to the total spectrum power. The subsequent computation of the spectral entropy in the identified frequency sub-band provides a solution to the problem of search for the response patterns in noised electrograms under the above conditions. **Practical significance:** the results confirm the effectiveness of the developed functions whose use is limited by the requirement that the electrogram should be recorded on more than one lead.

Keywords: Electrogram · High-frequency pattern · Shannon function
Spectral entropy · Local frequency sub-band

1 Introduction

The studies of reactions of living organisms to external stimuli solve the problems of analysis of response electrograms and search for patterns containing useful information and their identification in such electrograms [1]. The analysis of different entropy functions [2–6] is sufficiently effective. Spectral entropy (SE) is the Shannon function computed from the components of the power spectrum [7]. SE has been widely used in

© Springer Nature Switzerland AG 2019
A. Abraham et al. (Eds.): IITI 2018, AISC 875, pp. 138–147, 2019.
https://doi.org/10.1007/978-3-030-01821-4_15

biomedical research, for example, in problems of identifying the stages of anaesthesia based on recorded electrograms [8, 9], for assessing chaotic heart rhythm [10], diagnosing epilepsy [11–14] and other application.

Similar practical problems, when entropy uses in the context of pattern classification and information technology, have a great appliance also in technologies for industry. The SE employed as a quantitative measure for monitoring the output noised signals of a various technical dynamical systems and provided a useful tool to identify their models in physics [15], in electric-power industry [16], in communication technique [17], in problems, related to the non-destructive control [18], and others.

At the same time, the problems of using SE in certain special cases still exist. It is still unclear how to choose the boundaries of the analysed frequency sub-band for computing SE [19–22]. No problem arises when these boundaries are a priori known with sufficient accuracy. But the feature extraction of high-frequency pattern from noised electrograms using spectral entropy is often impossible. This can be explained by the fact that the contribution of low-frequency components to the total spectrum power of the entire signal is significant, which accounts for the sensitivity of the SE cost function being sufficiently high to describe the reactivity effects. The same fact accounts for the low sensitivity of the SE cost function, if the desired patterns are high-frequent and the boundaries of their energy spectra are a priori unknown

In addition, the sampled electrograms may be corrupted by various noises, such as baseline wander, electrode motion, power line interference, motion artifact and so on [1]. This does not allow obtaining detailed statistical data on the properties of noise components, on high-frequency artifacts of the sampled electrograms. At the same time, the energy spectra of noises, high-frequency artifacts and patterns in question may overlap or be rather close. E.g., in the course of the experimental studies, the worst results of applying the SE method were observed in the problems of searching through electrograms for patterns with a relatively low amplitude and narrowband frequency spectra within 70..150 Hz.

To overcome the above problems, there has been developed an additional cost function, which makes it possible to search for the frequency range that would most closely correlate with the frequencies of the spectral components of the patterns in question maximizing the total electrogram spectrum power. This increases the sensitivity of the SE cost function in the problems of feature extraction of high-frequency patterns, especially when the energy spectrum boundaries of these patterns are a priori unknown. The use of the developed additional cost function is limited by the requirement that the sampled electrogram should be observed on several leads. The results of its practical application are presented.

2 Search for and Temporal Localization of SE Response Patterns in Electrograms

There still exist certain problems of using SE in studies of reactions of living organisms to external stimuli.

It is still unclear how to choose boundaries of the analysed frequency sub-band.

Suppose that, simultaneously for time moments $t \in [0; \ T]$ on N leads of a multidimensional electrogram (EG), it is possible to register $X(t) = (x_1(t) \ \ x_2(t) \ \ \ldots \ \ x_N(t))^T$, where T is the sign of transposition; N is the number of sensors. Due to the discreteness of the procedure of measuring EG with a constant time increment $\Delta t = 1$, each J-th component of the EG $x_J(t)$ $(J = \overline{1, N})$ of the n number size is written as

$$t_k = t_0 + (k-1)\Delta t : x_{J,k} = x_J(t)|_{t=t_k}, \ k = 1, 2, \ldots, n. \tag{1}$$

It is assumed that the EG (1) contains the pattern of response bioelectric activity that exists on a short time interval $\Delta T_p \in [0; \ T]$, $\Delta T_p < \ <T$. The temporal position of the pattern in the EG (1) is unknown.

The time series (1) is traditionally divided into $L > 1$ temporal epochs and an epoch most likely to contain the pattern of a bioelectrical response is searched for [2]. Through consecutive application of the Fourier transformation to each l-th sample from the series (1), where $l = 1, 2, \ldots, L$ is the number of the epoch, corresponding power spectrum density (PSD) functions $P_l(f)$ are identified and normalized [10]:

$$\sum_{f_i} \tilde{P}_l(f_i) = C_l \sum_{f_i} P_l(f_i) = 1; f_i \in \left[f^{\min}; f^{\max}\right], \tag{2}$$

Where $f^{\min}, f^{\max}$ are the boundaries of the analysed local frequency sub-band; $\tilde{P}_l(f_i)$ is the normalized PSD function for each l-th epoch; C_l are the normalization coefficients. Further SEs are calculated

$$H_l\left[f^{\min}; f^{\max}\right] = -\sum_{f_i} \tilde{P}_l(f_i) \log\left(\tilde{P}_l(f_i)\right), \tag{3}$$

associated with a particular functional, search for whose extremum reflects the essence of the solution of the problem formulated above. The criterion [2] is frequently used

$$\max_l \left\{ \Gamma_l = \frac{H_l - H}{H} \right\}, \tag{4}$$

where Γ_l is a variation of the SE H_l of the l-th epoch relative to the SE H computed similarly to (2), but for the entire time series (1). The larger the value of the cost function is, the more the "event-related" electrogram in this epoch differs from the entire time series (1), the more likely the presence of a pattern of the response bioelectrical activity is.

It is obvious that in formulas (2)–(4) the method of choosing the boundary values of $f^{\min}$, $f^{\max}$ of the examined frequency band remains unclear. In the course of the experiments, there was found a significant dependence of the sensitivity of the cost functions SE (2)–(4) on the accuracy of the choice of the boundaries $f^{\min}$, $f^{\max}$.

Suppose that the sensitivity of the SE cost functions can be increased by using for computation not an extended frequency band $[f^{\min};f^{\max}]$ but a narrower sub-band

$$[\hat{f}^{\min};\hat{f}^{\max}]:\left(\hat{f}^{\max}-\hat{f}^{\min}\right)<<\left(f^{\max}-f^{\min}\right). \tag{5}$$

It is obvious that the choice of estimates $\hat{f}^{\max},\hat{f}^{\min}$ must be made based on the condition that it is the spectral components of the patterns in question that make the maximum contribution to the total spectrum power in the narrow frequency sub-band (5). It is known that the maximum values of entropy are reached for the signal representing white noise, since in this case the spectral components are evenly distributed over the whole band of frequencies involved and the relative energies are practically uniform at all resolution levels. The relative energy contribution of the pattern concentrates around a certain frequency sub-band. The spectral components are distributed unevenly over frequencies and some of them, depending on the properties of the pattern, make the maximum contribution to the total electrogram spectrum power. Consequently, in those epochs in which the pattern is present, a burst of certain spectral components of the PSD should be observed. Therefore, the entropy of the distribution over time epochs of any individual spectral component from the extended frequency band will be all the higher as the distribution is more even. If a burst of any spectral component is observed at a certain time epoch, which may indicate a "contribution" of the pattern, then the entropy of the distribution of this particular spectral component over the time epochs will be low.

Proceeding from the above arguments, we formulate the cost function

$$h_{f_i}=-\sum_{l}\tilde{P}_l(f_i)\log\left(\tilde{P}_l(f_i)\right), \tag{6}$$

where h_{f_i} is the entropy of the epoch distribution of each spectrum component of the frequency f_i from the extended frequency band $f_i\in[f^{\min};f^{\max}]$.

Function (6) will have a minimum value for those i-th spectral components that are maximally unevenly distributed over the time epochs, which will indicate the presence of characteristic local features in the EG at the frequency f_i.

It is clear that, at any frequencies from the extended frequency band and for various reasons, the noised EG may have various spontaneous oscillations – local features that are in no way related to the desired patterns. These local oscillator structures will lead to the same effects as the desired patterns – that is minimizing function (6). Therefore, we will further take into account the fact that the EG is registered simultaneously on N leads. It is clear that the temporal and frequency characteristics – the shape, the average amplitude of significant oscillations, the width of the spectrum, the duration and temporal location of each spontaneous local feature in each J-th component $(J=\overline{1,N})$ of the EG – will be different. The same characteristics belonging to the response patterns registered on all N leads, will, on the contrary, coincide within small values. Therefore, averaging over all N leads of the same values of each J-th function (6) will allow finding the narrow frequency sub-band (5), which corresponds to the maximum contribution of the spectral components of the response patterns to the total EG spectrum power.

Therefore, we specify the criterion for determining the estimates $\hat{f}^{\max}, \hat{f}^{\min}$ of the narrow frequency sub-band corresponding to the maximum contribution to the total spectrum power of exactly the spectral components of the sought-for response bio-electrical activity patterns as follows

$$\min_{f_i} \left\{ \tilde{h}_{f_i} = \frac{1}{N} \sum_{J=1}^{N} h_{J,f_i} \right\}, \tag{7}$$

where $h_{J,f_i} = -\sum_l \tilde{P}_{J,l}(f_i) \log\left(\tilde{P}_{J,l}(f_i)\right)$ is the entropy of the distribution over the time epochs of the spectral component of frequency $f_i \in \left[f^{\min}; f^{\max}\right]$ for the J-th component of the EG; $P_{J,l}(f_i)$, $\tilde{P}_{J,l}(f_i)$ are respectively, the PSD and the normalized PSD of the l-th sample of the J-th component of the EG,

$$\sum_{f_i} \tilde{P}_{J,l}(f_i) = C_{J,l} \sum_{f_i} P_{J,l}(f_i) = 1; \; l = 1, 2, \ldots, L,$$

$C_{J,l}$ are corresponding normalization coefficients.

We should also note the robustness of the function $\tilde{h}_{f_i}$ of criterion (7) to the inevitable temporal variations in the time of observation of response patterns and their shapes on different leads. For example, it would be incorrect to average over N leads of the same k^{th} values of a multivariate EG

$$\tilde{x}_k = \frac{1}{N} \sum_{J=1}^{N} x_{J,k}, \tag{8}$$

with the subsequent computation of function (6) for the averaged EG $\tilde{x}_k$ precisely because of the presence of temporal shifts in the response patterns over the leads. Such temporal shifts can lead to the effect which in radio-communication is called fading of a radio signal due to fluctuations of amplitudes, phases, and angles of arrival of patterns in each lead. Accordingly, in some cases the patterns in different leads can be observed, for example, in antiphase and, thus, averaging over the EG components (8), as opposed to (7), will not lead to the expected effect of EG recording on multiple leads.

We further rewrite formulae (3) and (4) as follows:

$$H_{J,l}\left[\hat{f}^{\min}; \hat{f}^{\max}\right] = -\sum_{f_i} \tilde{S}_{J,l}(f_i) \log\left(\tilde{S}_{J,l}(f_i)\right); \tag{9}$$

$$\max_{l} \left\{ \tilde{\Gamma}_l = \frac{\tilde{H}_l - \tilde{H}}{\tilde{H}} \right\}, \tag{10}$$

Where $\tilde{H}_l = \frac{1}{N}\sum_{J=1}^{N} H_{J,l}$, $\tilde{H} = \frac{1}{N}\sum_{J=1}^{N} H_J$ are the corresponding SE values averaged over N leads; $f_i \in \left[\hat{f}^{\min}; \hat{f}^{\max}\right]$.

The effectiveness of cost functions (7), (9), (10) was studied on the same practical examples.

3 Numerical Studies of the Proposed Cost Functions

We studied the cases when additive high-frequency patterns of low intensity were encapsulated in the multidimensional EG $\left\{z_{J,k}\right\}_{k=1}^{n}$, $N = 8$:

$$\left\{x_{J,k}\right\}_{k=1}^{n} = \left\{z_{J,k} + y_k^{(q)}\right\}_{k=1}^{n}, \tag{11}$$

where $y_k(q)$ is a simulated additive pattern of the qth shape.

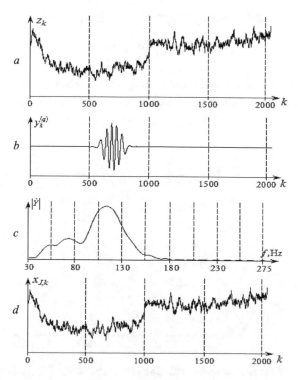

Fig. 1. The examples of the J-th EG component without pattern $z_{J,k}$(a), the pattern of the q-th type $y_k^{(q)}$(b) and its spectrum (c), as well as model EG $x_{J,k}$ (d).

The analysis involved model discrete samples $\left\{x_{J,k}\right\}_{k=1}^{2048}$, $J = 1, 2, \ldots, 8$, with the size of $n = 2048$ readouts divided into $L = 7$ epochs of 512 readouts with overlap of 256 readouts. In order to improve the quality of the comparative analysis, there were used about $q = 30$ types of pattern models characteristic of biomedical applications differing in shape, mean amplitude of significant oscillations, spectral width, duration ΔT_p and location in the EG. In each J-th lead, random variations of shapes and time parameters of the encapsulated patterns were specified. The maximum shift of the encapsulation moments of the patterns in the EG $z_{J,k}$ was ± 25 readouts, and the variations of the mean amplitude of the significant oscillations were approximately 10%. All the characteristics of the patterns were considered "a priori unknown" for the purpose of numerical computation.

The examples of the J-th EG component without pattern $z_{J,k}$(a), the pattern of the q-th type $y_k^{(q)}$(b) and its spectrum (c), as well as model EG $x_{J,k}$ (11) (d) are shown in Fig. 1(a, b, c, d).

The entropy analysis was to clearly show the presence of the pattern $y_k^{(q)}$ in the model EG (11) in the $l = 3$ time epoch corresponding to the time interval of [512, 1024] readouts.

The extended boundary values $f^{\min}$, $f^{\max}$ were set equal to

$$f^{\min} = 50\,\text{Hz}, f^{\max} = 250\,\text{Hz}. \tag{12}$$

Figure 2 shows the graphs of the function $\tilde{h}_{f_i}$ of criterion (7) computed for EG $z_{J,k}$ $\left(J = \overline{1,8}\right)$, that did not include model patterns, and for the model EG $x_{J,k}$ (11) with the pattern $y_k^{(q)}$.

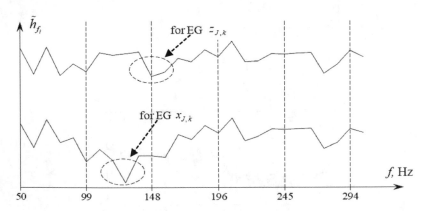

Fig. 2. The graphs of the function $\tilde{h}_{f_i}$ of criterion (7) computed for EG $z_{J,k}$ $\left(J = \overline{1,8}\right)$, that did not include model patterns, and for the model EG $x_{J,k}$ (11) with the pattern $y_k^{(q)}$.

The following estimates were selected from the graphs for subsequent computation of functions (9) and (10)

$$\hat{f}^{\min} = 110\,\mathrm{Hz}, \hat{f}^{\max} = 135\,\mathrm{Hz} \tag{13}$$

The comparison of estimates (13) with the spectrum of the model pattern $y_k^{(q)}$ (see Fig. 2) confirms the correctness of the reasoning.

Figure 3 shows the graph of the function $\tilde{\Gamma}_l$ (10) calculated for the model EG $x_{J,k}$ (11) with the pattern $y_k^{(q)}$.

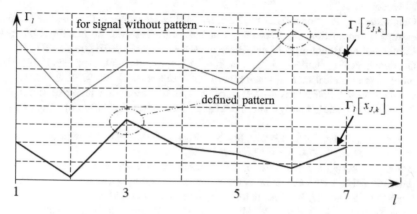

Fig. 3. The graph of the function $\tilde{\Gamma}_l$ (10) calculated for the model EG $x_{J,k}$ (11) with the pattern $y_k^{(q)}$.

The cost function $\tilde{\Gamma}_l$ clearly reflects the fact of the presence of the event-related pattern in the EG and allows us to correctly carry out its temporary localization.

4 Conclusion

The results of the numerical studies confirmed the conclusion that it is possible to increase the sensitivity of the SE cost functions through selection for analysis of a frequency sub-band corresponding to the maximum contribution of the components of bioelectric response patterns in question to the total spectrum power. The developed cost functions are sufficiently sensitive for searching in noised electrograms for the patterns of low-intensity rhythms g1...g3 with a priori unknown frequency boundaries.

The developed method was used in research of the rodent olfaction system. It allowed to extract from the noised electro-olfactograms (EOG) the time-domain potentials, evoked by experimental measurement hardware. The amplitude of the founded time-domain evoked potentials was less than 10% of the mean of EOG-amplitude and their frequency parameters were matched with frequency parameters of the biological noises. Although existence of this time-domain evoked potentials in EOG was not suggested and their visual identification was impossible. Obtained results made it possible to modify the experimental measurement hardware and to exclude this negative effects.

The limitation of the developed approach is the fact that the electrogram where the patterns are to be temporally localized should be recorded on more than one lead.

References

1. Rangayyan, R.M.: Biomedical Signal Analysis: A Case-Study Approach. Wiley-IEEE Press, New York (2007)
2. Giannakakis, G.A., Tsiaparas, N.N., Xenikou, M.S., Papageorgiou, C., Nikita, K.S.: Wavelet entropy differentiations of event related potentials in Dyslexia. In: 8th IEEE International Conference on Bioinformatics and Bioengineering, pp. 1–6. IEEE Greece (2008)
3. Gorshkov, A.A., Osadchi, A.E., Fradkov, A.L.: Regularization of EEG/MEG inverse problem with a local cortical wave pattern. Inf. Control. Syst. 5(90), 12–20 (2017)
4. Cvetkov, O.V.: Entropiinyi analiz dannykh v fizike, biologii i tekhnike (Entropy Data Analysis in Physics, Biology and Technique). SPbGETU "LETI" Publ., Saint-Petersburg (2015)
5. Zunino, L., Perez, D.G., Garavaglia, M., Rosso, O.A.: Wavelet entropy of stochastic processes. Phys. A 379, 503–512 (2007)
6. Hong, H., Yonghong, T., Yuexia, W.: Optimal base wavelet selection for ECG noise reduction using a comprehensive entropy criterion. Entropy 17, 6093–6109 (2015)
7. Inouve, T., Shinosaki, K., Sakamoto, H., Inouye, T., Shinosaki, K., Sakamoto, H., Toi, S., Ukai, S., Iyama, A., Katsuda, Y., Hirano, M.: Quantification of EEG irregularity by use of the entropy of the power spectrum. Electroencephalogr. Clin. Neurophysiol. 3(79), 204–210 (1991)
8. Viertio-Oja, H., Maja, V., Sarkela, M., Talja, P., Tenkanen, N., Tolvanen-Laakso, H., Paloheimo, M., Vakkuri, A., Yli-Hankala, A., Merilainen, P.: Description of the entropy algorithm as applied in the Datex-Ohmeda entropy module. Acta Anaesthesiol. Scand. 48, 154–161 (2004)
9. Kekovic, G., Stojadinovic, G., Martac, L., Podgorac, J., Sekulic, S., Culic, M.: Spectral and fractal measures of cerebellar and cerebral activity in various types of Anesthesia. Acta Neurobiol. Exp. 70, 67–75 (2010)
10. Ostanin, S.A., Filatova, E.V.: A virtual instrument for spectral entropy estimation of heart rate. Izvestiia Altaiskogo gosudarstvennogo universiteta 1, 45–51 (2016)
11. Mirzaei, A., Ayatollahi, A., Gifani, P., Salehi, L.: Spectral entropy for epileptic seizures detection. In: 2nd International Conference on Computational Intelligence, pp. 301–307. Communication Systems and Networks, UK (2010)
12. Polat, K., Gunes, S.: Classification of epileptiform EEG using a hybrid system based on decision tree classifier and fast Fourier transform. Appl. Math. Comput. 187(2), 1017–1026 (2007)
13. Kannathal, N., Choo, M.L., Acharya, U.R.: Entropies for detection of epilepsy in EEG. Comput. Methods Programs Biomed. 80, 187–194 (2005)
14. Zhang, A., Yang, B., Huang, L.: Feature extraction of EEG signals using power spectral entropy. In: Internal Conference on BioMedical Engineering and Informatics, pp. 435–439. IEEE, China (2008)

15. Graham, D.J.: On the spectral entropy of thermodynamic paths for elementary systems. Entropy **11**, 1025–1041 (2009)
16. Misrihanov, A.M.: Primenenie metodov vejvlet-preobrazovanija v jelektro-jenergetike (Appliance of Wavelet-Transform Methods in Electrical Power Industry). Autom. Remote. Control. **5**, 5–23 (2006)
17. Cyplihin, A.I., Sorokin, V.N.: Segmentacija rechi na kardinal'nye jelementy (Speech signal segmentation on cardinal elements). Informacionnye processy **6**(3), 177–207 (2006)
18. Zahezin, A.M.: Metod nerazrushajushhego kontrolja dlja opredelenija zarozhdajushhihsja defektov pri pomoshhi Fourier i vejvlet-analiza vibracionnogo signala (Method of the non-destructive control for defects identification based on Fourier and wavelet analysis of vibration-signal), Vestnik Juzhno-Ural'skogo gosudarstvennogo universiteta (J. South. Ural. State Univ.) **13**(2), 28–33 (2013)
19. Toh, A.M., Togneri, R., Nordholm, S.: Spectral entropy as speech features for speech recognition. In: Electrical Engineering and Computer Science, PEECS, Australia, pp. 22–25 (2005)
20. Jia, C., Xu, B.: An improved entropy-based endpoint detection algorithm. In: International Symposium on Chinese Spoken Language Processing, ISCSLP, Taiwan, pp. 96–97 (2002)
21. Zhang, Y., Ding, Y.: Voice activity detection algorithm based on spectral entropy analysis of sub-frequency band. BioTechnology Indian J. **10**(20), 12342–12348 (2014)
22. Misra, H., Ikbal, S., Bourlard, H., Hermansky H.: Spectral entropy based feature for robust ASR. In: International Conference on Acoustics, Speech, and Signal Processing, pp. 2–6. IEEE, Canada (2004)

Results of Using Neural Networks to Automatically Creation Musical Compositions Based on Color Image

Vladimir Rozaliev[✉], Nikita Nikitin, Yulia Orlova,
and Alla Zaboleeva-Zotova

Volgograd State Technical University, 28 Lenin Avenue, Volgograd 400005,
Russia
vladimir.rozaliev@gmail.com

Abstract. In this work we show the results of development and experiments with the program for automated sound generation based on image color spectrum with using the neural network. The work contains a description of the transition between color and music characteristics, the rationale for choosing and the description of a recurrent neural network. The choices of the neural network implementation technology as well as the results of the experiment are described.

Keywords: Sound generation · Emotion · Image analysis
Recurrent neural network · Sampling

1 Introduction

Making music is the creative process. Now there are almost no automatic music creation systems. Understanding of emotionality through musical sounds and images is a task that remains relevant for many years [1].

The process of making music at the moment does not give a clear formalization, although it is based on clearly defined music rules. Since the music began to be recorded on paper in the form of musical notation, the original "ways" of its composition began to appear. One of the first methods of algorithmic composition was the method of composing music invented by Mozart - "The Musical Game of the Dice". The first computer musical composition - "Illiac Suite for String Quartet" - was created in 1956 by the pioneers of using computers in music - Lejaren Hiller and Leonard Isaacson [2]. In this work, almost all the main methods of algorithmic musical composition are used: probability theory, Markov chains and generative grammars.

The development of computer music, including the sound generation by image, in the last century was severely limited by computing resources - only large universities and laboratories could afford to buy and hold powerful computers, and the first personal computers lacked computing power. However, in the 21st century, almost everyone can study computer music.

Now, computer music can be used in many industries: creating music for computer games, advertising and films. Now, to create background music compositions in

A. Abraham et al. (Eds.): IITI 2018, AISC 875, pp. 148–157, 2019.
https://doi.org/10.1007/978-3-030-01821-4_16

computer games and advertising, companies hire professional composers or buy rights to already written musical works. However, in this genre the requirements for musical composition are low, which means that this process can be automated, which will allow companies to reduce the cost of composing songs. Also, the generation of sounds based on image can be applied in the educational process [3]. The development of musical perception in preschool children can be in the form of integrated educational activities, which is based on combinations of similar elements in music and arts (the similarity of their mood, style and genre) [4].

The greatest success of the theory of automation of the process of writing and creating music made up relatively recently (at the end of XX century), but mostly associated with the study and repetition of different musical styles [5].

Since the process of creating music is difficult to formalize, artificial neural networks are best suited for automated sound generation – they allow identifying connections that people do not see [6]. In addition, to reduce the user role in the generation of music, it was decided to take some of the musical characteristics from the image. Thus, the purpose of this work is to increase the harmony and melodicity of sound generation based on image colour spectrum through the use of neural networks.

2 From Color to Musical Characteristics

To reduce the user role in the generation of music, some of the musical characteristics are obtained by analysing the colour scale of the image. Thus, the character of the output musical composition will correspond to the input image. This feature makes possible to use this approach for creating background music in computer games, advertising and films.

The key characteristics of a musical work are its tonality and tempo. These parameters are determined by analysing the color scheme of the image. The tonality of a composition is determined by two colour characteristics - a hue and a colour group, the tempo by brightness and saturation. To determine the tempo of composition, it's necessary to get the brightness and saturation of predominant color, and calculate the tempo, according to these parameters.

3 The Choice of a Neural Network to Generate Musical Compositions

An important feature of feedforward neural networks is that, this neural network has a common limitation: both input and output data have a fixed, pre-designated size, for example, a picture of 100×100 pixels or a sequence of 256 bits. A neural network, from a mathematical point of view, behaves like an ordinary function, albeit very complexly arranged: it has a pre-defined number of arguments, as well as a designated format in which it gives the answer.

The above features are not very difficult if we are talking about the pictures or pre-defined sequences of symbols. But for the processing of any conditionally infinite sequence in which it is important not only the content but also the order of information, for example, text or music, neural networks with feedback should be used - recurrent neural networks (RNN). In recurrent neural networks neurons exchange information among themselves: for example, in addition to a new piece of incoming data the neuron also receives some information about the previous state of the network. Thus, the network realizes a "memory" which fundamentally changes the nature of its work and allows to analyse any data sequences in which it is important the order of information [7].

However, the great complexity of RNN networks is the problem of explosive gradient, which consists in the rapid loss of information over time. Of course, this only affects the weights, not the states of the neurons, but it is in them information accumulates. Networks with long-short term memory (LSTM) try to solve this problem through the using filters and an explicitly specified memory cell. Each neuron has a memory cell and three filters: input, output and forgetting. The purpose of these filters is to protect information. The input filter determines how much information from the previous layer will be stored in the cell. The output filter determines how much information the following layers will receive. Such networks are able to learn how to create complex structures, for example, compose texts in the style of a certain author or compose simple music, but they consume a large amount of resources [8].

Thus, to implement the program for automated sound generation based on image colour spectrum, it is necessary to use recurrent neural networks with long short-term memory - RNN LSTM. This kind of neural networks is used to generate musical compositions in various programs such as Magenta. Magenta is an open source music project from Google. Also, RNN LSTM is used in BachBot. This is the program that creates the musical composition in the Bach style. And this kind of neural network is used in DeepJaz - the system that allows to generate jazz compositions based on analysis of midi files [9].

4   Description of the Used Neural Network

Recurrent neural network (RNN) has recurrent connections which allow the network to store information related to the inputs. These relationships can be considered similar to memory. RNN is especially useful for the study of sequential data, such as music.

In TensorFlow, the repeated connections on the graph are deployed into an equivalent feedforward neural network. Then this network is trained using the technique of gradient descent, called backpropagation through time (BPT).

There are a number of ways in which RNN can connect to itself with cyclic compounds. The most common are networks with long-short term memory (LSTM) and gated recurrent units (GRU). In both cases, networks have multiplicative neurons that protect their internal memory from overwriting, allowing neural networks to process longer sequences. In this work, LSTM is used. All recurrent neural networks have the form of a chain of repeating modules. In standard RNNs, this repeating module will have a very simple structure, for example, one layer of tanh. LSTMs also have this chain, but the repeating module has a more complex structure. Instead of having one layer of the neural network, there are four interacting with each other in a special way [10].

The first step in LSTM is to decide what information we are going to throw out of the cell state. This decision is taken by the sigmoid layer. This layer looks at the value of ht-1 output and xt input, calculates a value in the range from 0 to 1 for each Ct-1 state. If the layer returned 1, this means that this value should be left (remember), if 0 - removed from the state of the cell. For example, in the state of a cell, the characteristics of the current measure can be stored - if the measure is not yet complete, then it is necessary to leave the characteristics in memory, if work is already in progress, then new parameters must be memorized.

The next step is to decide what new information we are going to store in the state of the cell. To do this, firstly the sigmoid layer decides what values we will update. Next, the tanh layer creates a vector of new candidate values, Ct, that can be added to the state.

The next step is to update the old Ct-1 state in the new Ct state. To do this, it is necessary to multiply the old ft state, thus deleting the information from the state. Then, it's necessary to add resulting value and it * Ct. Thus, we get new candidate values, scaled by the update coefficient value of each state value.

At the last step, we need to decide what will output this layer. This output will be based on the state of the cell. First, we pass the input value through the sigmoid layer, which decides which parts of the cell state should be output. Then, we process the state of the cell using tanh (to shift the value between −1 and 1), and multiply it by the output of the sigmoid layer.

The behaviour of a neural network is determined by the set of weights and displacements that each node has. Therefore, for the properly work of neural network we need to configure them to some correct value. First, it is necessary to determine how good or bad any output is according to the input value. This value is called cost. Once the cost is received, we need to use the backpropagation method. In fact, it reduces to calculating the cost gradient relative to the weights (differential of the cost for each weight for each node in each layer), and then it is necessary to use the optimization method to adjust the weights to reduce the cost. In this work, we will use the method of gradient descent.

For the training of a neural network, it is proposed to feed a vector that contains the following parts [11]:

- Note name: MIDI representation of current note. Used to represent the pitch of a note.
- Time when note on.
- Time when note off.
- The velocity of the note playback.

To determine the correct output according to the input, it is suggested to transform the vector as follows: let there be a vector of notes {c, d, e, f, g, a, h}, then the learning vector will be {{c, d}, {d, e}, {e, f}, {f, g}, {g, a}, {a, h}}. This method of learning a neural network is used, for example, to predict time series [12].

5   Sound Synthesis

In the process of studying the methods of sound synthesis, four most popular methods of synthesis were considered: additive synthesis, FM synthesis, phase modulation and sampling.

Additive synthesis is very difficult to implement, due to the need for separate control of the volume and height of each harmonic, which even a simple timbre consists of dozens [13].

FM - synthesis is well - suited for synthesizing the sound of percussion instruments, the synthesis of other musical instruments sounds too artificial. The main disadvantage of FM synthesis is the inability to fully simulate acoustic instruments [14].

Phase modulation gives a good enough sound, but is very limited, so it's rarely used in practice.

Sampling is used in most modern synthesizers, since it gives the most realistic sound and is fairly simple to implement [15].

Each of the methods has its advantages and disadvantages, but Sampling was chosen as the most suitable method for sound generating based on image colour spectrum. This method gives the most realistic sound of instruments, which is an important characteristic for the program, and this method is relatively easy to implement. The disadvantage of sampling is its limitation, but for the implementation of the program it is not essential, since for the program needs it is not required possibilities of changing the ready-made presets

6   The Experiments

To test the developed program, two experiments were conducted:

- Analysis the dependence of the quality of generated sounds on the size of the training set.
- Analysis the dependence of the quality of generated sounds with using neural networks and without neural networks.

Since there are no automated ways of assessing the quality of musical works, experts (people with a musical education) were used to evaluate the quality of compositions. For all experiments, the set of ten images were used. For all ten images, output musical compositions were prepared and stored (can be found here: https://github.com/NikitaNikitinVSTU/ImageSoundGeneration). Experts evaluated the compositions according to the following criteria:

- Matching character of image (on a five-point scale).
- Realistic sound of an instrument (piano or guitar).
- Melodiousness of the composition.
- The quality of harmony (accompaniment).
- The pleasantness of the melody for the perception.
- Integrity of the composition.
- Realism/artificiality of the composition.

An example of an abstract image is shown in Fig. 1, an example of a landscape is shown in Fig. 2 and an example of a city is shown in Fig. 3

Fig. 1. An example of an abstract image.

Fig. 2. An example of a landscape.

Fig. 3. An example of city image.

6.1 The Experiment of Analysis the Dependence of the Quality of Generated Sounds on the Size of the Training Set

For this experiment, a training sample consisting of 4295 pieces of music in.midi format was taken. Five models were trained at different sample sizes. For the first model, 120 classical works were taken, for the second model - 500, for the third - 1000 classical works. For the fourth and fifth models, 2000 and 4,295 (the maximum number of music pieces in the .midi format that are available for learning) of works of a different genre were taken. For all models, 3 songs were generated, and sent to experts for analysis. The results of the experiment are shown in Table 1.

Table 1. Average values for all test for the first experiment

Criterion	Average value for all tests
120	3.5
500	3.7
1000	4
2000	4.5
4925	4.1

Thus, we can conclude that the best model from the point of view of the quality of the generated musical composition is the model, trained for 2000 classical works.

It can be seen from the Table 1 that the last model, trained on 4295 tracks, generates sounds worse than the previous model. This is due to the fact that the compositions of a different genre were taken for the training of the last model, that at the output gives a work consisting of a mixture of different genres - such works were judged by experts worse than just classical musical works.

6.2 The Experiment of Analysis the Dependence of the Quality of Generated Sounds with Using Neural Networks and Without Neural Networks

In this experiment, 5 musical compositions were created using artificial neural networks, and 5 without their use. These compositions were sent to experts for analysis. The expert evaluated each pair of works of one of three evaluations:

- −1 - The sounds without the use of neural networks are better than with their use.
- 0 - The sounds without the use of neural networks are like sounds with their use.
- 1 - The sounds with using the neural networks are better than without using them.

The results of expert's evaluation of pairs of products are presented in Table 2.

Table 2. Average values for all tests for the third experiment

Criterion	Average value for all tests
Matching character of image	0
Realistic sound of an instrument	0
Melodiousness of the composition	0.8
The quality of harmony	1
The pleasantness of the melody for the perception	1
Integrity of the composition	0.8
Realism/artificiality of the composition	0.8

Thus, it can be seen from the Table 2 that the criteria "Matching character of image" and "Realistic sound of an instrument" did not change when the method of generating sounds was changed (using artificial neural networks and without). This is due to the fact that the parameter "Matching character of image" is determined by the chosen tonality, which does not depend on the use of neural networks. And the parameter "Realistic sound of an instrument" depends on the synthesizer module of sounds, which was not affected either by changing the approach to generating tracks.

The most important parameters for assessing the achievement of the research goal - "Melodiousness of the composition", "The quality of harmony" and "The pleasantness of the melody for the perception" were noted by experts as improved. This is due to the fact that without the use of neural networks a "naive" approach to the generation of compositions was used, while neural networks allow us to identify such connections that are not visible to a person, which affects the quality of the generated musical compositions.

The parameters "Integrity of the composition" and "Realism/artificiality of the composition" have improved slightly with the use of neural networks. This is due to the fact that these criteria depend on the previous three criteria, as they have improved, so the use of neural networks has positively affected these criteria.

7 Conclusion

In this work, the scheme of correlation of color and musical characteristics was determined, an overview of the types of neural networks was made and the most suitable type for the generation of musical compositions was selected. Also, the used neural network was described in detail, the technology of implementing the neural network was chosen and the method of synthesis of sounds was chosen. To evaluate the effectiveness of the proposed algorithms, an experiment was conducted to assess the harmony and melodiousness of the output musical compositions.

Analysis of various types and architectures of ANNs concluded that the most suitable network for processing musical information is the recurrent neural networks with long short-term memory (RNN LSTM) [16].

During the description of the used neural network, it was determined that for learning the network it is supposed to input a vector that contains the following parts: MIDI representation of current note, time when note on, time when note off and the velocity of the note playback.

In analyzing the libraries for implementing a neural network in the Python programming language, it was discovered that the Keras library should be used to develop a recurrent neural network, since this library allows to work based on Theano and TensorFlow, taking advantage of them, while the development of neural networks using this library simple and fast, which allows to create prototypes for rapid experimentation.

As a result of the experiment, the model (neural network) on Beethoven's compositions was trained, and compositions of 10 images were generated. These compositions were sent for analysis to experts. As a result of the analysis of expert assessments, it can be concluded that the program generates quite melodic compositions, but it appears that the model was trained on a small number of compositions by only one author.

Acknowledgments. The work is partially supported by the Russian Foundation for Basic Research (16-47-340320 and 17-07-01601 projects).

References

1. Rozaliev, V., Zaboleeva-Zotova, A.: Methods and Models for Identifying Human Emotions by Recognition Gestures and Motion. Atlantis Press, Singapore (2013)
2. Ariza, C.: Two pioneering projects from the early history of computer-aided algorithmic composition. Comput. Music. J. **35**(3), 40–56 (2012)
3. Chereshniuk, I.: Algorithmic composition and its role in modern musical education. Art Educ. (3), 65–68 (2015)
4. Vygotsky, L.: Imagination and creativity in childhood. J. Russ. East Eur. Psychol. **42**(1), 7–97 (2004)
5. Cope, D.: Computer Models of Musical Creativity. MIT Press, Cambridge (2005)
6. Mazurowski, L.: Computer models for algorithmic music composition. In: Proceedings of the Federated Conference on Computer Science and Information Systems, pp. 733–737 (2012)

7. Sak, H., Senior, A., Beaufays, F.: Long short-term memory based recurrent neural network architectures for large vocabulary speech recognition. ArXiv e-prints (2014)
8. Doornbusch, P., Nierhaus, G.: Algorithmic composition paradigms of automated music generation. Comput. Music J. **34**(3), 70–74 (2014)
9. Brinkkemper, F.: Analyzing six deep learning tools for music generation. http://www.asimovinstitute.org/analyzing-deep-learning-tools-music/. Last Accessed 05 May 2018
10. Mikolov, T.: Recurrent neural network based language model. In: Proceedings of Interspeech International Speech Communication Association, vol. 2010, no. 9, pp. 1045–1048 (2010)
11. Kim, H.K., Ao, S.I., Mahyar, A.: Transactions on Engineering Technologies. Special Issue of the World Congress on Engineering and Computer Science, p. 796. Springer Publishing Company, New York (2013)
12. Fernandez, J.D., Vico, F.: AI methods in algorithmic composition. a comprehensive survey. J. Artif. Intell. Res. **48**, 513–582 (2013)
13. Korvel, G., Simonyte, V., Slivinskas, V.: A modified additive synthesis method using source-filter model. J. Audio Eng. Soc. **63**(6), 443–450 (2015)
14. Lazzarini, V., Timoney, J.: Theory and practice of modified frequency modulation synthesis. J. Audio Eng. Soc. **58**, 459–471 (2012)
15. Russ, M.: Sound Synthesis and Sampling, p. 568. Taylor & Francis Group, London (2012)
16. Nikitin, N.A., Rozaliev, V.L. Orlova, Y.A.: Program for sound generation based on image color spectrum with using the recurrent neural network. In: Proceedings of the IV International research conference «Information technologies in Science, Management, Social sphere and Medicine» (ITSMSSM 2017), vol. 72, pp. 227–232. Atlantis Press (2017)

Hybrid Expert Systems and Intelligent Decision Support Systems in Design and Engineering

Intelligent Integrated System
for Computer-Aided Design and Complex
Technical Objects' Training

Alexander Afanasyev, Nikolay Voit[✉], and Andrey Igonin

Ulyanovsk State Technical University, Ulyanovsk, Russia
{a.afanasev,n.voit}@ulstu.ru, igonin@ritg.ru

Abstract. This paper is focused on the generalized structure, mathematical and software of the complex system of design and training that is based on author's models and methods. The system includes a block of training and block of recommendations. The domain model consists of the scheme level, presentation level and practical level, and is based on the ontology of objects of the problem area. Experiments on teaching student showed an increase in the average level of skill of trainees.

Keywords: Computer-aided design · Intelligent · Training

1 Introduction

The paper presents the author's methods and techniques of workflow processing and complex technical objects' design and using. We deal with two classes of complex technical objects: complex automated systems (CAS) and complex mechanical products (CMP).

The key problem of CAS' design and development is to create a successful project. According to the Standish Group data, only 40% of software projects are successful projects (i.e., the projects completed in duly time and budget, with all the specified features and functions) today. And the especially significant role in achievement of the CAS development's success is given to the diagrammatic models in visual forms of business process artifacts, particularly at the concept phase of CAS design. For this purpose, the visual languages (UML, IDEF, ER, DFD, eEPC, SDL, BPMN, etc.) were developed, and they are widely used in practice. Such models' use sufficiently increases an effectiveness of the design process and a quality of the design solutions, through the unification of interaction language of the CAS development participants, the strict documentation of the project-architectural, functional solutions, and the formal control of diagram notation correctness.

In recent years the large industrial companies and enterprises actively use distributed dynamic flows of designing and manufacturing activities. For example, according to [1], the first generation of statistical management systems of product lifecycle and project workflow can no longer meet the requirements of many companies. The approach and automated tools of the first generation of project workflow standardization have already exhausted its resources, and, as a result, there are poorly

© Springer Nature Switzerland AG 2019
A. Abraham et al. (Eds.): IITI 2018, AISC 875, pp. 161–170, 2019.
https://doi.org/10.1007/978-3-030-01821-4_17

formalized processes (often containing semantic errors) increasing the growth of expenses for their development and improvement.

However, in theory and practice of corporate use of diagrammatic models there are no effective methods and tools for monitoring diagrammatic representations of dynamic distributed workflows of CAS, that results in the serious design errors.

Thus, the analysis, monitoring and processing of distributed dynamic workflows in CASs' design and operation, presented via their diagrammatic models, is an important scientific and technical task.

In frames of creating and interpreting of training flows in automated training systems, the world practice is currently dominated by an adaptive approach associated with the formation of individual learning paths. A formation of the custom-oriented (designed for a specific employer) competences, based, in particular, on the development of the employer's experience and best practices, is nowadays a key task of implementing the software engineer's training. Even such a concept as the World of Work has appeared. However, the modern training systems do not take into account the unique features of project activities' training, there is no integration of such systems with CAD software packages and project repositories, and the project activities of designers are not assessed. Formation of the necessary competences and recommendations for a designer will improve an efficiency of his/her work. And an important element of the modern training systems is a virtual environment (within the project's context) in a form of training systems, virtual worlds, including virtual work places, work stations, workshops, and enterprises in total. However, the task of trainee's action assessment is not automated in these environments (as usual, the experts manually make an expert analysis of trainees' action reports or an expert visual real-time analysis). That is why a development of the approach that allows the experts to evaluate automatically their trainees' actions in virtual environments and to generate recommendations, including the same for enhancing their skills, has a great fundamental and practical significance.

Thus, the goals of this research work are: to extend the class of diagnostic errors within the process of CAS design and operation - through the development and implementation of methods and tools for analysis and control of dynamic diagrammatic workflows' models, as well as to improve the designers' competences - through the development and implementation of methods, models and tools for the analysis of project solutions and the formation of personally-oriented training on basis of a uniform intelligent project repository.

The paper has the following structure. In the Sect. 2, the general structure of a complex system is discussed. In the Sect. 3, methods of distributed diagrammatic workflows' analysis and control at CAS's design are described. The Sect. 4 deals with personified training software for CAS and CMP designers, as well as the method of actions' analysis and recommendations' formation for CMP designers. Findings and further directions of the research are presented in the conclusion.

2 The Generalized Structure of an End-to-End System of Design Automation and Training of Difficult Technical Objects

The system's structure is presented in Fig. 1.

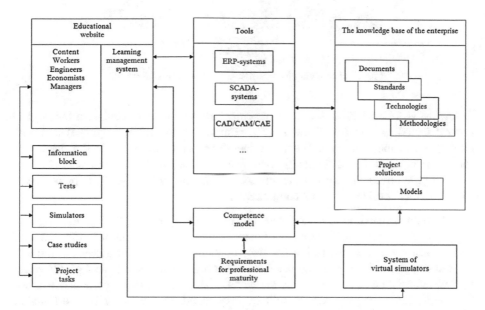

Fig. 1. The structure of the intelligent training environment.

The project part of the system is represented by a computer-aided design system and an electronic workflow system, and the training part is represented by a training portal.

The data storage in the system is based on the data bus development (unified service-oriented framework) that connects the tools for developing / designing artifacts of the enterprise's design and production process and an intelligent repository built on a dynamic ontological model and physical database (Fig. 2).

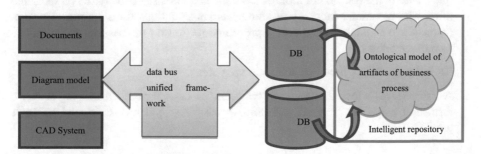

Fig. 2. Database organization.

The distributed workflows are characterized by the diverse usage of visual language facilities (for example, the scheme in Fig. 3).

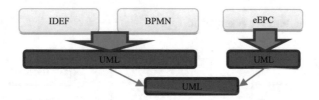

Fig. 3. The example of a distributed workflow's scheme.

At the same time the task of such schemes' analysis is connected with topological control including complex diagrams (for example, in the UML language) and with control of semantic approval. The above-mentioned problems are offered to be solved via the author's device of the multi-level syntactic oriented RVM grammars [11].

The generalized structure of the automated training system (ATS) is presented in Fig. 4. Let us consider the ATS components.

1. The training block is organized to train (materials' selection, testing, training scenario's development). The subject domain stores a set of training materials and links between them. The test and task's base contains both tests for assessing trainee's knowledge-level and practical tasks for assessing skills. The trainee's profile stores the information concerned with the level of trainee's current competences, test results, etc. A training path is built on tests based on data analysis of a trainee's profile and subject domain information. A training scenario is the sequence of trainee's training materials and tests' selection. Monitoring is a set of tests for assessing new knowledge the trainees gained from learning materials.
2. The recommendation block is responsible for recommendation development and trainee's skill-level diagnostics. The operation's generator tracks trainee's operations and codes them for the further analysis. The state's generator forms a project's state on the basis of operations. The project's state stores a history of project's states throughout the work in a CAD. The facts box is a base of facts received from a set of operations and project's states. The rules' generator fills rules for the expert system (ES) by the certain algorithms. Expert is an expert in CAD who fills the rule's base in the ES. ES is a tool for recommendation's selection based on facts and rules. The recommendations box stores recommendations for a trainee. The recommendations' analysis is a user's profile correction on the basis of the provided recommendations.

Training process is based on the following steps.

1. A trainee takes a test, the results of which is the basis for current training scenario formation with data from subject domain base, test and task base, and trainee's profile.

2. The work in CAD gives the operations' generator an opportunity to process user's operations, and this information is the basis for the facts' formation for the ES.
3. The ES creates recommendations based on laid down rules and received facts and provides trainees with them.
4. The trainee's profile is corrected based on performed actions and recommendations.
5. After trainee's profile correction a new training scenario is formed.

The domain model is based on the ontology and consists of three levels: the scheme level, presentation level and practical level.

The scheme level describes the domain knowledge structure and relations among the knowledge units. At this stage the order of studying the elements is determined.

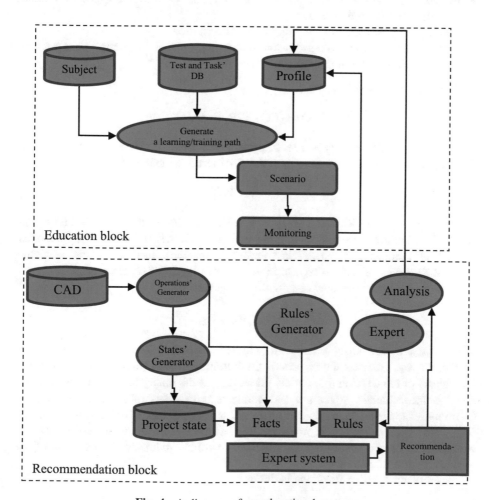

Fig. 4. A diagram of an educational system.

The presentation level is a set of training materials and reference books given as hypertext. The reference books are additional materials presented as hyperlinks for optional review. There are three types of supplementary training materials: reference books, glossary, and similar training materials.

The practical level is a level for designer's skill testing: the ability to perform design activities based on the acquired knowledge of automated design and design practice. This level is represented by a set of project solutions.

In subject domain there are the following classes:

- knowledge atoms;
- terms/concepts are the atoms grouped by any principle;
- training materials are the materials for studying. One of the types of training materials can be reference books.
- training purposes group atoms that should be studied;
- competences group the atoms that are included into this competence. Competences are a subclass of training purposes.

The domain model is given as:

$$O = (PSL, T, R, F, Ax), \tag{1}$$

where $PSL = \{psl_i | i = 1..x\}$ is a set of project solutions, T is the terms of the applied domain, which the ontology describes. A set of terms is defined as:

$$T = \{C, In\}, \tag{2}$$

where $C = \{A, P, D, GOAL, COMP\}$ is a set of ontology classes (A is knowledge atoms, P is terms/concepts, D is training materials, GOAL is the training objectives, COMP is competences). The class «training material» has a slot «is a Reference» with «true» or «false» values, In is a set of objects of ontology classes.

R is a set of relations between ontology objects:

$$R = \{R_{learn}, R_{part}, R_{next}\}, \tag{3}$$

where R_{learn} is the binary relation «is_studied_in» which has the semantics «connected_to» and connects the objects of the ontology classes («Atom», «Concept») to the objects of the «Training Material» class, R_{part} is the binary relation «is_a_part_of» that has the semantics «part_of» and connects the objects of the ontology classes («Atom», «Concept») to the objects of the «Concept», «Purpose of Learning» classes, R_{next} is the binary relation «is_trained_after», which has the semantics «after_of» and connects the objects of the ontology classes («Atom», «Concept») to objects of the «Concept» and «Atom» classes.

A set of interpretive functions is defined as:

$$F = \{Fatom_{op}, Fpsl_a, Fedu, Fdefine, Fsimilar, T\}, \tag{4}$$

where Fatom$_{op}$: A → {Operation} is a function of the «Atom» class object mapping into a set of project solution operations, Fpsl_a : PSL → {A} is a function of the project solution mapping into a set of the «Atom» class objects, Fedu : {A} → {D} is a function for constructing an ordered set of training materials for studying certain knowledge atoms, Fdefine : P → {D} is a function of finding training materials that describe a certain concept, Fsimilar : D → {D} is a function of finding the most similar training materials, T : D → Q+ is the didactic material complexity.

A set of axioms is defined as:

$$Ax = \{AxAHP, AxAHL, AxAHD, AxPAfP, AxPAfA, AxAAfP\}, \qquad (5)$$

where AxAHP is «atoms are part of the terms/concepts» , if the atom Y is related to the concept of X, which is related to the concept of Z, then the atom Y is related to the concept of Z as Semantic Web Rule Language (SWRL): Term/Concept(?x) ∧ Atom(?y) ∧ Term/Concept(?z) ∧ be_a part of (?y, ?x) ∧ is_a part of (?y, ?z) → be_a_part of (?y, ?z).

AxAHL is «atoms are related to training purposes» , if the atom Y is related to the concept of X, which is related to the training purpose of Z, then the atom Y is related to the training purpose of Z as SWRL: Term/Concept(?x) ∧ Atom(?y) ∧ the_training_objective (?z) ∧ be_a_part of (?y, ?x) ∧ be_a_part of (?y, ?z) → be_a_ part of (?y, ?z).

AxAHD is «atoms are related to training materials» , if the atom Y is related to the concept of X, which is studied in the training material Z, then the atom Y is studied in the training material Z as SWRL: Term/Concept(?x) ∧ Atom(?y) ∧ Training Material(?z) ∧ be_a_part of (?y, ?x) ∧ be_ studied_in(?y, ?z) → be_ studied_in (?y, ?z).

AxPAfP is «atoms are studied after atoms» , if the atom Y is related to the concept of X, which is studied after the concept of Z, and the atom C is related to Z, then the atom Y is studied after C as SWRL: Atom(?y) ∧ Term/Concept(?x) ∧ Term/Concept(?z) ∧ Atom(?c) ∧ be_ a_part_of (?y, ?x) ∧ be_a_part_of (?c, ?z) ∧ be_ studied_after(?x, ?z) → be studied after(?y, ?c).

AxPAfA is «concepts are studied after atoms» , if the atom Y is related to the concept of X, which is studied after the atom C, then the atom Y is studied after C as SWRL: Atom(?y) ∧ Term/Concept(?x) ∧ Atom(?c) ∧ is a part_of (?y, ?x) ∧ be_ studied_after(?x, ?c) → be_studied_after(?y, ?c).

AxAAfP is «atoms are studied after concepts» , if the atom Y is studied after the concept of X and the atom C is related to X, then the atom Y is studied after C as SWRL: Atom(?y) ∧ Term/Concept(?x) ∧ Atom(?c) ∧ is a part_of (?c, ?x) ∧ be_studied_after(?y, ?x) → be_studied_after(?y, ?c).

Mathematical basis and representation of domain models, a trainee, test design tasks are given in [12–17].

3   Experiment

The experiment in training was performed among 160 students of the Ulyanovsk State Technical University. In order to assess the initial skills' level, these students passed a pre-test, as a result of which two groups of 80 and 80 students were organized. Their average skill's level was 0.5. The groups were trained on the materials of CAD KOMPAS-3D [15]. The first group was trained based on a linear scenario; the second group was trained in accordance with an adaptive scenario. After training, the task was to build an assembly «Pump cover» of 471 elements. The experiment's results are shown in Table 1. When applying the adaptive training scenario, the average level of students' skills is 20% higher compared with the linear training scenario due to the more detailed training. This resulted to 17% increase in time required for preparation.

Table 1. Comparison of linear and adaptive learning scenarios.

	The average level of skills before training	Average time of training, hours	The average level of skills after training	Average time of creation of assembly, hours	The general spent time, hours
Linear script	0,5	1,65	0,745	2,8	4,5
Adaptive script	0,5	2	0,9	1,9	3,9
Prize		−17%	20%	50%	20%

4   Realization

The training block was built according to the classical three-tier architecture: a client, an application server and a database server. Java Platform, Standard Edition technology is used as a development platform. This technology supports built-in client-server applications (RMI and ISOAP technologies). MySQL is used as the database server. The system is scalable, allows for functionality expansion by adding components (plugins). Plugins allow client and server's components to be extended by providing a client and server API. In order to implement the client-server technology at Java SE, the technology of Web Services is chosen. The SOAP protocol is used to exchange messages between Web services. The recommendation block is built on .NET Framework. As an example, the computer-aided design system called KOMPAS-3D V16 was chosen. In order to process events, it is used the Automation technology that is implemented as a library at C#.

5   Conclusion and Future Work

A generalized structure of an integrated design and training system for a large design and manufacturing enterprise is proposed. This system is based on the intelligent repository of design and production artifacts. Methods for the analysis and control of distributed workflows, represented by diagrammatic models in various visual languages, are proposed.

We have developed ATS that uses a new ontological domain model and method for forming an individual training scenario. The method for recommendation formation is developed on basis of the proposed modules. A trainee's profile is corrected based on designer's work tasks performance, which allowing improve the effectiveness of practical training.

The main quantitative positive indicator of this research work is 20–25% increase of designer's productivity. Further research directions are the development of temporal RV-grammars that take into account the dynamic and temporal nature of the development of project workflows; the study of this finite-state grammars class; development of a universal expert system for forming recommendations for designers.

Acknowledgement. The reported study was funded by RFBR and Government of Ulyanovsk Region according to the research project No. 16-47-732152. The research is supported by a grant from the Ministry of Education and Science of the Russian Federation, project No. 2.1615.2017/4.6.

References

1. A global Swiss company offering advanced intelligent application software for multiple business sectors. http://whitestein.com
2. Dorca, F.A., Lima, L.V., Fernandes, M.A., Lopes, C.R.: Comparing strategies for modeling students learning styles through reinforcement learning in adaptive and intelligent educational systems: an experimental analysis. Expert Syst. Appl. **40**, 2092–2101 (2013)
3. Graf, S., Kinshuk, Liu, TC.: Identifying learning styles in learning management systems by using indications from students' behaviour. In: Proceedings of the 8th IEEE International Conference on Advanced Learning Technologies (ICALT 2008), pp. 482–486 (2008). https://doi.org/10.1109/icalt.2008.84
4. Ozpolat, E., Akar, G.B.: Automatic detection of learning styles for an e-learning system. An Int. J. Comput. Educ. **53**, 355–367 (2009). https://doi.org/10.1016/j.compedu.2009.02.018
5. Chang, Y.C., Kao, W.Y., Chu, C.P., Chiu, C.H.: A learning style classification mechanism for e-learning. An Int. J. Comput. Educ. **53**, 273–285 (2009)
6. Simsek, O., Atman, N., Inveigle, M.M., Arian, Y.D.: Diagnosis of learning styles based on active or reflective dimension of felder and silverman's learning style model in a learning management system. In: Proceedings of the ICCSA 2010, Part II, LNCS, vol. 6017, pp. 544–555. Springer, Heidelberg (2010)
7. Darwesh, M.G., Rashad, M.Z., Hamada, A.K.: Behavioural analysis in a learning environment to identify the suitable learning style. Int. J. Comput. Sci. & Inf. Technol. (IJCSIT) **3**(2), 48–59 (2011). https://doi.org/10.5121/ijcsit.2011.3204

8. Deborah, L.J., Baskaran, R., Kannan, A.: Learning styles assessment and theoretical origin in an E-Learning scenario: a survey. Artif. Intell. Rev. **42**, 801–819 (2012). https://doi.org/10.1007/s10462-012-9344-0

9. Dung, P.Q., Florea, A.M.: An approach for detecting learning styles in learning management systems based on learners' behaviors. In: Proceedings of the International Conference on Education and Management Innovation (IPEDR 2012), vol. 30, pp. 171–177 IACSIT Press, Singapore (2012)

10. Jyothi, N., Bhan, K., Mothukuri, U., Jain, S., Jain, D.: A recommender system assisting instructor in building learning path for Personalized Learning System. In: IEEE Fourth International Conference on Technology for Education on Proceedings, pp. 228–230 https://doi.org/10.1109/t4e.2012.51

11. Afanasyev, A., Voit, N., Gaynullin, R.: The analysis of diagrammatic models of workflows in design of the complex automated systems. In: Proceedings of the First International Scientific Conference «Intelligent Information Technologies for Industry» (IITI 2016), vol. 450, pp. 227–236 (2016). https://doi.org/10.1007/978-3-319-33609-1_20

12. Afanasyev, A.N., Voit, N.N., Kanev, D.S.: Development of intelligent learning system based on the ontological approach. In: Proceedings of the 10th IEEE International Conference on Application of Information and Communication Technologies (AICT), pp. 690–694 (2016)

13. Afanasyev, A.N., Voit, N.N.: Intelligent learning environments for corporations. In: Proceedings of the 9th IEEE International conference on Application of Information and Communication Technologies, AICT, pp. 107–112 (2015)

14. Afanasyev, A., Voit, N., Kanev, D., Afanaseva, T.: Organization, development and implementation of intelligent learning environments. In: Proceedings of the 10th International Technology, Education and Development Conference, INTED-2016, pp. 2232–2242 (2016). https://library.iated.org/view/AFANASYEV2016ORG. Last Accessed 14 May 2018

15. Afanasyev, A., Voit, N.: Intelligent agent system to analysis manufacturing process models. In: Proceedings of the First International Scientific Conference «Intelligent Information Technologies for Industry» (IITI 2016), vol. 451, pp. 395–403 (2016). https://doi.org/10.1007/978-3-319-33816-3_39

16. Afanasyev, A.N., Voit, N.N., Kanev, D.S.: Development of intelligent learning system based on the ontological approach. In: Proceedings of the 10th IEEE International Conference on Application of Information and Communication Technologies (AICT), pp. 690–694 (2016). https://doi.org/10.1109/icaict.2016.7991794

17. Afanasyev, A., Voit, N., Kanev, D.: Intelligent and virtual training environment. In: Proceedings of the International conference on Fuzzy Logic and Intelligent Technologies in Nuclear Science, FLINS2016, pp. 246–251 (2016). https://doi.org/10.1142/9789813146976_0041

Methods of Conceptual Modeling of Intelligent Decision Support Systems for Managing Complex Objects at All Stages of Its Life Cycle

Aleksey D. Bakhmut[1], Vladislav N. Koromyslichenko[2],
Aleksey V. Krylov[3(✉)], Michael Yu. Okhtilev[3], Pavel A. Okhtilev[3],
Boris V. Sokolov[3], Anton V. Ustinov[1], and Alexander E. Zyanchurin[3]

[1] St.Petersburg State University of Aerospace Instrumentation,
Saint-Petersburg, Russia
adbakhmut@gmail.com
[2] Scientific Center « Petrokometa», Saint-Petersburg, Russia
koromislichenko@petrocometa.ru
[3] St.Petersburg Institute for Informatics and Automation of the Russian
Academy of Sciences (SPIIRAS), Saint-Petersburg, Russia
pavel.oxt@mail.ru

Abstract. The article suggests a set of methods for conceptual modeling of the main components of specialized intelligent decision support systems: models and methods for integration and analysis of data, applications, a complex of distributed data and knowledge bases, the design of user interfaces and the organization of distributed computations as part of an intelligent computer-aided engineering system of specialized software. The detailed substantiation of application of a new intelligent information technology and its methods for information modeling of functioning of complex objects in various subject areas of state and industrial purpose is given.

Keywords: Decision support systems · Artificial intelligence
Computer-Aided software engineering

1 Introduction

The experience of implementation of various automated systems (AS) over the past decades shows that the correct management of information flows through the automation of processes in organizations and enterprises can significantly improve the efficiency of their operation and ensure compliance with the requirements for them. Such AS include various systems for monitoring the state, supporting decision-making (DSS), management, information and analytical (IAS), information and reference systems (AIS), etc., as well as their counterparts – Supervisory Control And Data Acquisition (SCADA), Enterprise Resource Planning (ERP), Manufacturing Execution System (MES), Computer-Aided Design (CAD), Continuous Acquisition and Lifecycle Support (CALS), Product Lifecycle Management (PLM), Online Analytical Processing (OLAP), etc. [1]. At the same time, the specificity of many subject areas (SA) implies fixing the final decisions for the person - the decision-maker. In this regard nowadays

© Springer Nature Switzerland AG 2019
A. Abraham et al. (Eds.): IITI 2018, AISC 875, pp. 171–180, 2019.
https://doi.org/10.1007/978-3-030-01821-4_18

DSS plays a special role in such SA as systems that provide information and analytical support to decision-makers, and also offer solution alternatives. However, nowadays there are no unified requirements, a unified approach to design, and the strict life cycle (LC) of the DSS development is not defined, and the composition of the main components is not defined also. It is known that empirical experience of the specialists of the SA is often enough to provide the required level of quality of functioning of organizations. This feature, from the point of view of the scientific method, determines the possibility of using heuristic models and methods limited by the scope of their application and at the same time being relatively simple in designing in relation to their analytical "analogs". In this regard, the authors of this work suggest that the DSS should be based on principles and methods that allow the formalization of models to extract and present knowledge of experts (SA experts) and to organize the computing process on the basis of this knowledge. In other words, the development of modern DSS lies in the field of the theory of Artificial Intelligence (expert systems) and deals with the issues of the formation of knowledge bases (KB) and data bases (DB), solver and communication systems for the designer (knowledge engineer) and program user (PU).

2 Development Lifecycle and Proposed Methodology for Designing Decision Support Systems

Since, ultimately, the DSS is realized as a specialized software (SS), an important issue is the formation of a strictly ordered sequence of LC stages in the development of the SS to improve the efficiency of this process and the quality of its result [2]. At the same time, increasing requirements for the functioning of organizations and enterprises, the multifaceted complexity of Complex Organizational and Technical Objects (COTO) requires, in turn, the availability of more special software complex (SC) of DSS, which must be supported, scaled, modified. Under these conditions, the greatest effect can be achieved if each LC stage of the design and development process of the SS and the transitions between them are automated in such a way that it is possible to isolate on each of them formalized knowledge in machine-interpetable languages of varying degrees of abstraction, starting from the formalization of goals, tasks, requirements and SA description of the software and up to the schemes of organizing the computing process. Currently, software development often uses the international standard of software engineering ISO/IEC 12207, based on the SWEBOK standard of the Software Engineering Coordinating Committee [3, 4]. According to this standard, the technical development stage is presented in the left column of Table 1.

On the basis of the above statements, the authors propose a methodology for the design of a DSS based on a system of interrelated knowledge representation models (KRM) and an integrated solver, the implementation of which can be called the intelligent computer-aided engineering system (ICAES) of SS. This system is aimed at automating the development of SS at all stages of its LC in order to increase the efficiency of the development process, ensure the predicted quality of the SS and the results of its operation, provide the ability to design and replicate complex KB without special skills in programming in high-level languages, the approximation of knowledge representation language (KRL) to natural languages for the design in terms of SA, automatic synthesis of SS on the basis of the forms created for the SA KB. The main components of ICAES of SS and their relationship with the LC of the SS development, as well as the requirements for the DSS and the design process are presented in Table 1.

Let us consider in more detail the components of ICAES of SS associated with the conceptual modeling of computational tasks, DB and data manipulation requests, interfaces and distribution of computations.

3   Modeling the Subject Area (Specification of Requirements and Architecture)

At the heart of the simulation phase of the SA is the provision that the LC of the SS development begins with the formation of the vocabulary [5]. So ontologies are included in the ICAES of SS as a means of conceptualizing the SA. Ontologies allow to annotate the conceptual models of the SA; limit the created entities and their relationships. This property allows to verify models in automatic mode from the point of view of their ontological expressiveness [6, 7]. ICAES of SS uses an ontological system that allows the creation of an interrelated extensible system of models describing the various aspects of the SA [8]. As a top-level ontology, a modification of the Bunge-Wand-Weber ontology was developed (BWW-ontology, [7]) the design of information and analytical systems, which defines the concepts associated with the formation of conceptual models of:

- information entities and actors (top-level SA models). KRM, focused on the design of these models, uses formalizations of conceptual schema design ("entity-relation") as a unified modification of UML notation, as well as a representation of the multi-agent system (MAS), based on the intelligent agent model (IA) of K. Cetnorovich, later developed by Rybina [9]. Designing the model of information entities and actors allows, in a unified view, to clearly identify the entity structure in the SA, the automated processes, goals and tasks on the basis of information about the actors;
- business processes. Using BPMN notation allows you to describe threaded parallel asynchronous processes, represented with the help of events, tasks and their interrelationships [10]. Each track of the BPMN-model is a description of the functioning of the intelligent agent.
- Each track of the BPMN-model is a description of the functioning of computing tasks. Used for a given constraint on the assumption that any task in the BPMN-model is computational in the sense that it can be represented by a computational model in terms of Tyugu E.Kh. As the KRM of computational problems, G-models are used.

Thus, such a system of interconnected KRM allows to extract and formalize a variety of knowledge of SA experts at all stages of the software development, which greatly facilitates the process of development, scaling and modification of SS, increasing its quality, accuracy and reliability of the results of operation.

Table 1. Correlation of the stages of the LC of the development of the DSS, the components of the ICAES of SS and the requirements for ICAES of SS

Stages of the life cycle of the development of SS and the tasks within their framework	Components of ICAES of SS	Requirements for the ICAES of SS when designing the DSS
Identification of requirements: Formulation of the goal; Assessment of requirements; Harmonization of requirements; Registration of requirements.	Ontological system: • Metaontology (the ontology of the design of information and analytical systems), • Subject ontology, • Ontology of tasks, • Applied ontology of SS. KRM about the composition and structure of information entities and actors (automated processes, intelligent agents);	Ensuring the availability of a formalized general special vocabulary (thesaurus) of the SA, which is oriented towards a uniquely interpreted system of interrelated concepts in the considered SA. Ensuring that ICAES has an explicit tree of goals and tasks related to the design of the SS.
Specification requirements: Allocation of functional requirements; Allocation of requirements to interfaces; Allocation of requirements for restrictions; Identification of requirements for data models.	KRM of business processes; KRM of computing tasks; Interaction board Model of intelligent agents; KRM of visualization (interfaces); KRM of scenarios of interface behavior; Models of data manipulation requests; Algorithm for ontology synthesis according to the existing conceptual database scheme; KRM of the composition and structure of technical means; A model for managing the distribution of computational processes.	Consideration of COTO as a dynamic system due to the need to assess the state of COTO by time-consuming processes throughout its LC; Use of the model-oriented approach to the design of SS; Polymodel representation of COTO due to its decentralization, heterogeneity of processes and information; The use of methods of simulation-analytical, structural-functional modeling of computational problems due to the impossibility of their implementing strictly analytically; Use of well-known notations in the design of SS models in order to facilitate the process of extracting and presenting expert knowledge of the SA; Providing ICAES of SS with language tools with a visual notation that allows to design interface models and data marts that are oriented towards representing the cognitive image of an object of the SA, taking into account the information context.
Creating a system architecture: Definition of technical means; Definition of software; The distribution (decomposition) of components and the distribution of requirements for these components; Verification of architecture.		The possibility of creating an integrated KB for the subsequent replication of knowledge in the tasks of other SA; Accounting for the characteristics of the technical components of the hardware-software complex of DSS at the level of designing models of computing tasks; The possibility of designing a distributed system of software components on the basis of a unified model presentation for decentralized COTO; The possibility of interaction of the SS with a distributed database system; Providing an explicitly defined order of interaction between interface programs and programs that solve computational tasks.

(continued)

Table 1. *(continued)*

Stages of the life cycle of the development of SS and the tasks within their framework	Components of ICAES of SS	Requirements for the ICAES of SS when designing the DSS
Implementation: Definition of functions; Definition of interfaces; Definition of restrictions; Definition of data models.	An integrated solver based on the use of formal grammars for the synthesis of schemes of computational task programs, interface behavior scripts, data manipulation requests, distribution of computations with formalities of colored Petri nets; Integrated KB.	Possibility of synthesis of DB schemes, schemes of programs for solving computing tasks, interfaces, data manipulation requests, distribution of calculations; Using the principle of "programming without programming" for the automatic synthesis of software program schemes; Ensuring the receipt of the result of functioning in conditions of undetermined and incomplete information; Finding the result of analysis of the state of COTO for the forecasted time (functioning of the ICAES executive system in real time mode); Use of the principle of data management, streaming and parallel computing to ensure the invariance of the state of the object of analysis in the SA, characterized by natural parallelism, and the state of the computing process.
Integration (complexing): Integration of realized components into a single system; Verification of the integrated system.	It is provided by conceptual models formed in the previous stages in according to: KRM of the composition and structure of information entities and actors (automated processes); Models of data manipulation requests; Algorithm of ontology synthesis according to the existing conceptual database scheme; A model for managing the distribution of computing processes.	The possibility of forming an unified information space on the basis of an interconnected complex of structural and functional models with the property of interoperability, interrelated concepts of the vocabulary of vocational education and the common design environment; The possibility of modeling decentralized COTO in a unified view; The possibility of dynamic distribution of computing processes by technical components of the hardware and software complex in accordance with the semantics of tasks in the SA. Provision of a unified representation of heterogeneous information resources in the SA, in particular, data from databases executed on various technologies; At the execution stage, the ability to adapt the data marts to the decision-maker categories and the conditions of the tasks to be solved in accordance with the data management principle to simplify the process of developing a variety of data marts that characterize the state of the subsystems and the results of solving tasks related to COTO.

(continued)

Table 1. (*continued*)

Stages of the life cycle of the development of SS and the tasks within their framework	Components of ICAES of SS	Requirements for the ICAES of SS when designing the DSS
Testing: Verification; Validation.	Algorithms for checking the completeness and consistency of generated ontologies; Algorithms for checking for ontological expressiveness of conceptual models of information entities and actors of SA, business processes. Algorithms for checking the completeness and consistency of schemes of computational tasks programs, interfaces behavior scripts, data manipulation requests, computation distributions.	Limitation of the representation languages of the knowledge of the developed conceptual models by the concepts introduced in the vocabulary of the SA; Ensuring the correctness and uniqueness of the vocabulary concepts; Providing automation of model checking for completeness, closure and consistency for the purpose of automating and simplifying the DSS testing phase of DSS.
Documentation: Description of the results of the development process at each stage of the I.C.	It is partially provided with conceptual models formed at the previous stages with the help of KRL with a visual notation.	Providing the possibility of automatically forming part of the documentation (technical tasks, programs and test methods) associated with demonstrating the architecture of the software, models of interaction with the end user, models for solving computing problems, models of decomposition of the system of hardware and software.
Installation: Installation of SS on automated workplaces.	Executive system of ICAES of SS at the automated workplace.	Installing a software product that meets the specified requirements, in the target application environment.
Maintenance of the SS	It is provided by the system of interrelated conceptual models formed at previous stages and the presence of an integrated KB.	If you need to scale or modify the SS (and hardware), the design methodology should allow you to redesign without sacrificing performance and without attracting additional human resources.
Termination of the use of SS	–	–

4   Conceptual Modeling of DSS Interaction with a Distributed Database Complex

Modern COTO often have in its composition a number of information systems (IS), each of which is related to the availability of its database. The aggregate of such databases stores information characterizing the state of COTO. In this regard, since in designing the DSS it is necessary to take into account all this information for the formation of solution alternatives, the ICAES software model-algorithmic complex should include some KRM, which allows the DSS designer to solve the problem of interaction of the SS with the above-mentioned DB, i.e. task of data manipulation. To solve this task using conceptual modeling methods, it is necessary to propose such a KRL for database queries that would, depending on the features of the SA, database access technologies and used data models, project database queries in terms available to the expert.

Given the task, the use of such traditional approaches to solving this problem as ORM, ODBC, etc. is impossible [11, 12]. In this regard, to solve this problem, the authors propose to use ontologies as a means of annotating conceptual models of SA [13]. An important property of ontological modeling is the possibility of the formation of the UIS, within the framework of which it seems feasible to provide ICAES SS with a single view when forming requests to the DB. To implement this approach, algorithms must be implemented that allow them to create ontologies according to the existing conceptual schemas of the database in accordance with known data models (relational, network, hierarchical, etc.). The development of the language of ontological query language (OQL) will solve the problem of designing conceptual models of data manipulation. OQL provides the ability to formulate declarative requests to the ontology in the form of a set of terms SA and given constraints to the requested data, to automatically verify the model by estimating the consistency of a given constraint system. In addition, it should be possible to automatically synthesize the scheme of the query execution program [12, 14].

Thus, based on the use of KRM for query generation and ontological modeling, a methodology for designing ontology-driven DB and queries to them can be specified.

5 Conceptual Modeling of Distribution Processes of Calculations in DSS

Modern COTO, as a rule, decentralized, and decision-maker in such COTO are faced with the need to decompose the tasks to be solved on their subsystems, as well as their redistribution. The situation arises when, in the process of automation, it turns out that the tasks are connected, on the one hand, in the sense of the order of their solution in the SA, and on the other hand, that they must be divided into components of the hardware-software complex. This controversial situation requires the development of the ICAES SS component of the software to provide the ability to design a distributed system of software components on the basis of a single model representation with the ability to dynamically distribute computational processes on the hardware and software components in accordance with the semantics of tasks in the SA. This component is based on the model of managing the distribution of computing processes (MDCP).

The MDCP model describes the dependence of the hardware components of the hardware and software complex on the models of computational problems. To implement such a description, the model of the description of the structure and characteristics of the ICAES SS. This model allows the SA expert to specify characteristics describing the logical structure, as well as the technical characteristics of the components (computing power, communication bandwidth, etc.) at the design stage of the SS. Based on the description of the structure and characteristics of technical components and metric characteristics of program schemes synthesized from models of computational problems, it is necessary to perform a search for a minimal partition of graphs of program schemes, taking into account the structure of problems [15].

The introduction of these models into the ICAES SS will allow to distribute and synchronize the computational processes to solve the problems in accordance with their computational complexity on the basis of a single representation, thereby ensuring more efficient functioning of the distributed DSS.

6   Conceptual Modeling of DSS User Interfaces

One of the actual tasks in the design of the DSS is the development of user interfaces (UI), reflecting the results of the evaluation of the state of COTO and the recommendations formulated for the decision maker. In connection with the specifics of modern COTO, the state of which is characterized by large volumes of heterogeneous information and, as a consequence, it is difficult to perceive the decision maker, the task arises of creating tools for presentation (visualization) of generalized, relevant and current information about the state of COTO with the aim of improving the quality of decisionmaker decisions on the current situation. In solving the problem of designing a UI, it is proposed to take as a basis a model-oriented approach, the advantages of which were outlined above [16].

The basis of the model-oriented approach to the design of the UI should be laid out with a visual notation KRL, focused on the formation of interface models and interface behavior scenarios, through which it is possible to design the UI by an expert of the SA, as well as the possibility of adapting the UI at the stage functioning of a DSS under a finite set of data showcase (DS).

In order to adapt the UI state to various PU tasks and the parameter values appearing in the streaming mode, it is necessary to provide the possibility of designing a scenario KB in which the schemes of the UI reconfiguration programs are determined based on a given set of mathematical transformation algorithms oriented to interface programming. Scenarios are described by "states" and "actions", where the "state" in the case under consideration is associated with graphical elements and is represented by a set of parameters, and the actions are represented by operators that reflect the relationships between the parameters. The development of scenarios should be carried out at the KRL, close to natural languages [17, 18].

It is necessary to synthesize the schemes of the UI programs, for which the formal apparatus of the G-models can be taken as a basis in terms of the use of formal grammars and in the present case the principle of invariance of the state of computation of problems and the state of the UI [19].

7   Conclusion

Thus, in the article the problem of modern complex DSS design and modeling of its main components is considered. It was revealed that there are no well-established common requirements for the DSS and its development process. There is no any unified approach to the design of the DSS and its strictly defined LC. So, a new methodology for the design of the DSS is proposed, aimed at the use of a unified approach to the design of systems using methods of an Artificial Intelligence and, as a consequence, a system of interconnected KRM. Such KRM system allows designing top-level conceptual models that define the structure of information entities, goals and objectives, business process models, computational tasks, interfaces, data manipulation and distribution of computations. The set of models is determined in accordance with the requirements for DSS in modern COTO. The use of heterogeneous KRM determined the possibility of forming an integrated KB containing knowledge of various aspects of the functioning of COTO and the formation of alternative solutions.

Acknowledgments. The research described in this paper is partially supported by the Russian Foundation for Basic Research (grants **16-07-00779, 16-08-00510, 16-08-01277, 16-29-09482-ofi-i, 17-08-00797, 17-06-00108, 17-01-00139, 17-20-01214, 17-29-07073-ofi-i, 18-07-01272, 18-08-01505**), grant **074-U01** (ITMO University), state order of the Ministry of Education and Science of the Russian Federation №2.3135.2017/4.6, state research **0073–2018–0003**, International project ERASMUS +, Capacity building in higher education, № 73751-EPP-1-2016-1-DE-EPPKA2-CBHE-JP, Innovative teaching and learning strategies in open modelling and simulation environment for student-centered engineering education.

References

1. Pogorelov, V.I.: System and Its Life Cycle: An Introduction to CALS-Technology: A Tutorial Balt. Gos. Tehn. Un-tPubl., St. Petersburg (2010). 182 p. ISBN 978-5-855-46-581-5 (in Russian)
2. Orlov, S.A., Cil'ker, B.J.: Software Development Technologies: A Textbook for Universities. 4th edn. The Standard of the Third Generation. Piter Publ, St. Petersburg (2012). 608 p. ISBN 978-5-459-011 01-2 (in Russian)
3. GOSTRISO/MJeK 12207-2010. Information technology. System and software engineering. Software life cycle processes (in Russian)
4. ISO/IEC 12207:2008. System and software Engineering - Software life cycle processes
5. Kogalovskij, M.R.: Data access systems based on ontologies. In: Programming Institute for Market Problems, №4, pp. 55–77. RAS Publ. (2012). (in Russian)
6. Fedorov, I.G.: Adaptation of the bunge-wanda-weber ontology to the description of executable models of business processes. Applied Informatics Publ. -T. 10, №4, pp. 82–92 (58) (2015). (in Russian)
7. Gehlert, A., Pfeiffer, D., Becker, J.: The BWW-model as method engineering theory. In: Americas Conference on Information Systems (AMCIS), AMCIS 2007 Proceedings, 83, (2007). http://aisel.aisnet.org. Date of the application 29 January 2018
8. Gavrilova, T.A., Kudrjavcev, D.V., Muromcev, D.I.: Knowledge Engineering. Models and Methods: Textbook. Lan Publ, St. Petersburg (2016). 324 p. ISBN 978-5-8114-2128-2 (in Russian)
9. Rybina, G.V., Parondzhanov, S.S.: Modeling the processes of interaction of intelligent agents in multi-agent systems. In: Artificial Intelligence and Decision Making Publ., №3, pp. 3–15 (2008). (in Russian)
10. Business Process Model and Notation (BPMN). Version 2.0.2. OMG Document Number: formal/2013-12-09. www.omg.org/spec/BPMN. Date of the application 31 October 2017, C. 532
11. Halpin, T.: Information Modeling and Relational Databases: From Conceptual Analysis to Logical Design. Morgan Kaufmann (2001). 792 p

12. Date, C.J.: An Introduction to Database Systems, 8th edn. Pearson/Addison Wesley Publ. (2004). 1005 p. ISBN 5-8459-0788-8
13. Borgest, N.M.: Ontology of Design: Theoretical Foundations. Part 1. Concepts and Principles. Tutorial. SGAI Publ, Samara (2010). 92 p. (in Russian)
14. Gray, P.M.D.: Logic, Algebra, and Databases. Wiley, New York (1984). 294 p. ISBN 0-321-18956-6
15. Schloegel, K.: Graph Partitioninng for High Performance Scientific Simulations. Minneapolis, Minnesota: Department of Computer Science and Engineering, University of Minnesota (2000). 39 p
16. Narinyani, A.S.: Model or Algorithm: A New Paradigm Of Information Technology, № 4, pp. 11–16. Information Technology Publ. (1997). (in Russian)
17. Malyshev, A.V., Gorodeckij, V.I., Karsaev, O.V., Samojlov, V.V., Tihomirov, V.V., Man'kov, E.V.: Multi-agent system for resolving conflict situations in airspace. In: SPIIRAS Proceedings, №3, t. 1. Science Publ, St. Petersburg (2006). (in Russian)
18. Gorodeckij, V.I., Troickij, D.V.: Scenario model and knowledge description language for assessing and predicting situations. In: SPIIRAS Proceedings, № 8, pp. 93–127 (2009). SPIIRAS Publ., St. Petersburg (in Russian)
19. Okhtilev, M.Y.: Basics of the Theory of Automated Analysis of Measurement Information in Real Time. Synthesis of the Analysis System. VIKU them. Mozhaisky Publ., St. Petersburg (1999). 161 p. (in Russian)

About the Integration of Learning and Decision-Making Models in Intelligent Systems of Real-Time

Alexander P. Eremeev and Alexander A. Kozhukhov[(✉)]

Institute of Automatics and Computer Engineering,
Moscow Power Engineering Institute, Moscow, Russia
eremeev@appmat.ru, saaanchezzz@yandex.ru

Abstract. The paper considers integrated tools consist of multi-agent temporal differences reinforcement learning, statistical and main analysis modules. Deep reinforcement learning approach were analyzed to improve performance of reinforcement learning algorithms under time constraints. The possibilities of including anytime algorithm, particularly milestone method, into the forecasting subsystem type of intelligent decision support system of real-time for improving performance and reducing response and execution time were proposed. The work was supported by RFBR projects 17-07-00553, 18-51-00007.

Keywords: Artificial intelligence · Intelligent system · Real time
Reinforcement learning · Deep learning · Forecasting · Decision support
Anytime algorithm

1 Introduction

Reinforcement learning (RL) methods [1], based on the using large amount of information for learning in arbitrary environment, are the most rapidly developing areas of artificial intelligence, related with the development of advanced intelligent systems of real-time (IS RT) typical example of which is an intelligent decision support system of real-time (IDSS RT) [2]. One of the most promising in terms of use in IDSS RT and central in RL is Temporal Difference (TD) learning [1]. TD-learning process is based directly on experience with TD-error, bootstrapping, in a model-free, online, and fully incremental way. Therefore, process do not require knowledge of the environment model with its rewards and the probability distribution of the next states. The fact that TD-methods are adaptive is very important for the IS of semiotic type able to adapt to changes in the controlled object and environment [3].

Using the multi-agent approach contains of groups of autonomous interacting entities (agents) having a common integration environment and capable to receive, store, process and transmit information in order to address their own and corporate (common to the group of agents) analysis tasks and synthesis information is the fastest growing and promising approach for dynamic distributed control systems and data mining systems, including IDSS RT. Multi-agent systems could be characterized by the possibility of parallel computing, exchange of experience between the agents, resiliency, scalability, etc. [4].

Usually data encountered by an online RL-agent is non-stationary, and online RL updates are strongly correlated. Deep reinforcement learning (DRL) approach provide rich representations that can enable RL-algorithms to perform effectively and enables automatic feature engineering and end-to-end learning through gradient descent, so that reliance on environment is significantly reduced or even removed. Common idea of DRL is storing the agent's data in an experience replay memory where the data can be batched or randomly sampled from different time-steps. Aggregating over memory in this way reduces non-stationarity and decorrelates updates, but at the same time limits the methods to off-policy RL-algorithms [5].

When modern IDSS RT are developing, important consideration should be given to means of forecasting the situation at the object, consequences of decisions, expert methods and learning tools [6]. In addition, attention should be given to optimal using of system available resources and ability to work in the environment with restrictions in time. These resources are necessary for modification and adaptation of IDSS RT regarding changes in object and external environment and for enhancing the application field and improving system performance.

For solving these problems, developed tool using parallel algorithms for deep reinforcement learning and statistical methods and possibilities of implementation of anytime algorithm that can receive acceptable information within the limited time interval were considered.

2 Reinforcement Learning Outlines

We assume, that the uncertainty of information entering the IDSS RT database about the problem area current state of the object and environment, mainly associated with the erroneous operation of the sensor (sensors) or errors from dispatching personnel. The functions of the RL includes of the non-Markov decision models adaptation to the situation by analyzing the history of decision and improving their quality [1, 7]. RL decision-making module adjusting the decision-making strategy by interacting with the environment and the analyzing the evaluation function (payment function) called the agent. Agent target is to find an optimal (for a Markov process) or acceptable (for not-Markov process) decision-making strategies, also called the policies, in the process of learning. Intelligent agents must be able to support multiple learning paths and adapt to experience changes in the environment. The target of RL is to maximize the expected benefits R_t, which is determined as a function defined on the sequence of rewards:

$$R_t = r_{t+1} + r_{t+2} + \ldots r_{t+T}$$

where T - the final time step, r_t - reward at the time step t. This approach can be used in applications where the final step can be defined by natural way, from the kind of the problem, i.e., when the interaction of the agent-environment can be divided into a sequence, called episodes.

The main problem of RL is to find compromise between learning and application by agent. Agent must prefer actions that he had already applied and found that they are effective in terms of getting more reward. On the other hand, to detect such actions, the agent must try to perform actions that were not previously performed. An important characteristic of RL is getting delayed compensations that occur in complex dynamic systems. This means that the action produced by the agent can affect not only the current award, but also all subsequent [8].

In terms of the use in IDSS RT, TD-methods can solve several problems: the problem of prediction the values of certain variables within a few time steps and management problem based on the how RL-agent's learning affects the environment.

In practice, the number of states Q (s, a) can be huge. In this case, it is necessary to represent states $s \in S$ and actions $a \in A$ in the form of certain characteristic attributes, and the function $Q(s, a)$ in the form of a parametric learning model $Q(s, a; \theta)$, describing s and a. Then we assume the function $Q(s, a; \theta)$ such as a complex function of attributes that represent real number, i.e. the expected gain, or its probability and its parameters θ, can be taught by learning methods.

Thus, each step of learning will look like this: the agent makes a move a from the state s, goes to the new state s', receiving a direct reward r and then does one step of learning function $Q(s, a; \theta)$ with the input (s, a) and the output $max_{a'} Q(s', a'; \theta) + r$, where in most cases $r = 0$, because of the reward can only be obtained at the end of the learning episode [9].

In addition, the agent can also take such steps in relation to the previous positions, updating the weights not only for the last entry, but also for several previous ones.

3   Deep Reinforcement Learning

For many machine learning algorithms, e.g., linear regression, logistic regression, support vector machines, decision trees and boosting, we have input layer and output layer where the inputs may be transformed with manual feature engineering before learning. In deep learning and particularly in DRL, between input and output layers, we have one or more hidden layers. At each layer except input layer, we compute the input to each unit, as the weighted sum of units from the previous layer. Then we usually use nonlinear transformation or activation function, such as logistic or rectified linear unit to apply to the input of a unit, to obtain a new representation of the input from previous layer. We have weights on links between units from layer to layer. After computations flow forward from input to output, at output layer and each hidden layer, we can compute error derivatives backward and back propagate gradients towards the input layer, so that weights can be updated to optimize some loss function.

We obtain DRL-methods to approximate any of the following component of reinforcement learning: value function, $\tilde{v}(s; \theta)$ or $q(s, a; \theta)$, policy $\pi(a \mid s; \theta)$ and model (state transition function and reward function), where parameters θ are the weights parameters. The main difference between DRL and simple RL is what function approximator is used. For example, we can utilize stochastic gradient descent to update weight parameters in DRL [10].

Learning on successive episodes, where adjacent frames are too similar to each other and strongly correlate is not the best way. Eventually, their distribution shifts depending on the state of the environment but remains localized. It hinders effective learning, because of the usual formulation of the learning problem, where we assume that the learning data is independent, and the distribution of data does not change over time.

Thus, using DRL we can make some improvements, relative to the standard scheme described above. DRL-approach first accumulates some experience, preserving its actions and their results for some time and then selects from this experience a random mini-batch of individual examples for learning taken in a random order. For the accumulation of learning experience, we can use ε-greedy strategy.

Formally speaking, at each stage t of learning system:

- choose the following action a_t (in ε-greedy strategy we choose a random action with probability ε and $a_t = arg\,maxQ(s_t, a; \theta)$ otherwise);
- do this action, getting the reward r_t and the next state s_{t+1};
- new unit of experience (s_t, a_t, r_t, s_{t+1}) is write into memory;
- select from memory a random mini-batch of such "units of experience" for learning, for example (s_j, a_j, r_j, s_{j+1});
- calculate the output of the network y_j and make one step of the gradient descent for the error function:

$$L = (y_j - Q(s_j, a_j; \theta))^2.$$

This means that we shift the weights to:

$$\nabla_\theta L = 2(y_j - Q(s_j, a_j; \theta))\nabla_\theta Q(s, a; \theta).$$

For correctness, it is necessary that the network does not immediately use the updated version in the objective function and learning for a long time on the old samples before making a full-fledged global update of the parameters $y_j = r_j$ if episode is final and $y_j = r_j + \gamma\,max_{a'}Q(s, a; \theta_0)$ otherwise. Where θ_0 are certain fixed weights that do not change after every test case, while only θ changes. At some point, usually once in one or more learning episodes, we need to return to these weights of the objective function and assign $\theta_0 := \theta$. Finally, we have two parallel networks: one determines the behavior and second the objective function. They have the same structure and they learn identically on the same data, but one network gradually lags behind the other from time to time jump to "catching up" the first.

Thus, one part of the network is train to assess the current state, and the other is to predict how useful the various actions in this state will be [11, 12].

4   Analysis of Anytime Algorithms

When agent interacting in complex dynamic real-time system, where available time for planning is highly limited, generating optimal solutions can be infeasible. In such situations, the agent must be satisfied with the best solution that can be generate within the available computation time and within the range of tolerance of error. A useful set of algorithms for generating such solutions are known as anytime algorithms. Typically, these start out by computing an initial, potentially highly suboptimal solution and then improve this solution as time allows [13].

Anytime algorithms allow making a tradeoff between computation time and solution quality, making it possible to compute approximate solutions to complex problems under time constraints. They also need to have settings that can adjust the flexibility of finding a tradeoff. They can be represented as sampling rates or iterative improvement functions that affect the quality in terms of accuracy, coverage, certainty and level of detail. A tradeoff between quality and time can be achieve by several methods:

- *milestone method* that executed in the minimum period of time and made subsequent evaluation of progress at control points. Based on the remaining time, the algorithm can decide to perform both mandatory and optional operations or simply mandatory operations;
- *sieve functions* that allows to skip the calculation steps. So, the minimum useful selection can be reached in a shorter period of time;
- *multiple versions of the same algorithm* in which intensive calculations can be replaced with faster but less accurate versions of the same algorithm;

For each of the above methods, it is necessary that the different implementation approaches have the ability to measure the quality of explicit metrics in the current state.

At any time, the way to execute the algorithm depends on several factors such as: the quality of the available solution, the prospects for further improving the quality of the solution, current time, cost of delaying the actions taken, current state of the environment and prospects for further environmental change that could be determined only in runtime. Thus, it is necessary to determine the path that provides the most optimal result relative to the current state of the environment [14].

Anytime algorithms should be able to be interrupt at any time or at predetermined control points, to output an intermediate result and be able to continue working using intermediate and incomplete results.

Anytime algorithms are increasingly used in a number of practical areas including: planning, deep confidence networks, evaluation of impact diagrams, processing queries to databases, monitoring and collecting information, etc. This approach can make of decisive importance for complex IDSS RT, with a large number of sensors capable of analysis, and large numerical complexity of the scheduling algorithms to obtain optimal solutions in a limited time and can significantly improve the system's productivity and efficiency.

5 Improvements of Reinforcement Learning Tool for the Forecasting Subsystem

The architecture and algorithm of emulator work for forecasting subsystem consist of forecasting module (includes statistical and expert sub-modules), the multi-agent RL-module (includes agents with various learning algorithms, such as TD(0), TD(λ),

SARSA, Q-learning) and the module of analysis and making decisions (which collect data from other modules, analyze received information and form final action on the environment) were proposed at [15]. For the developed subsystem it was decided to make several modifications on behalf of improving system performance.

5.1   Anytime Algorithm Implementation

Additional sub-module consists of an anytime algorithm that can obtain results under time constraints was included in subsystem. In the ideal version of the algorithm, if sufficient time and resources are available, parallel forecasting with a combined statistical method will be performed, then system learning with RL-algorithms by various methods. After that, analysis of the obtained results and the organization the optimal solutions will be done. Under conditions of severe time constraints, the milestone method [16] is apply to the system. In this method algorithm chooses which of the paths is the most promising, relative to the accuracy of the forecast and the execution time, and calculates the result only by methods capable of obtaining the necessary optimal results at the current moment. In this case, all other steps can also be executed in the background and could be included in the analysis in the next steps.

The introduction of the developed flexible algorithm into the prediction subsystem has the following advantages:

1. flexibility of the system: depending on the state of the environment, various parallel algorithms for learning and forecasting and their combinations can be used;
2. the system can always calculate the intended action, regardless of the available time and resources;
3. the possibility of immediate issue of a solution and continuation of calculations in the background mode;
4. identify the most effective algorithms for the current environment and determine the possible optimal solutions;

5.2   Upgrading of Reinforcement Learning Module

For the reinforcement learning sub-module developed in [15], it is planned to make a modification based on the DRL-approach [17]:

- each sub-module of a multi-agent network, that learning in parallel by various algorithms are divided into two networks also learning in parallel - one determines the behavior and second the objective function;
- several additional intermediate hidden learning layers are created between the input and output layer;
- storing the separate agent's data in an experience replay memory, where the data can be batched or randomly sampled from different time-steps;
- after the end of the episode, the gradient descent is calculated, and the network is completely updated;

Modification of the sub-module with using of DRL-scheme is shown on Fig. 1.

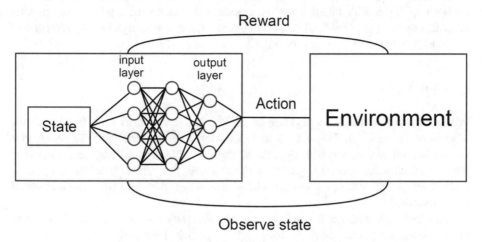

Fig. 1. Deep reinforcement learning sub-module scheme

At the end, we propose architecture of complex IDSS RT, consist of prediction sub-module and reinforcement learning sub-module with different TD-learning agents that all learning and computing in parallel. Inside each DRL-agent, we have hidden layers and replay memory. Over all system, we have sub-module working by milestone

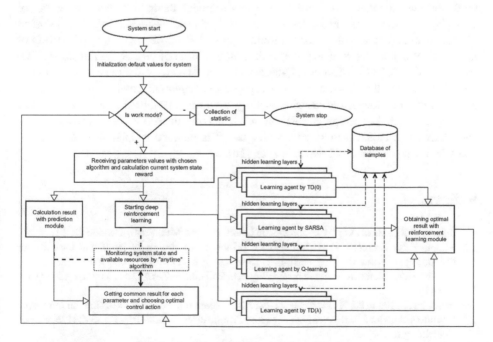

Fig. 2. Architecture of forecasting subsystem

anytime algorithm that evaluate system state and could interrupt in any module work, make fast decisions in situations with time constraints and limited resources that often happens in IDSS RT. Finally, we have main sub-module of analysis and making decisions that can perform final result and operate the entire subsystem. Architecture of forecasting subsystem is shown on Fig. 2.

6 Conclusion

In the paper, in terms of application in IS RT (the systems of type IDSS RT) the scheme of RL based on TD-methods and DRL-approach was analyzed. The basic idea and main methods for anytime algorithms, capable of finding a compromise between the computation time and the quality of the solution, that allows calculating approximate solutions of complex problems under time constraints and their basic methods were described.

The main directions that can improve the efficiency and productivity of the prediction forecasting subsystems by using DRL-approach, including in each sub-module of multi-agent RL-module of additional hidden layers and additional memory are proposed. The analysis of performance and time characteristics of the system after the inclusion of the anytime and DRL algorithms in the forecasting subsystem is carried out and the prospects of their use for IDSS RT are estimated.

The approach based on the integration of learning and decision-making modules was applied in the development of a prototype of IDSS RT for monitoring and control of one of the subsystems of the nuclear power plant unit [6]. Were implemented: the emulator that simulates the state of the environment using different algorithms of changing system parameters in the operational database; the forecasting module based on statistical and expert methods; the multi-agent RL-module, consisting of a group of independent agents, each of that is learning on the basis of one of the developed TD-methods (TD(0), TD(λ), SARSA, Q-learning), as well as used for the accumulation of knowledge about the environment and capable of adaptation, modification and accumulation of knowledge; the decision-making module is designed to analyze the data coming from the forecasting and RL modules, making decisions on follow-up actions and methods to adjust management strategies. It is planned to include a module that takes into account temporal dependencies [18].

References

1. Sutton, R.S., Barto, A.G.: Reinforcement Learning. The MIT Press, London (2012)
2. Vagin, V.N., Eremeev, A.P.: Some basic principles of design of intelligent systems for supporting real-time decision making. J. Comput. Syst. Sci. Int. **6**, 953–961 (2001)
3. Osipov, G.S.: Methods of Artificial Intelligence, 2nd edn. FIZMATLIT, Moscow (2015). (in Russian)
4. Busoniu, L., Babuska, R., De Schutter, B.: Multi-agent reinforcement learning: an overview. In: Innovations in Multi-Agent Systems and Applications, vol. 310, pp. 183–221. Springer, Heidelberg (2010)

5. Mnih, V., Badia, A.P., Mirza, M., Graves, A., Harley, T., et al.: Asynchronous methods for deep reinforcement learning. In: Proceedings of the 33rd International Conference on Machine Learning (PMLR 48), pp. 1928–1937 (2016)
6. Eremeev, A.P., Kozhukhov, A.A.: Implementation of reinforcement learning methods based on temporal differences and a multi-agent approach for real-time intelligent systems (in Russian). J. Softw. Syst. **1**, 28–33 (2017)
7. Sort, J., Singh, S., Lewis, R.L.: Variance-based rewards for approximate Bayesian reinforcement learning. In: Proceedings of Uncertainty in Artificial Intelligence, pp. 564–571 (2010)
8. Wiering, M., Otterlo, M.: Reinforcement Learning: State-of-the-Art (Adaptation, Learning, and Optimization). Springer (2012)
9. Mnih, V., Kavukcuoglu, K., Silver, D., Rusu, A.A., Veness, J., Bellemare, M.G., et al.: Human-level control through deep reinforcement learning. Nature **518**, 529–533 (2015)
10. Li, Y.: Deep reinforcement learning: an overview, arXiv (2017). http://arxiv.org/abs/1701.07274
11. Nikolenko, S., Kadurin, A., Archangelskaya, E.: Deep Learning. Immersion in the world of neural networks. PITER, St. Petersburg (2017). (in Russian)
12. Guo, H.: Generating text with deep reinforcement learning, arXiv (2015). http://arxiv.org/abs/1510.09292
13. Hansen, E.A., Zilberstein, S.: Monitoring and control of anytime algorithms: a dynamic programming approach. J. Artif. Intell. **126**, 139–157 (2001)
14. Mangharam, R., Saba, A.: Anytime algorithms for GPU architectures. In: IEEE Real-Time Systems Symposium (2011)
15. Eremeev, A.P., Kozhukhov, A.A.: Methods and program tools based on prediction and reinforcement learning for the intelligent decision support systems of real-time. In: Proceedings of the Second International Scientific Conference "Intelligent Information Technologies for Industry" (IITI 2017), vol. 1, pp. 74–83. Springer (2017)
16. Likhachev, M., Ferguson, D., Gordon, G., Stentz, A., Thrun, S.: Anytime dynamic A*: an anytime, replanning algorithm. In: Proceedings of the International Conference on Automated Planning and Scheduling (ICAPS), pp. 262–271 (2005)
17. Nair, A., Srinivasan, P., Blackwell, S., et al.: Massively parallel methods for deep reinforcement learning. In: Deep Learning Workshop, International Conference on Machine Learning, Lille, France (2015)
18. Vagin, V.N., Eremeev, A.P., Guliakina, N.A.: About the temporal reasoning formalization in the intelligent systems. In: OSTIS-2016, Minsk, pp. 275–282 (2016)

Intelligent Planning and Control of Integrated Expert System Construction

Galina V. Rybina$^{(\boxtimes)}$, Yury M. Blokhin, and Levon S. Tarakchyan

National Research Nuclear University MEPhI (Moscow Engineering Physics
Institute), Kashiskoe sh. 31, Moscow 115409, Russian Federation
galina@ailab.mephi.ru

Abstract. Intelligent planning and control of construction processes of
integrated expert system is the main subject of this paper. Problem-
oriented methodology of integrated expert system construction is used
as a theoretical basis. The development of theoretical and software
implementation of intelligent software environment is discussed, which
is performed by adding special operational component - intelligent planner which performs planning and control of construction process. Planning method is described in details and software implementation details
are briefly given. Intelligent planner is implemented as a part of AT-
TECHNOLOGY workbench.

Keywords: Integrated expert system
Problem-oriented methodology · Intelligent software environment
Intelligent planning · AT-TECHNOLOGY workbench
Intelligent planner · Applied IES prototyping

1 Introduction

Intelligent systems and technologies are currently the basis of key technologies
of the 21st century. They are used widely in all industrial and socially significant
areas of human activity, because they allow to increase the efficiency of using
computer technology significantly in traditional spheres of its application by
solving new classes of problems using methods and means of artificial intelligence.

The most popular class of intelligent systems (in comparison with others, like
[1–3]) is *integrated expert systems* (IES) class [4]. An analysis of the experience
of developing foreign and domestic IES, including those based on a problem-
oriented methodology for IES development (authored by G.V. Rybina) and the
AT-TECHNOLOGY workbench supporting this methodology [4,5] showed that
the greatest complexity is still on the stages of design and implementation of
IES, and the specificity of the specific problem area and the human factor play
a significant role. Here you can note several main problems that show the need
to create effective tools for automated support for the development of IES [4,5]:

- knowledge engineers (users) are not able to fully determine the requirements
  for the systems being developed, the lack of a reliable method for assessing

© Springer Nature Switzerland AG 2019
A. Abraham et al. (Eds.): IITI 2018, AISC 875, pp. 190–197, 2019.
https://doi.org/10.1007/978-3-030-01821-4_20

the quality of verification and validation of the IES, the inapplicability of traditional tracing technology to the knowledge base (KB) of the IES;
- there are a lot of intermediate stages and iterations in the life cycle models of the construction of individual components of the IES;
- if not taking in account AT-TECHNOLOGY workbench, there is not enough specialized tools which provide: automated design of software and information support for applied IES at all stages of the life cycle; reuse of individual software and informational components; presence of internal integrability of tools; support for a convenient cognitive-graphic interface; openness and portability of tools.

The most of commercial tools like G2 (Gensym Corp.), RTWorks (Talarian Corp.), SHINE (NASA/JPL), RTXPS (Environmental Software & Services GmbH), etc. do not "know" what is being designed and developed with their help by a knowledge engineer so the effectiveness of the application is completely determined by the art of developers.

In the context of solving the above-mentioned problems, certain results were obtained in the framework of approaches such as KBSA (Knowledge Based Systems Assistant) and KBSE (Knowledge Base Software Engineering), which usually integrate capabilities of means to support the development of traditional expert systems (ES) and CASE-tools, and in some cases providing "intellectualization" of the processes of developing intellectual systems [4,5].

The AT-TECHNOLOGY workbench [4,5] is the most widely known new generation workbench, where KBSE approach is implemented. The workbench provides automated support of the processes of building the IES based on the task-oriented methodology, and includes intellectual assisting (planning) the actions of knowledge engineers for account of the use of technological knowledge on standard design procedures (SDP) and reusable components (RUC) of previous projects that are basic components of the model of the intellectual software environment, the concept of which is described in detail in [4] and other works.

Major research step in intelligent software environment of ATTECHNOLOGY workbench development was performed by significantly improvement of role and functions of intelligent planner. It provides effective assistance to knowledge engineers, but also planning and managing the development process as a whole. Analysis of the current versions of the intelligent planner of the AT-TECHNOLOGY workbench [4,5] showed that since the complication of the IES architectures and the appearance of a large number of SDP and RUC in the technological KB, the time for finding solutions has significantly increased. Therefore, the search became quite labor-intensive, and the negative effect from the non-optimal choice of solutions became more significant. Hence, there was a need to improve the methods and algorithms of a planning used by the intelligent planner. The results of the system analysis of modern methods of intelligent planning and conducted experimental studies [6,7] have shown the expediency of using a fairly well-known approach related to planning in the state space.

Below some of the results of further development of the basic components of the intellectual software environment of the AT-TECHNOLOGY workbench will be discussed. They are used for automating and intellectualisation of the processes of development IES with different architectural typologies on the basis of a problem-oriented methodology (detailed description is given in [7–10]).

2   Aspects of Processes of the Prototyping of Applied IES

As a theoretical basis for current research the *problem-oriented methodology* was taken. Its conceptual basis is a multilevel model of integration processes in the IES, modeling of specific types of problems, relevant technologies of traditional ES, methods and methods of building the software architecture of IES and its components at each level of integration, etc. A detailed description of the methods for constructing various IES with a broad architectural typology developed and tested in practice is contained in [6–9], as well as in other works, and therefore we will focus here on the general formulation of the problem of intelligent planning [7–10] with reference to models, methods and means of the intellectual software environment of the AT-TECHNOLOGY workbench.

The main goal of the problem-oriented methodology and its supporting tools is the intellectualization of complex and time-consuming processes of applied IES prototyping for various IES architectures during all the life cycle (from system analysis of the problem area to the creation of a series of prototypes of IES). To reduce the intellectual load on knowledge engineers and to minimize possible erroneous actions and time risks in the prototyping of IES it is envisaged to use a technological KB containing a significant number of SDP and RUC reflecting the expertise of knowledge engineers in the development of applied IES (according to [4]).

Therefore, the formal definition of the problem of intelligent planning of the prototyping processes of IES is considered in the context of the model of prototyping processes of IES in the following form:

$$M_{proto} = \langle T, S, Pr, Val, A_{IES}, PlanTask_{IES} \rangle, \qquad (1)$$

where T is the set of problem domains for which applied IES are created; S is the set of prototyping strategies; Pr is the set of created prototypes of IES based on a problem-oriented methodology; Val - the function of expert validation of the prototype of the IES, determining the need and/or the possibility of creating subsequent IES prototypes for a particular problem domain; A_{IES} - the set of all possible actions of knowledge engineers in the prototyping process; $PlanTask_{IES}$ is the function of planning knowledge engineer's actions to obtain the current prototype of IES for a particular problem domain. In [10], a detailed description of all the components of M_{proto} is given, as well as the specification of some basic components of the model of the intellectual software environment.

Modern methods of intelligent planning in the state space have been researched [11–15] for involving into efficient implementation of the component $PlanTask_{IES}$ of the M_{proto} model. Experimental software research has shown that the best results are achieved if the search space is formed by modeling the actions of the knowledge engineer while constructing fragments of the architecture model [4,5] of the prototype of IES using the appropriate SDPs (for a formal description of this process, graph theory can be used by reducing the problem to the problem of covering of the model of the architecture of the IES prototype, presented as a labeled graph, with SDP fragments in the form of corresponding subgraphs).

A significant role in the implementation of the $PlanTask_{IES}$ component is played by the plan for constructing a specific coverage (i.e., the sequence of applied SDP fragments), which can be uniquely transformed into a plan for constructing a prototype of IES. In this case, the plan for constructing the prototype of the IES can be represented like [10]:

$$Plan = \langle A_G, A_{atom}, R_{prec}, R_{detail}, PR \rangle, \tag{2}$$

where A_G is the set of global tasks decomposable into subtasks); A_{atom} is the set of planned (atomic) tasks, the execution of which is necessary for the development of the IES prototype; R_{prec} is the function that determines the predecessor relation between the planned tasks; R_{detail} is the relation showing the affiliation of the planned task to the global tasks; PR is the representation of the plan, convenient for the knowledge engineer.

The relation R_{prec} and the sets A_G and A_{atom} are used to form two task networks - enlarged and detailed. The enlarged network of tasks obtained with R_{prec} and A_G is called the global plan (the relation of the precedence between the elements of the A_G is obtained on the basis of the R_{detail} detail relation), and the detailed network of tasks obtained on the basis of R_{prec} and A_{atom} is called the detailed plan, with each planned task associated with specific function by a function of a specific operational RUC.

Certain limitations are imposed about the structure of any SDP which are based on the analyzing of the experience of developing applied IES. The most valuable limitation is that at least one compulsory fragment and an arbitrary number of optional fragments associated with the mandatory fragment of at least one data flow. As a rule, the first element of the set of fragments includes an non-formalized operation (NF-operation) [4], which plays the main role in the construction of the architecture of the current prototype of IES.

3   General Definition of the Problem of Generation of Plans for the Development of Prototypes of IES

Now let us consider the general definition of the problem of generation of plans for developing prototypes of IES. The initial data is the model of the architecture of the prototype of the IES, described using the hierarchy of extended data flow diagrams (EDFD [4]); The technological KB, containing a lot of SDP and RUC. In addition, a number of restrictions and working definitions were introduced.

- The so-called *generalized EDFD* is a special oriented graph, which is obtained from the composition of the architecture model M_{IES}, built at the stage of

the analysis of system requirements. It is obtained by taking in account only a set of elements and a set of data flows represented in the form of a marked oriented graph $G_R IL$, where the labels determine the relationship between the elements the hierarchy of the EDFD and the nodes and arcs of the graph.
- Any arbitrary connected subgraph contained in G_{RIL} is called as *the fragment* of the generalized EDFD. A SDP instance is an aggregate of TPP and a fragment of a generalized EDFD satisfying the conditions of applicability (the C component of the TPP model) of the corresponding SDP. The cover ($Cover$) of a generalized RDPA is the set of instances of SDP with mutually disjoint fragments containing all the nodes of G_{RIL} (or cover the entire G_{RIL}).
- A coarse coverage of a generalized EDFD is a EDFD coverage in which all instances of the SDP contain only mandatory fragments. The fine coverage is an extension of the coarse coverage by including optional fragments in the coverage.
- The inclusion of each fragment of the SDP in the cover is compared with a certain value determined by an expert evaluation based on technological experience, which conceptually corresponds to the average costs of human resources for the implementation of the corresponding fragment of the model of the architecture of the IES.

Taking into account the initial data and imposed constraints, the problem of generating a plan for developing a prototype of IES, can be defined in terms of states and transitions in the form of the following model:

$$PlanTask_{IES} = <S_{IES}, A_{IES}, \gamma, Cost, s_0, G_{IES}, F_{COVER}>, \qquad (3)$$

where S_{IES} is the set of states of the graph G_{RIL} that describe the current coverage $Cover$; A_{IES} - the set of possible actions over G_{RIL}, which are the addition of fragments of specific instances of SDP to the cover (the total set is formed in the aggregate WKB and G_{RIL}); γ is the transition function between states; $Cost$ - a function that determines the cost of a sequence of transitions; s_0 is the initial state describing the empty coverage; G_{IES} - the function of determining whether the state belongs to the target state; F_{COVER} - the function of generating a development plan ($Plan$) from the coverage ($Cover$).

$Plan$ is partially ordered set of actions of the knowledge engineer and it is the solution of $PlanTask_{IES}$. The plan should be optimal, thats why it is necessary to develop an admissible heuristic function for guiding heuristic search algorithms.

An original method was developed [10] to solve the defined problem. This method can be decomposed into four following stages: obtaining a generalized EDFD (G_{RIL}) from the architecture model M_{IES}; generating an exact $Cover$ cover using heuristic search; generation of the plan of actions of the knowledge engineer ($Plan$) on the basis of the obtained fine coverage ($Cover$); generating a plan view (PR) based on coverage ($Cover$). Necessary algorithms were developed to implement each stage.

4 Aspects of Software Implementation of Basic Components of the Intellectual Software Environment

In lots of references like [7–10] the features of the design and software implementation of the intelligent planner and other components of the intelligent software environment were repeatedly described. Therefore, we give here only a brief description of the composition and structure of the software tools of the intelligent software environment, including the kernel, the user interface subsystem and the extension library, which implements interaction with operational RUCs. The subsystem of the user interface has a convenient graphical interface, on the basis of which the RUC interacts with the knowledge engineer using screen forms. Software architecture of the most valuable component - intelligent planner is given in the Fig. 1.

The technological KB is conventionally divided into an extension library that stores operational knowledge in the form of plug-ins that implement the relevant operational RUCs and the declarative part. The kernel implements all the basic functionality of automated support for the development of prototype IES, project file management, extension management, etc.

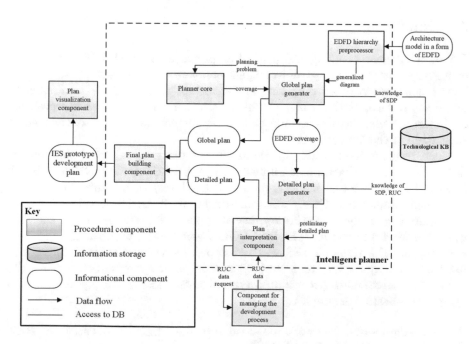

Fig. 1. Software architecture of intelligent planner

The intelligent planner is a part of the core and implements functionality related to planning the prototyping processes of IES. With the help of the preprocessor of the hierarchy of the EDFD, the pre-processing of the hierarchy of the

EDFD is carried out by converting it into one generalized diagram of maximum detailing.

The task of coverage the detailed EDFD with the available SDP is implemented with the help of a *global plan generator* (see Fig. 1) that, on the basis of the technological KB and the constructed generalized EDFD, provides the fulfillment of the task using the method described above, as a result of which a fine coverage is constructed, which is later transformed into a global plan of development.

Generator of the detailed plan on the basis of the given coverage of the EDFD and the technological KB performs a detalization of each element of the coverage, thus forming a preliminary detailed plan. Then, based on the analysis of available RUC and their versions (data about which are requested from the development process control component), a detailed plan is formed in the plan interpretation component, where each task is related to a specific RUC and can be performed by a knowledge engineer. With the help of the final plan building component, the required plan representation is generated for its use by other components of the intelligent software environment (the component of the plan visualization, etc.).

5   Conclusion

For more efficient planning and control of processes of applied IES prototypes construction new technology of intelligent software environment was developed. This technology is based on giving an important role to special component - intelligent planner, with the help of which risks of erroneous actions of knowledge engineers are minimized, cost and time development indicators are reduced at each stage of the life cycle, etc.

As a result of joint testing of the intelligent planner and other components of the intelligent software environment of the AT-TECHNOLOGY workbench it was obtained a lot of numeric data which allowed to measure efficiency of developed components. Thus it was concluded that all the developed tools can be used quite effectively in the prototyping of applied IES. The developed software was integrated into the AT-TECHNOLOGY workbench, with the use of which a prototype of a dynamic IES was developed to manage medical forces and facilities for major road accidents [10]; prototype of dynamic IES for satellite network resources management; prototype of static IES for diagnostics of blanks for electron-beam lithography.

Acknowledgements. The work was done with the Russian Foundation for Basic Research support (project №18-01-00457).

References

1. Giarratano, J.C., Riley, G.D.: Expert Systems: Principles and Programming, Fourth Edition. Course Technology, Boston (2004)
2. Meystel, A.M., Albus, J.S.: Intelligent Systems: Architecture, Design, and Control, 1st edn. Wiley, New York (2000)
3. Schalkoff, R.: Intelligent Systems: Principles, Paradigms, and Pragmatics. Jones & Bartlett Learning (2011). https://books.google.ru/books?id=80FXUtF5kRoC
4. Rybina, G.V.: Theory and Technology of Construction of Integrated Expert Systems. Monography. Nauchtehlitizdat, Moscow (2008)
5. Rybina, G.V.: Intelligent systems: from a to z. monography series in 3 books. In: Knowledge-Based Systems. Integrated Expert Systems, vol. 1. Nauchtehlitizdat, Moscow (2014)
6. Rybina, G.V., Blokhin, Y.M.: Methods and means of intellectual planning: implementation of the management of process control in the construction of an integrated expert system. Sci. Technical Inf. Process. **42**(6), 432–447 (2015)
7. Rybina, G.V., Blokhin, Y.M.: Use of intelligent planning for integrated expert systems development. In: 8th IEEE International Conference on Intelligent Systems, IS 2016, Sofia, Bulgaria, 4–6 September 2016, pp. 295–300 (2016). http://dx.doi.org/10.1109/IS.2016.7737437
8. Rybina, G.V., Rybin, V.M., Blokhin, Y.M., Sergienko, E.S.: Intelligent technology for integrated expert systems construction. In: Advances in Intelligent Systems and Computing, vol. 451, pp. 187–197 (2016)
9. Rybina, G.V., Blokhin, Y.M., Parondzhanov, S.S.: Intelligent planning methods and features of their usage for development automation of dynamic integrated expert systems. In: Advances in Intelligent Systems and Computing, vol. 636, pp. 151–156 (2018)
10. Rybina, G.V., Blokhin, Y.M.: Methods and software of intelligent planning for integrated expert systems development. Artif. Intell. Decis. Making **1**, 12–28 (2018)
11. Nau, D.S.: Current trends in automated planning. AI Mag. **28**(4), 43–58 (2007)
12. Ghallab, M., Nau, D.S., Traverso, P.: Automated Planning - Theory and Practice. Elsevier, San Francisco (2004)
13. Ghallab, M., Nau, D., Traverso, P.: Automated Planning and Acting, 1st edn. Cambridge University Press, New York (2016)
14. Geffner, H., Bonet, B.: A concise introduction to models and methods for automated planning. IEEE (Institute of Electrical and Electronics Engineers) IEEE Morgan & Claypool Synthesis eBooks Library, Morgan & Claypool (2013). https://books.google.ru/books?id=_KInlAEACAAJ
15. Korf, R.E., Taylor, L.A.: Finding optimal solutions to the twenty-four puzzle. In: Proceedings of the Thirteenth National Conference on Artificial Intelligence and Eighth Innovative Applications of Artificial Intelligence Conference, AAAI 96, IAAI 96, Portland, Oregon, 4–8 August 1996, vol. 2, pp. 1202–1207 (1996)

The Matrix Data Recognition Tool in the Input Files for the Computing Applications in an Expert System

Simon Barkovskii[1], Larisa Tselykh[2]([⊠]), and Alexander Tselykh[1]

[1] Department of Information and Analytical Security Systems,
Institute of Computer Technologies and Information Safety,
Southern Federal University, Nekrasovskii, 44, 347922 Taganrog, Russia
kharitonov.simon@yandex.ru, ant@sfedu.ru
[2] Chekhov Taganrog Institute, Rostov State University of Economics,
Initsiativnaya, 48, 347936 Taganrog, Russia
l.tselykh58@gmail.com

Abstract. This study proposes a simple and inexpensive tool for automating matrix recognition in heterogeneous input/output data from various computational applications. The recognition of matrix data is done at the file system level and is based on the development of recognition rules using data structure sample. Our contribution is as follows: we propose an automated mechanism for recognizing matrices written by different styles in input/output files; this procedure does not require special query and any participation from the end-user; no special skills are required for the end-user. The proposed recognition mechanism is an effective solution for the development of an automated system for plugging in and switching math modules integrated into a system of sequential computations without resorting to third-party developers.

Keywords: Expert systems · Matrix parsing · Matrix recognition
Text processing

1 Introduction

Modern intelligent systems, including expert systems, often contain various computational algorithms. The output of these algorithms is used internally. The development of such systems requires updating versions of computational algorithms or their complete substitution, or the addition of new computing applications. This raises the problem of integrating new input/output data structures after they are plugged-in.

As a rule, the system accepts the results of computations on the logical level in the relational database. Therefore, even small changes in the output data (format, form, structure, etc.) require reprogramming of their indexing and query support. This makes the users seek assistance of professional programmers and domain experts, which entails an increase in the cost of using the system.

There is an alternative to solving the problem of plugging computational applications – managing input/output files at the physical level in the file system. In this case, there is a need to transform the input/output files into files with a certain type and

A. Abraham et al. (Eds.): IITI 2018, AISC 875, pp. 198–208, 2019.
https://doi.org/10.1007/978-3-030-01821-4_21

structure required by the system and/or other plugged-in computational applications. High specificity of the output files presents a considerable difficulty. This kind of barrier might be bridged through developing special tools for particular types of data. For decision support systems (for example, for clustering programs, for finding influences, fuzzy cognitive modeling, etc.), the required data are in the form of matrices. The process of "detection-recognition-transformation" will be referred to as post-processing of data. The solution to the post-processing task should be done automatically, without the involvement of end-users.

In this article, we present the main phase of data post-processing – recognition, to be used in applications that require the input/output data in matrix form. The proposed solution of matrix data recognition is implemented at the level of the file system and is based on the development of generation rules using samples of data structures. Our contribution is as follows: an automated mechanism for recognizing matrices written by different styles in input/output files; this procedure does not require special query and any participation from the end-user; no special skills are required for the end-user.

2 Related Work

The existing data post-processing solutions offer the following approaches:

- Development of programs for extracting and indexing the contents of files designated for specific programs, for example, scientific computing packages, such as the Gaussian program [1] or for the web [2];
- Based on special programming language [3, 4];
- At the logical level [5–8].

The MATHSCOUT package uses keywords to find data from named lists and tables using special command coding methods. Since this package is intended for the Gaussian program, it is geared to use a specific structure and type of the input/output files as a list of numerical characteristics. The console application TabbyXL developed by [2] uses recognition of tagged documents to exclusively bring tables with a set of certain attributes in the canonicalization process to their canonical relational form. The approach using special languages such as OIL [3], VisualTPL [4], means that they are employed by the end-users who have the knowledge of both programming and mathematics, and of relevant subject areas. The approach proposed by [8] performs data processing and control in MatCloud and is based on sharing the file system and NoSQL database at the logical level which requires predefined characteristics. Therefore, this tool is aimed at software developers. The article [5] proposes a solution through an extraction function using the R2RML mapping language, based on the combination of Semantic Web and Data Warehouse technologies.

The philosophy behind the proposed software is to integrate the existing software packages as easily as possible, without turning for help to professional developers.

3 Data Recognition Process

In this section, we show the data post-processing to the document contained the matrix at the file system level for documents containing matrices. The program is written in C#, for the Windows; the software platform is .NET Framework. In this realization of the program, it is assumed that the matrices can be written in three different types: (i) type 1 as an $n \times n$ table, where n is the dimension of the matrix; (ii) type 2 as a single line of numbers that are elements of the matrix, including zero ones; or (iii) type 3 in the form of a table with three columns in which the row number, the column number and the numerical value of the nonzero element of the matrix are set.

The proposed recognizer is designed for computational programs and is not related to a particular program. The diagram of the program components is shown in Fig. 1.

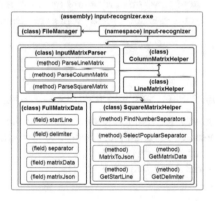

Fig. 1. Deployment diagram for the Input-Recognizer

Our recognizer is brought into play by "input-recognizer.exe" (manual) or from the wrapper-program session. The program takes the document with a sample of input data and notation (if any). The document and notation can be supplied to the program: (i) directly by the user through the command prompt of the program and (ii) automatically from the controlling wrapper-program in case of using the Input-Recognizer as a library. Input-Recognizer contains six classes:

1. FileManager – downloads and writes the files.
2. InputMatrixParser – initializes matrix recognition.
3. ColumnMatrixHelper – recognizes the 3rd type matrix.
4. LineMatrixHelper – recognizes the 2nd type matrix.
5. SquareMatrixHelper – recognizes the 1st type matrix.
6. FullMatrixData – configures the output data.

The `InputMatrixParser` class includes three functions –
`ParseSquareMatrix(inputMatrix, fileName)`,
`ParseLineMatrix(inputMatrix, fileName)`,
`ParseColumnMatrix(inputMatrix, fileName)` –
which recognize matrices written in three types.

Input: `ParseSquareMatrix(inputMatrix, fileName)` takes (1) the text of the input data from the file and (2) the name of this file. Each tool of the Input-MatrixParser class has six main functions:

1. `FindNumberSeparators(inputMatrix)` – identifies decimal separator types.
2. `SelectPopularSeparator(separators)` – searches and selects the predominant type of decimal separator (creation of the 1st generation rule "Separator").
3. `GetDelimiter(separator, inputMatrix)` – searches and selects the prevailing matrix delimiter (creation of the 2nd generation rule "Delimiter").
4. `GetMatrixData(inputMatrix, delimiter)` – creates the list of matrix cells.
5. `GetStartLine(matrixData, delimiter, inputMatrix)` – searches the initial line of the matrix entry in the document (creation of the 3rd generation rule "StartDrawMatrixAtLine").
6. `MatrixToJson (matrixData, fileName)` – rules-based transformation of the recognized matrix into *json*-format.

After the completion of all the functions above, we obtain a set of generation rules (3 types) and the recognized matrix in *json*-format. We determined that each matrix has the following mandatory attributes, which must be identified in order to recognize the matrix image correctly:

1. Decimal separator – `Separator`.
2. Matrix cell delimiter – `Delimiter`.
3. The initial line of the matrix entry in the document – `StartDrawMatrixAtLine`.
4. Notation with the command description for getting the values of special cells (if any).

Consider the recognition process using the example of the `ParseSquareMatrix` tool to recognize the input data file from the `Fortran` program for computing the influence nodes. The input matrix (`inputMatrix`) of type 1 is shown in Fig. 2.

As shown in Fig. 2, this matrix feed has the following distinctive features:

– Apart from the matrix itself, the document contains text comments ("noise"), which are not directly related to the matrix and interfere with the recognition of the matrix;
– In the first line there is an additional value that is an indicator of the dimension of the matrix necessary for the correct operation of the math algorithm. This value must be specified by the author of the sample file in the notation, otherwise it is recognized as "noise";
– The decimal separator is " . ";
– The last two values of the matrix have a different decimal separator – " , ";
– There are matrix elements with the minus sign;

Two elements of the matrix, unlike the others, do not have decimal separators.

Fig. 2. Primary data of type 1 matrix

Therefore, in the recognition process, the following tasks must be solved:

1. Remove the "noise" and recognize matrix elements.
2. Recognize the elements of the matrix that have different decimal separators, and the elements that do not have decimal separators.
3. Create the generation rules regarding mandatory attributes (Separator, Delimiter, StartDrawMatrixAtLine).
4. Recognize the author's notation (if any), and add to the existing rules the author's additional rules necessary to generate an input file to be transferred to another math algorithm.
5. Transform the recognized matrix into *json*-format for further work with the matrix data inside the program.

The flowchart of the recognition mechanism is shown in Fig. 3.

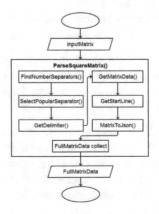

Fig. 3. The flowchart of the recognition mechanism

Following this algorithm, for the case under consideration we obtain the following results. In the function FindNumberSeparators(inputMatrix), the command

```
List<char> separators =
SquareMatrixHelper.FindNumberSeparators(inputMatrix)
```

creates a list of all detected decimal separators, which is passed to the next stage of matrix recognition.

The function SelectPopularSeparator(separators) is used to create the first generation rule – definition of the Separator type. The selection of separator type is made using the statistical approach. The function groups the values in the decimal separator list and selects the maximum frequency index from the types of separators that occur in the document matrix. The command

```
char separator =
SquareMatrixHelper.SelectPopularSeparator(separators)
```

selects the "." separator as the most frequent and stores it in RAM as the 1st type generation rule (Fig. 4), as well as for further use in the control system.

Fig. 4. Visualization of matrix attributes

Then, taking into account the chosen decimal separator, the function GetDelimiter(separator, inputMatrix), searches all possible cell delimiters of the matrix. Next, these delimiters are grouped by value, and the most common type of delimiter of the cells of the matrix is selected. In this way, the 2nd generation rule - the Delimiter type definition - is created. The delimiter represents a sequence of zero or more Unicode characters. Data are generated by the command

```
string delimiter =
SquareMatrixHelper.GetDelimiter(separator, inputMatrix)
```

and are stored in RAM as the 2nd type generation rule (Fig. 4), as well as for further use in the control system.

In the fourth step, on the basis of the gotten delimiters of the matrix cells, the function `GetMatrixData(inputMatrix, delimiter)` performs the markup of the matrix cells and all possible strings of the matrix, and counts the groups by string length (the number of cells in each row of the matrix). At the end of the process, all strings in the group with the lowest repetition rate in the document are cut off. The command

```
List<string[]> matrixData =
SquareMatrixHelper.GetMatrixData(inputMatrix, delimiter)
```

forms a list of all matrix cells with the attributes Separator and Delimiter, without "noise" (Fig. 4).

Fig. 5. Visualization of the recognized matrix without "noise"

As a result, the function collects all the received cells of the matrix with one type of separators and without "noises", and makes a correctly written matrix (Fig. 5).

In the fifth step, the function `GetStartLine(matrixData, delimiter, inputMatrix)` creates the 3rd generation rule `StartDrawMatrixAtLine` – the line number at which the recording of the matrix starts in the document. The command

```
int startLine =
SquareMatrixHelper.GetStartLine(matrixData, delimiter,
inputMatrix)
```

initializes the search process and stores the result in RAM as the 3rd generation rule, as well as for further use in the control system.

The numbering of the lines of `startLine` values starts from zero. The function loops over the lines of the recognized document and finds a line that corresponds to the result of the concatenation of the values of the first list of values returned from `GetMatrixData (inputMatrix, delimiter)` and the Delimiter gotten earlier, as shown in Fig. 4. In the example in Fig. 6, the 3rd generation rule is written as `StartDrawMatrixLine`: "2".

At the final step, the function MatrixToJson(matrixData, fileName) transforms the recognized matrix in *json* format. The object in the *json*-format matrix (MatrixJson) has the following structure:

```
public class MatrixJson
    {
            [JsonProperty("name")]
            public string name;
            [JsonProperty("n")]
            public int n;
            [JsonProperty("connections")]
            public List<ConnectionsJson> connections;
    }
public class ConnectionsJson
    {
            public int from;
            public int to;
            public double value;
    }
```

where "name" is the name of the matrix (corresponds to the name of the document storing the matrix), "n" – the dimension of the matrix, "connections" – list of ConnectionsJson classes characterizing connections inside the matrix that have the following values: "from" – id of the output vertex, "to" – id of the input vertex, and "value" – the weight of the connection between these vertices.

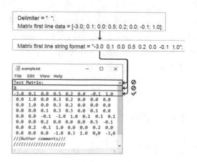

Fig. 6. Concatenating the first line of the matrix and searching it in a document

At the end of the program's work, we receive the following data in the report file (Fig. 7).

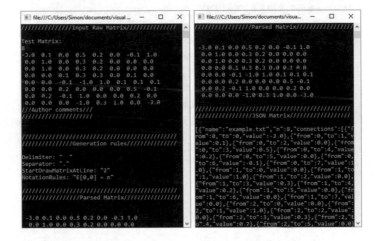

Fig. 7. The Report file

Thus, as a result of the work of the program, we have:

1. Generation rules:
   (a) The 1st rule: `Separator`: "." – decimal separator.
   (b) The 2nd rule: `Delimiter`: " " – the delimiter of the matrix cells, equal to two spaces in this case.
   (c) The 3rd rule: `StartDrawMatrixLine`: "2" – the number of the line from which filling of the matrix starts (the count starts from zero).
   (d) The 4th rule: `NotationRules`: "E[0,0] = n" – an additional rule to write a mandatory elements gotten from notation.
2. *JSON*-matrix having the following view:

```
[{
  "name" :  "example.txt",
  "n" : "8",
  "connections" : [{"from" : 0, "to" : 0, "value" :    -
3.0}, {…}, …]
}]
```

4   Discussion

In this article, we proposed the main part of the automation tool for plugging math algorithms, designed to recognize heterogeneous data of input/output files. The program recognizes 3 types of matrices (in tabular and linear form), eliminates "noise" that is not related to the matrix, and constructs the correct matrix entry in *json*-format. The main advantage of the approach used in the program is the control of input/output data at the physical level. The input-recognizer does not require a predefined data representation structure; it automatically determines the data in the form of matrices. The proposed approach frees users from the routine procedure of searching for matrices themselves and entering them into math applications manually.

The essential task was to provide the possibility of working with the program to users who do not have special skills in programming. This task was achieved for the developed module. All procedures are automated and do not require the involvement of end-users. Because the Input-recognizer reads and writes files in small data packets, it does not consume a lot of system resources and does not impose restrictions on the size of files that need to be recognized. Another advantage of this approach is that it does not require any modifications of the system and its existing applications to connect to data. The Input-Recognizer has certain limitations. In the case of loading user matrices with errors (for example, extra and/or missing matrix values, random symbols in the matrix, etc.), the development of additional module functionality is required. However, this does not present significant difficulty. The architecture of the program is scalable and allows, if necessary, to add a new set of classes and functions. In the case of a new way of writing for matrix, the program can be improved.

5   Conclusion

In this work, we demonstrated a simple and inexpensive tool that encapsulates the functions of automated recognition of input files matrices of math programs. Our solution is as follows:

1. Recognition of the output data at the physical level;
2. Generation of rules based on pattern recognition of data structures in the input data file;
3. The recognition process does not require end-user intervention.

We also developed a prototype of the proposed solution for matrix recognition, a console application "input-recognizer", which can be used either separately by manual controls or as a module in other programs by class library.

Acknowledgements. This work was supported by the Russian Foundation for Basic Research [grant number № 17-01-00076].

References

1. Barnett, M., Capitani, J.: The MATHSCOUT Mathematica package to postprocess the output of other scientific programs. Comput. Phys. Commun. **177**, 944–950 (2007). https://doi.org/10.1016/j.cpc.2007.07.009
2. Shigarov, A., Mikhailov, A.: Rule-based spreadsheet data transformation from arbitrary to relational tables. Inf. Syst. **71**, 123–136 (2017). https://doi.org/10.1016/j.is.2017.08.004
3. Khan, K., Akhter, G., Ahmad, Z.: OIL – output input language for data connectivity between geoscientific software applications. Comput. Geosci. **36**, 687–697 (2010). https://doi.org/10.1016/j.cageo.2009.09.005
4. Chen, W.-K., Tu, P.-Y.: VisualTPL: a visual dataflow language for report data transformation. J. Vis. Lang. Comput. **25**, 210–226 (2014). https://doi.org/10.1016/j.jvlc.2013.11.003
5. Nath, R., Hose, K., Pedersen, T., Romero, O.: SETL: a programmable semantic extract-transform-load framework for semantic data warehouses. Inf. Syst. **68**, 17–43 (2017). https://doi.org/10.1016/j.is.2017.01.005
6. Silva, V., Leite, J., Camata, J., de Oliveira, D., Coutinho, A., Valduriez, P., Mattoso, M.: Raw data queries during data-intensive parallel workflow execution. Future Gener. Comput. Syst. **75**, 402–422 (2017). https://doi.org/10.1016/j.future.2017.01.016
7. Silva, V., Oliveira, D.d., Valduriez, P., Mattoso, M.: Analyzing related raw data files through dataflows. Concurr. Comput. Pract. Exp. **28**, 2528–2545 (2016). https://doi.org/10.1002/cpe.3616
8. Yang, X., Wang, Z., Zhao, X., Song, J., Zhang, M., Liu, H.: MatCloud: a high-throughput computational infrastructure for integrated management of materials simulation, data and resources. Comput. Mater. Sci. **146**, 319–333 (2018). https://doi.org/10.1016/j.commatsci.2018.01.039
9. Tselykh, A., Tselykh, L., Vasilev, V., Barkovskii, S.: Expert system with extended knowledge acquisition module for decision making support. Adv. Intell. Syst. Comput. **680**(2), 21–31 (2017). https://doi.org/10.1007/978-3-319-68324-9

Modern Approaches to Risk Situation Modeling in Creation of Complex Technical Systems

Anna E. Kolodenkova[1]([⊠]), Evgenia R. Muntyan[2], and Vladimir V. Korobkin[3]

[1] Samara State Technical University, Samara, Russia
anna82_42@mail.ru

[2] Scientific Research Institute of the Multiprocessor Computing Systems of Southern Federal University, Taganrog, Russia
evgenia_muntyan@mail.ru

[3] Southern Federal University, Taganrog, Russia
vvk@niimvs.ru

Abstract. Creation of complex technical systems (CTS) is complicated iterative process, which is connected with considerable expenses of material, labor and financial resources together with many arising risk situations caused by constructive defects, industrial releases of unproven technologies, staff faults, inadequate skill level, etc. It can lead to sufficient backlog or fall of CTS creation project. Therefore modeling of the risk situations arising during creation of complex technical systems in the conditions of fuzzy input data is the relevant. Two modern approaches are proposed to detect and predict risk situations: fuzzy cognitive modeling and situation modeling. The convolution algorithm is presented for situation graph generalization. The construction and impulse modeling for fuzzy cognitive model of risk detection in nuclear industry is considered. The results of modeled scenarios of possible risk evolution and their analysis are shown. Proposed approaches allow to identify and analyze the facts impacting on risk situation, obtain possible scenarios of emergence, find the decision ways in modeled situations. It can be used a basis in the production of scientifically proven management actions.

Keywords: Risk situations · Complex technical system
Fuzzy cognitive modelling · Situation modelling · Fuzzy data

1 Introduction

Creation of the complex technical systems (CTS) is the difficult iterative process consisting of the number of consecutive stages. It is characterized by the wide set of uncertainties connected by fuzzy initial data (fuzzy intervals, fuzzy triangle and trapezoidal values and verbal descriptions), large capital investments, big labor input, considerable expenses of resources, risk situations, the large number

The work was supported by RFBR grants No. 17-08-00402-a.

of execution of various documentation [1, 2]. Even the insignificant mistakes made on any of stages of creation of CTS can lead to the fact that the behavior of system does not meet stated purposes and requirements to CTS that, in turn, leads to necessity of urgent adaptation of already implemented system.

In this connection, the present paper proposes the following approaches for the risk situation detection and prediction for CTS creation for nuclear stations:

1. Fuzzy cognitive modeling applying fuzzy cognitive models (FCM) for pre-project researches.
2. Situation modeling based on fuzzy situation graph (FSG).

2   Principles of Risk Situation Detection

Risk situations are the events caused by the reasons and risk factors connected with scheduling errors, staff provision, changes of initial requirements, adoption of design decisions, terms and volumes of the allocated resources on creation of CTS, which can lead to negative or positive effects.

The following principles are used for development and implementation of certain approaches and techniques for risk situation researching, modeling and analyzing:

– Multifarious principle assumes consideration of creation of CTS from the different points of view (i.e. performance of executor, number of executor's errors, time and cost of CTS creation, staff qualification, etc.);
– Plurality principle considers that various FCM and FSG has the same right to exist for certain risk situation modeling;
– Principle of factor omnipotence considers the factors, which does not impacts on risk situation in past, but can impact on this in future;
– Principle of model limitation assumes that a FCM or FSG cannot fully reflect risk situations for CTS creation;
– Principle of information sufficiency considers that complete absence of information about CTS creation excludes FCM and GSG creation;
– Simplicity and economy principle considers that constructed FCM and FSG must be enough simple, i.e. they should not be as complex as it is required for leader and executor. Information obtained on this basis must be available for leaders and executors in convenient form at all stages of CTS life cycle.

These principles allow to form various tasks connected with detection and prediction of risk situations during CTS creation using fuzzy cognitive and situation modeling and with decision ways on risk situations from the uniform theoretical point of view.

3   State of Art

Cognitive modeling (CM) is a tool for analysis, decision making and management in semi-formalized tasks proposed by R. Axelrod in 1975. CM is based on modeling of subjective expert (leaders or executors) ideas about the situation. CM includes knowledge representation model in form of cognitive map and techniques of situation analysis [3,4].

Sufficient contribution into CM development was made by the scientists: Abramova, Avdeeva, Atkin, Gorelova, Khoroshevskiy, Kosko, Kulba, Maksimov, Silov, Zhang [5–7].

Nowadays, the following systems for CM modeling are known: Decision Explorer and FCMapper oriented on analysis of CM structure, CM system developed under the leadership of G.V. Gorelova, "Kosmos" system developed under the leadership of V.B. Silov, system modeling labeled oriented graphs and labeled weighted graphs developed under the leadership of V.V. Kulba, "IGLA" of Bryansk state technical university developed under the leadership of A.G. Podvesovskiy, "Kanva" system developed under the leadership of O.P. Kuznetsov etc.

The main advantages of CM are follows: prediction of possible events, estimation of interaction of executed factors defining possible evolution scenarios of situations, which stronger affect on modeling object.

In 1986, Kosko proposed fuzzy cognitive models (FCM), which are theoretical basis of behavior description for any CTS in uncertainty conditions. These models were developed in number of works (Carvalho, Fedulov, Ginis, Groumpos, Kulinich, Kuznetsov, Lagerea, Papageorgiou, Podvesovsky, Stylios, etc.). Depending on certain task interpretation, the different modifications of FCM are considered [8–10].

However, despite on sufficient number of above mentioned works, the problem of initial data processing (values of vertices' parameters and their connections) is still opened. It is connected with those fact that the conventional approaches considers the parameters of vertices to be dimensionless with values at $[0, 1]$ and the connections to be values at $[-1, 1]$ [10,11]. It is obvious that these conditions are not satisfied in the real situation of CTS creation.

4   Fuzzy Cognitive Modeling for Detection and Analysis of Risk Situations at Pre-project Researches Stage

FCM is the fuzzy cognitive map, where vertices are factors and arcs are fuzzy causal relations among the factors [12]:

$$G_{fuzzy} = \langle X, W \rangle, \tag{1}$$

where $X = x_i$ is the set of vertices ($i \in [1, h]$, where h is the number of vertices), $W = w_{ij}$ is the set of arcs characterizing direction and power of affecting between i-th and j-th vertices.

It should be noted that FCM considers the most important factors, which number is between 10 and 15. If this number is less, FCM unlikely has a sense. If this number is greater, taking into account the number of all connections, the reading of FCM is rather complicated.

It should be also noted that while FCM is not considered as mathematical model the terms "factor", "concept" and "object" have to be used. Then, "vertex" must be operated.

To detect and predict risk situations, impulse modeling of FCM is proposed. As a result, various scenarios of risk situation prediction connected with CTS creation can be constructed with the purpose to decrease negative tendencies and/or increase positive ones.

To perform impulse modeling for FCM, it is necessary to check the dependencies $x_i(t)$, i.e. vertices' changes during time. Graph perturbation in time is characterized as follows:

$$x_i(n+1) = x_i(n) + \sum_j w_{ij}(n)P_j(n) + Q_i(n), \qquad (2)$$

where $x_i(n+1)$ and $x_i(n)$ are the states of vertices at the consequent time steps n and $n+1$, P characterizes the changes of x_i at n, $Q_i(n)$ is the vector of external impulses at n.

The scenarios generated by (2) give the answer for a question: "What is at $t = n + 1$, if ...?".

The algorithm of fuzzy input data processing [13] is proposed to perform impulse modeling of FCM because CTS creation assumes input data to be presented in form of intervals and fuzzy values. Novelty of this algorithm is concluded in normalization and structuring of factor values, which allows to apply fuzzy cognitive modeling of process of risk detection and prediction.

5   Situation Modeling and Risk Situation Detection at Operation Stage

The task of control over diverse components arises in CTS design. It is convenient to use situation graphs to model the situations during CTS operation [14–16]. In situation graphs, vertices are system states (situations) or set of factors affecting on risk situations. Such set of factors together with transitions form the situation, which is changeable in time depending on their combination. In total, factors of the modeled system can be divided into three general groups (input, output and external).

Situation graphs can be rather big because of multicomponent nature of CTS, which leads to the problems of their processing (i.e. representation, merging and division). It is proposed to use fuzzy situation graph to model the CTS execution, where the i-th vertex is characterized by $\langle x_i, \eta_i \rangle$ and the arc between i-th and j-th vertices is characterized by $\langle w_{ij}, \mu_{ij} \rangle$, where $\eta_i \in [0, 1]$ and $\mu_{ij} \in [0, 1]$ are the membership degree of fuzzy set [17, 18]. These degrees can be defined by experts or computed based on membership functions.

Situation graph can be constructed in advance or during monitoring process. It can be finite or conditionally infinite. "Conditionally" means that the size is unknown at the time of evaluation.

It is often that the number of possible following situation is rather big and estimation is very difficult. In this case, it is useful to consequently estimate surrounding possible situations than consequent movement through the graph beginning from the neighboring situation. It is proposed to estimate the complete situation, but not the consecutive events. To except the possible losses the reversibility is provided.

The convolution algorithm for graph is concluded in division of graph into clusters of difference levels in the certain way, representation of each cluster in form of arcs list and absorption of low-level cluster by high-level cluster. The convolution algorithm consists of eight steps and described below.

Let x_{init} be the initial vertex of graph, Cl_k be the cluster of level k, $Ns(x_i)$ be the set of neighbors of x_i, c be the number of neighbors, $Cat(Cl_i, Cl_j)$ be the function of concatenation of Cl_i and Cl_j, $hd(Cl_i)$ be the operator pointing to the head of cluster Cl_i, $tl(Cl_i)$ is the operator pointing to the tail of cluster Cl_i.

Algorithm 1. Convolution algorithm

1: Cluster level is set as $k = 0$
2: The initial vertex x_{init} is set;
3: Function of search for $Ns(x_{init})$ is performed
4: **if** $c \neq 0$ **then** $k = k + 1$ and Goto 7
5: **else** Goto 9
6: **end if**
7: Cluster of k-th level is defined as x_{init} and $Ns(x_{init})$. Here, cluster is defined as $Cl_k = cat(x_{init}, Ns(x_{init}))$ and $hd(Cl_k) = x_{init}$, $tl(Cl_k) = Ns(x_{init})$.
8: Cluster of $k_t h$ level is mapped into a vertex obtaining new x_{init}. Goto 4
9: End of algorithm

The detailed algorithm of convolution for generalization of graph is presented in [16]. Novelty of the proposed algorithm is that it can be used for planing and management of both single objects, and group of objects. In case of group, the construction of situation graph can be distributed between the objects with representation of absorbing vertices in the form of clusters. In [19] the possibility is shown and the example of implementation is considered for convolution algorithm in division of situation graph into separated parts.

It should be noted that the execution of this algorithm is real-time. Presented convolution algorithm shows economy of memory to 30% due to the reduction of computation redundancy, which is 20–30% (depending on graph completeness) faster than algorithms applying adjacency matrices or incidence ones together with consecutive search for vertices [12]. This algorithm is unique for conditionally infinite graph processing.

6 Risk Situation Modeling Based on Fuzzy Cognitive Modeling in Example of Nuclear Stations

The following factors are defined to detect and model risk situation. Let the arcs be estimated using five-point scale. As a result of implementation of fuzzy data processing algorithm the values of $w_{ij} \in [-1, 1]$ are obtained and illustrated in Fig. 1.

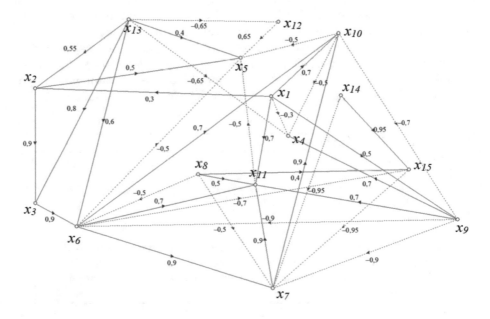

Fig. 1. FCM in example of nuclear station

In Fig. 1, x_1 is the number of tasks, x_2 is the speed of work execution (rate per our), x_3 is the number of feasibility estimates for creation of CTS, x_4 is the completion of the project, x_5 is the profitability, x_6 is the reliability of CTS, x_7 is the security and protection of CTS, x_8 is the external factors (fire or power supply), x_9 is the number of errors of executors, x_{10} is the time for CTS creation, x_{11} is the cost of CTS creation, x_{12} is the number of executors, x_{13} is the qualification of executors, x_{14} is the violation of normal operation, x_{15} is the emergency situation.

The analysis of the structural stability of FCM showed that the provided FCM is stable as degree of the structural stability equals to 0.53.

Figure 2 shows the fragments of impulse modeling and analysis is given. The axis of abscissas is labeled by iterations and the ordinates one is change rate of vertex in %.

Scenario 1. Impulse enters to two vertices. The following question arises: "What is obtained with CTS if x_3 is decreased by 10% and x_{11} is decreased by 10%" (Fig. 2a).

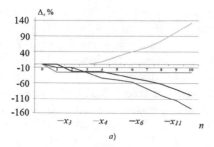
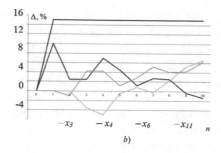

Fig. 2. Results of impulse modeling of risk situation evolution at FCM

Recommendations: decreasing of number of feasibility estimates and costs of CTS creation leads to the tendency of reliability reduction of CTS. It leads to sudden fall of project completion. It should be noted that reliability is not increased when finance resources are slightly increased.

Scenario 2. Impulse enters to two vertices. The following question arises: "What is obtained with CTS if x_6 is decreased by 10% and x_8 is decreased by 15%" (Fig. 2b).

Recommendations: increase of reliability and external factors leads to decrease of fall of project completion and emergency situations. However, slight decrease of CTS reliability leads to sudden fall of project completion, where observing of emergency situations is big.

Results of impulse modeling shows the key role of CTS reliability. The results of implementation do not contradict the practical observations and demonstrate adequacy of the proposed FCM. The obtained scenarios of risk situation evolution allow to predict possibility of problems connected with perturbation observing (i.e. increase of external factors affecting) and to model the probability of evolution to take technical, administrative and other actions for generation of reasonable management decisions. It should be noted that such factors as "Project completion", "External factors" and "Number of errors of executors" are obtained based on quiz and are formalized by means of specially developed scales [8] for the considered domain.

The analysis of structure of FCM [9] was performed to obtain information about hidden affects between factors of FCM. It shows that, firstly, the most positive affect has x_2 (0.19) and x_3 (0.10), change of which allows to "to turn" the whole system into positive side, secondly, x_1 (0.01) does not strongly affects on FCM, therefore can be excluded from the model, thirdly, x_{11} (0.2), x_3 (0.13), x_{10} (0.13), x_6 (0.08), therefore risk situations together with negative external impact can be reduced changing them.

7 Conclusions

Two modern approaches are proposed to modeling and analysis of risk situations arising at different stages of life cycle of CTS in conditions of fuzzy data. These approaches are fuzzy cognitive modeling, which is applied at stage of pre-project researches and fuzzy situation graph, which is applied at operation stage.

Performed analysis showed that the model is stable. It is confirmed BY presented fragments of analysis of risk situation scenarios. The results do not contradict practical observations and demonstrate adequacy of the developed FCM. The estimation of factors' affect was performed for creation of CTS. It allowed to identify, which factors has the greater influence and, vice versa, which factors are mostly affected by FCM. Modeling based on FSG using convolution algorithm and graph division allows to plan and control both single CTS and group of CTS.

Information obtained as a result of implementation of the proposed approaches is the basis of generation of scientifically reasonable management actions directed to detection and elimination of possible risk situations during CTS creation to perform a set of economical and technical decisions.

References

1. Yurkov, N.K.: System approach to the organization of life cycle of difficult technical systems. Reliab. Qual. Difficult Syst. Sci. Pract. Mag. **1**, 27–35 (2013). (in Russian)
2. Korobkin, V.V., Kolodenkova, A.E., Kukharenko, A.P.: Accounting of risk situations when modeling the designing process of complex managing systems on the basis of cognitive models. News of SFU. Technical science vol. 9, pp. 103–111 (2017). (in Russian)
3. Katalevskiy, D.Yu.: Fundamentals of simulation and system analysis in management: a tutorial (2011). (in Russian)
4. Kulinich, A.A.: Methodology of cognitive modeling of complex ill-defined situations. http://kk.docdat.com/docs/index-406893.html
5. Abramova, N.A., Avdeeva, Z.K.: Cognitive analysis and management of the development of situations: the problems of methodology, theory and practice (2008)
6. Dickerson, J., Kosko, B.: Virtual worlds as fuzzy cognitive maps. In: Virtual Reality Annual International Symposium, pp. 471–477 (1993)
7. Wang, C., Chen, S., Chen, K.: Using fuzzy cognitive map and structural equation model for market-oriented hotel and performance. Afr. J. Bus. Manag. **5**(28), 11358–11374 (2011)
8. Silov, V.B.: Making Strategic Decisions in a Fuzzy Environment. INPRO-RES, Moscow (1995) (in Russian)
9. Borisov, V.V., Kruglov, V.V., Fedulov, A.S.: Fuzzy models and networks. Hot line - Telecom, Moscow (2007). (in Russian)
10. Papageorgiou, E.I., Stylios, C.D., Groumpos, P.P.: Active Hebbian learning algorithm to train fuzzy cognitive maps Internet. Int. J. Approx. Reason. **37**, 219–249 (2004)
11. Carvalho, J.P.: On the semantics and the use of fuzzy cognitive maps and dynamic cognitive maps in social sciences. Fuzzy Sets Syst. **214**, 6–19 (2013)
12. Kosko, B.: Fuzzy cognitive maps. Int. J. Man Mach. Stud. **1**, 65–75 (1986)
13. Kolodenkova, A.E.: Modeling of process of feasibility of the project on creation of management information systems using fuzzy cognitive models. Mess. Comput. Inf. Technol. **6**(144), 10–17 (2016). (in Russian)

14. Nguyen, D., Fisher, D.C., Stephens, R.L.: A graph-based approach to situation assessment. http://web.cs.ucla.edu/~miryung/Publications/oopsla10-libsync.pdf. Accessed 13 July 2018

15. Gavgani, M.H., Eftekharnejad, S.: A graph model for enhancing situational awareness in power systems. https://ieeexplore.ieee.org/document/8071427/. Accessed 13 July 2018

16. Sergeev, N.E., Muntyan, E.R., Tselykh, A.A., Samoylov, A.N.: Situation graph generalization for situation awareness using a list-based folding algorithm. News of SFU. Technical science, vol. 3, pp. 111–121 (2017). (in Russian)

17. Cormen, T.H., Leiserson, C.E., Rivest, R.L., Stein, C.: Introduction to Algorithms, 3rd edn. The MIT Press, McGraw-Hill Book Company (2009)

18. Foster, J.M.: List Processing, p. 54. Macdonald, London (1968)

19. Sergeev, N.E., Muntyan, E.R.: Using convolution algorithm to separate a graph on the proportional subgraphs. Vestnik UGATU, vol. 22, no. 1(79), pp. 121–130 (2018). (in Russian)

Automated Quality Management System in Mechanical Engineering

Georgy Burdo$^{(\boxtimes)}$

Tver State Technical University, A. Nikitin Emb. 22, Tver, Russia
gbtms@yandex.ru

Abstract. The key feature of modern machinery production is a wide range of products of specified quality targeted at a specific consumer. The paper shows the relevance of creating automated quality management systems in machine-building industries. It also justifies the necessity of product quality management at all stages of a product life cycle. The study used the system analysis apparatus. There is a developed typical scheme of product life cycle steps and stages for engineering products. In order to reveal an information exchange and decision-making mechanism in an automated system of product quality management, its set-theoretical model was developed. The author considers the structures of input and output data for each operator subsystem. There are the principles defining the methodology of creating industrial quality management systems. Their implementation will allow developing sufficiently effective algorithms and software tools for making decisions within the framework of an automated system for product quality management in engineering.

Keywords: Automated product quality management system
Set-theoretic model · System analysis · Artificial intelligence

1 Introduction

The quality of manufactured products directly affects the development level of state economy in general. The advanced economies of developed countries repeatedly proved this statement [1, 2].

The key feature of modern machinery production is a wide range of specified quality products targeted at a specific consumer, short terms of production-support work, which allows it to be defined as multi-product manufacture.

Therefore, nowadays, under conditions of multi-product manufacture, in order to ensure product quality, it is necessary to pay more and more attention to the entire product life cycle (PLC) [3–5], i.e. to manage quality at each step and stage (Fig. 1).

In order to manage product quality at all steps of a product life cycle, it is necessary to create a quality management system (QMS) for products. Its operating principles should be determined by the ISO 9000 standards [6].

This work was supported by RFBR, project 17-01-00566.

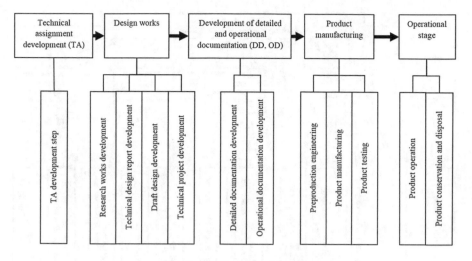

Fig. 1. PLC steps and stages

It is also important that quality management systems should reflect the ideas of Lean Production [7] and information integration of enterprise activity processes [8–10]. The necessity to adjust qualitative indicators effectively and a wide range of products in a modern machine-building enterprise proves that creation of an automated QMS is a relevant task.

2 A Set-Theoretical Model of a Quality Management System

In order to identify the mechanism of information exchange and decision-making in an automated QMS of products in a multi-product manufacture enterprise, its set-theoretic model has been developed [11].

From the control point of view, a QMS functioning at an enterprise is a complex system represented by subsystems $\{R_0\} = \{R_1^1, R_1^2, \ldots, R_1^5, R_2^1, R_2^2, \ldots, R_2^{12}\}$ (Fig. 2).

Here R_0 is a control subsystem of a QMS (top-level subsystem). Subsystems of the following level are as follows: R_1^1 subsystem performs quality management during development of technical assignment (TA), it has a subsystem R_2^1 quality management at the stage of technical assignment development (TA).

The subsystem R_1^2 performs quality management during design works, which includes four life cycle stages, its functions:

- R_2^2 manages quality at the stage of research work (R&D) development;
- R_2^3 manages quality at the stage of technical design report (TDR) development;
- R_2^4 manages quality at the stage of draft design (DD) development;
- R_2^5 manages quality at the stage of technical project (TP) development.

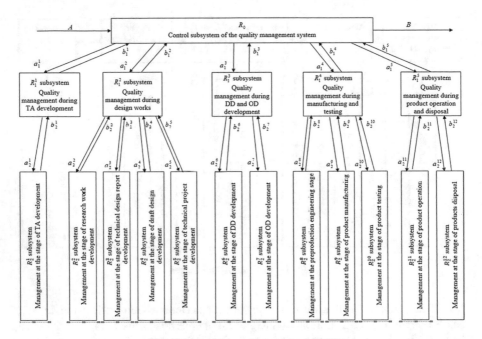

Fig. 2. A set-theoretic model of QMS

The R_1^3 subsystem performs quality management during development of detailed (DD) and operational documentation (OD). It manages two stages of the life cycle: R_2^6 is quality management at DD development stage, R_2^7 is quality management at OD development stage.

The subsystem R_1^4 performs quality management during product manufacturing and testing, which includes three life cycle stages. The functions are the following: R_2^8 is quality management at the preproduction engineering stage, R_2^9 is quality management at the stage of product manufacturing, R_2^{10} is quality management at the stage of product testing.

The subsystem R_1^5 performs quality management during operation and disposal of products, at two life cycle stages, its functions are the following: R_2^{11} is quality management at the stage of product operation, R_2^{12} is quality management at the stage of product disposal.

The control subsystem R_0 performs six control functions. The first function is managing the subsystem R_1^1, which is definition and adjustment of product quality parameters at the TA step: $R_{0_1} : A \times b_1^1 \rightarrow a_1^1$. In this formula A is a control signal, i.e. a set of requirements to product quality parameters at all PLC stages that are determined by an organization management system; b_1^1 is a set of values of actual quality parameters of the product at the TA stage; a_1^1 is a set of corrective actions on product quality parameters at the technical assignment step.

The second function is management of the subsystem R_2^1, which is determination and adjustment of product quality parameters at the design step: $R_{0_2} : A \times b_1^2 \to a_1^2$. Here b_1^2 is a set of values of product quality parameters at the design step; a_1^2 is a set of corrective actions on product quality parameters during design works.

The third function is R_3^1 subsystem control, i.e. determination and adjustment of product quality parameters during DD and OD development: $R_{0_3} : A \times b_1^3 \to a_1^3$. Here b_1^3 is a set of actual values of product quality parameters during DD and OD development; a_1^3 is a set of corrective actions on product quality parameters at the DD and OD development step.

The fourth function is R_4^1 subsystem control, i.e. determination and adjustment of product quality parameters during product manufacturing and testing: $R_{0_4} : A \times b_1^4 \to a_1^4$. Here b_1^4 is a set of actual values of product quality parameters during product manufacturing and testing; a_1^4 is a set of correcting influences on product quality parameters during product manufacturing and testing.

The fifth function is R_5^1 subsystem control, i.e. determination and adjustment of product quality parameters during operation and disposal; $R_{0_5} : A \times b_1^5 \to a_1^5$. Here b_1^5 is a set of actual values of product quality parameters during operation and disposal phase; a_1^5 is set of corrective influences on quality parameters during operation and disposal.

Finally, the sixth control function is determination of product quality parameters taking into account all PLC steps: $R_{0_6} : A \times b_1^1 \times b_1^2 \times b_1^3 \times b_1^4 \times b_1^5 \to B$, where B is an output signal, i.e. a set of quality parameters of a finished product.

The subsystem R_1^1 performs two control functions. The first is direct quality management at the stage of TA development: $R_{1_1}^1 : a_1^1 \times b_2^1 \to a_2^1$, where a_2^1 is a set of requirements to product quality parameters at the stage of the technical assignment; b_2^1 is a set of actual (current) values of product quality parameters at the TA stage.

The second function is transfer of information about the actual product quality at the TA stage to a superior control subsystem: $R_{1_2}^1 : b_2^1 \to b_1^1$.

The subsystem R_1^2 performs five control functions.

The first of them is direct product quality management at the stage of R&D development: $R_{1_1}^2 : a_1^2 \times b_2^2 \to a_2^2$, where a_2^2 is a set of requirements to product quality parameters at the R&D stage; b_2^2 is a set of actual (current) values of product quality parameters at the R&D stage.

The second function is product quality management at the TDR development stage: $R_{1_2}^2 : a_1^2 \times b_2^3 \to a_2^3$, where a_2^3 is a set of requirements to product quality parameters at the TDR stage; b_2^3 is a set of actual (current) values of product quality parameters at the TDR stage.

The third function is product quality management at the stage of draft design project development: $R_{1_3}^2 : a_1^2 \times b_2^4 \to a_2^4$, where a_2^4 is a set of requirements to product quality parameters at the draft design project stage; b_2^4 is set of actual (current) values of product quality parameters at the draft design project stage.

The fourth function is product quality management at the stage of TP development: $R^2_{1_4} : a^2_1 \times b^5_2 \rightarrow a^5_2$. Here a^5_2 is a set of requirements to product quality parameters at the TP stage; b^5_2 is a set of actual (current) values of product quality parameters at the TP stage.

The last, fifth function is a synthesis of information on the product quality state at the design step: $R^2_{1_5} : b^2_2 \times b^3_2 \times b^4_2 \times b^5_2 \rightarrow b^2_1$.

The R^3_1 subsystem is designed to perform three control functions.

The first function is direct quality management at the development stage of the DD: $R^3_{1_1} : a^3_1 \times b^6_2 \rightarrow a^6_2$, where a^6_2 is a set of requirements to product quality parameters at the DD stage; b^6_2 is a set of actual (current) values of product quality parameters at the DD stage.

The second function is product quality management at the stage of OD development: $R^3_{1_2} : a^3_1 \times b^7_2 \rightarrow a^7_2$, where a^7_2 is a set of requirements to product quality parameters at the stage of OD development; b^7_2 is a set of actual (current) values of product quality parameters at the stage of OD development.

The third function is related to a synthesis of information on the product quality state during DD and OD development: $R^3_{1_3} : b^6_2 \times b^7_2 \rightarrow b^3_1$.

The subsystem R^4_1 performs four control functions.

The first one is product quality management at the preproduction engineering stage: $R^4_{1_1} : a^4_1 \times b^8_2 \rightarrow a^8_2$. Here b^8_2 is a set of requirements to product quality parameters at the preproduction engineering stage; b^8_2 is a set of actual (current) values of product quality parameters at the preproduction engineering stage.

The second function is control of the product manufacturing stage: $R^4_{1_2} : a^4_1 \times b^9_2 \rightarrow a^9_2$. Here a^9_2 is a set of requirements to product quality parameters at the manufacturing stage; b^9_2 is a set of actual (current) values of product quality parameters at the manufacturing stage.

The third function is product testing phase control: $R^4_{1_3} : a^4_1 \times b^{10}_2 \rightarrow a^{10}_2$. Here a^{10}_2 is a set of requirements to products parameters during testing; b^{10}_2 is a set of actual (current) values of product parameters. The fourth function is a synthesis of information about the manufacturing and testing phase state: $R^4_{1_4} : b^8_2 \times b^9_2 \times b^{10}_2 \rightarrow b^4_1$.

The last R^5_1 subsystem performs three control functions. The first one is direct management of product quality at the stage of product operation: $R^5_{1_1} : a^5_1 \times b^{11}_2 \rightarrow a^{11}_2$. Here a^{11}_2 is a set of requirements to product quality parameters at the stage of product operation; b^{11}_2 is a set of actual (current) values of product quality parameters at the stage of product operation.

The second function is product quality management at the product disposal stage: $R^5_{1_2} : a^5_1 \times b^{12}_2 \rightarrow a^{12}_2$. Here a^{12}_2 is a set of requirements to product quality parameters at the stage of product disposal; b^{12}_2 is a set of actual (current) values of product quality parameters at the stage of product disposal. The last, third function is transferring information about the state of the product operation and disposal: $R^5_{1_3} : b^{11}_2 2^{11} \times b^{12}_2 \rightarrow b^5_1$.

All subsets of requirements for product quality parameters at PLC steps $\left(a_1^1, a_1^2, a_1^3, a_1^4, a_1^5\right)$ are determined by the requirements for product quality specified by an organization management system. The subsets $\left(b_1^1, b_1^2, b_1^3, b_1^4, b_1^5\right)$ form quality parameters of the finished product B. If quality is ensured, then the set of A parameters must include a number of B parameters.

Obviously, the information developed at each PLC step and stage is a subject of estimation. Taking into account the specifics of multi-product mechanical engineering and the conditions for design, technological preparation and organization of production, only the most significant information flows in the context of quality are analyzed.

3 The Principles of Creating Automated Quality Management Systems

The basis for automated system development should be the principles of its creation, which ensure: (a) implementation of decision-making procedures fitting the processes in a real production system adequately, (b) information integration with a higher-level organization management system, (c) integration of the system into automated product life cycle support systems, (d) compliance of decision-making procedures with the logic of human reasoning and algorithms of specialists' actions, (e) reduction of the total time (cycle) of sample product manufacturing. The principles are the following:

1. The principle of compliance of product sample output preparation steps to product life cycle steps.
2. The principle of iterative decision-making.
3. The system hierarchy.
4. Determinacy of system functioning.
5. Presence of artificial intelligence elements.
   Let us consider this principle in more detail. To develop possible solutions for quality, the production models of knowledge were used because of their simplicity. For the subsequent evaluation of possible solutions by experts, an algorithm based on fuzzy logic was implemented, which allows obtaining the resultant evaluation of experts using quantitative and qualitative criteria.
6. Combining the work step execution in time.
7. Criteria derivativeness from technical assignment and from the expected performance indicators of production in its industrial output.
8. The multivariance principle.
9. The principles of forming decision assessment criteria at the steps are as follows:
   (a) the principle of a criteria boundary type;
   (b) the principle of a criteria relative type;
   (c) the principle of criteria complexity.

These criteria were studied in detail in [4].

4 Conclusion

The presented set-theoretic model of an automated product quality management system and its creation principles are a methodological basis for their development.

Basically, the proposed apparatus is invariant with respect to products manufactured in machine-building production systems. It defines the rules of information exchange, its processing and assessment of synthesized solutions quality at the steps and stages of a product life cycle. Product quality assessment criteria must be linked to the technical assignment for a particular product and reflect its specificity.

It is also necessary to determine the methodology and decision-making methods for users of the quality management system. Obviously, some criteria will allow assessing decisions formally. Assessing in poorly formalized situations will require experts.

All this ultimately allowed us to develop sufficiently effective solutions for decision making and software tools within the framework of an automated product quality management system in a machine building industry. Currently, an experimental verification of the given system is carried out at one machine-building enterprise.

References

1. Conti, T., Watson, G.H., Kondo, Y. (eds.) Quality into the 21st Century: Perspectives on Quality and Competitiveness for Sustained Performance. Asq Pr, Milwaukee (2003)
2. Imai, M.: Gemba Kaizen: A Commonsense, Low-Cost Approach to Management. McGraw-Hill, New York (1997)
3. Kolchin, A.F., Ovsyannikov, M.V., Strekalov, A.F., Sumarkov, S.V.: Upravlenie zhiznennym tsiklom produktsii. Anarkhist, Moscow (2002)
4. Burdo, G.B., Stoyanova, O.V.: Avtomatizirovannaya sistema upravleniya protsessami sozdaniya naukoemkikh mashinostroitelnykh izdely. Programmnye produkty i sistemy 2 (106), 164–170 (2014)
5. Wenzel, B.A.: Data Integration Architecture for SC4. ISO TC184/SC4/WG10/N89 (1997)
6. GOST R ISO 9001-2008: Sistemy menedzhmenta kachestva. Osnovnye polozheniya i slovar. State standard, Standartinform, Moscow (2001)
7. Womack, J.P., Jones, D.T.: Lean Thinking: Banish Waste and Create Wealth in Your Corporation. Productivity Press, New York (2003)
8. Bernus, P., Nemes, L.: Modeling and Methodologies for Enterprise Entegration. Chapman and Hall, London (2006)
9. Duncan, W.R.: A Guide to the Project Management. Body of Knowledge. Project Management Institute, USA (2008)
10. Grady, J.O.: System Integration. CRC Press, Florida (2004)
11. Mesarovich, M., Takakhara, Ya.: Obshchaya teoriya sistem: matematicheskie osnovy. Mir, Moscow (1978)

Knowledge Representation Method
for Intelligent Situation Awareness System
Design

Maria A. Butakova, Andrey V. Chernov$^{(\boxtimes)}$, Alexander N. Guda,
Vladimir D. Vereskun, and Oleg O. Kartashov

Rostov State Transport University, Rostov-on-Don, Russia
{butakova,avcher,guda,vvd}@rgups.ru, okrstu@yandex.ru

Abstract. This work presents a novel method for knowledge representation which is adapted for the design of intelligent situation awareness system. Our idea behind the proposed method is to design the intelligent features of situation awareness system from a human-computer interaction point of view. This dictates the using soft and non-metric approaches for each stage of the proposed method. Our method is suited for distributed and dynamic systems design cycle, so the central part of the paper is devoted to the distributed dynamic logic based on the description logic. Knowledge representation architecture for the distributed case of the intelligent situation awareness system is presented. Detailed definitions, syntax and semantic constructors and axioms with use of the *SHOIN* description logic for dynamic distributed description logic have been developed.

Keywords: Knowledge representation · Situation awareness
Intelligent system · Description logic · Distributed dynamic description logic

1 Introduction

The design of situational awareness systems, which are, in fact, a class of decision support systems rises constant interest for researchers [1, 2]. Informally, but informatively enough for understanding, "situational awareness" (*SA*) itself can be defined as a process of perception, comprehension, and projection of current information about the situation, together with estimates and predictions of possible developments and outcomes in the future. At the same time, a wide variety of models and methods, mathematical approaches, and tools used in this process at its various stages is natural. This approach to the *SA* is mentioned quite often, since [3, 4]. However, for the modern development of information technologies, only a theoretical base is not enough, not to mention even the development of practical realization for the intelligent situation awareness system (*ISAS*).

The work was financially supported by Russian Foundation for Basic Research (projects 16-01-00597-a, 16-07-00888-a, 17-07-00620-a, 18-01-00402-a, 18-08-00549-a).

A. Abraham et al. (Eds.): IITI 2018, AISC 875, pp. 225–235, 2019.
https://doi.org/10.1007/978-3-030-01821-4_24

Intelligent technologies on the railway transportation require *ISAS* development also. In our previous papers [5–11] we have proposed various elements of *SA* applications for the multilevel intelligent control system. This work formulates a novel method for knowledge representation in *ISAS* presented from a human-computer interaction point of view. Our approach is entirely different from other same approaches to "system intelligence" and focuses on interaction aims mainly. We consider the "system intelligence" as a possibility the system to react adequately to groups and types of situations arose. In that regard, we underline four dual-contradicted properties of the *ISAS* design:

(1) individuality and collectivity;
(2) purposefulness and adjustability;
(3) integration and personalization;
(4) affectivity and expressiveness.

The individuality of *ISAS* is understood as the specialization of the system concerning types, groups, and varieties of emergent situations. Obviously, going into the detail of each particular situation, their total number tends to infinity. Specialists can compulsorily restrict the number, type of situations in different subject areas. However, the ability of the *ISAS* to establish the fact of the situation occurrence previously unknown and never arising before will be lost. In this case, one of the variants is the making a collective decision, which confirms the new situation occurrence.

The purposefulness of *ISAS* refers to the ability of the system to achieve a specific goal or goal of its functioning and to maintain conditions that contribute to a completely stable, consistent operation. Nevertheless, the intelligent features of the ISAS should also envisage its reorganization under the influence of newly aroused situations, at least from avoiding negative scenarios. There is a question about the reasoning, the result of which is the conclusion about the need to adjust the parameters, algorithms, operating regimes of the *ISAS*.

The integration in the *ISAS* is mainly due to technical reasons for the availability of heterogeneous information sources and is manifested in need for single information storage on situations. It is worth to note that in this case, we are neither talking about one format for storing data about situations nor about one single knowledge database. Thus, the *ISAS* should allow for the aggregation of similar situations, without recourse to differences in the ways of registering and documenting them, and then provide opportunities for transition to a personalized analysis and processing of some new situation.

The affectivity of the *ISAS* and the possibility of an emotional reaction within the framework of the technical system can be explained as follows. In complex information-control systems and networks, especially in systems that perform critical functions, the complete elimination of the human operator from the control loops of responsible technological processes has not yet occurred. Partly this is connected with the apparent need to monitor and control of the system in abnormal and various uncertain situations. It is clear that different situations, depending on the circumstances accompanying them and the expected consequences, have a different emotional coloring. In this case, the *ISAS* is confronted more not with the recognition of the emotional state of a person, but rather with the formation of emotions that correspond to the situation that has arisen. It should also be noted that for the formation of specific emotions in the technical system, measurements related to situational awareness are required.

Proper considering the intelligent features which is discussed above allows us to propose a novel method for knowledge-based designing of *ISAS*. Next, we will apply the distributed and dynamic ontology approach to the formal description of the system under design. The paper contains several sections to detail our approach. Section 2 presents stages of the proposed method with the using non-metric mathematical apparatus and from a human-computer interaction point of view. Section 3 reveals possible constructions of the *ISAS* distributed knowledge bases in the distributed and dynamic environments. Section 4 proposes the main definitions for distributed dynamic description logic for *ISAS* and ontology evolution rules also. Finally, we conclude the paper by summarizing our outcomes.

2   Proposed Method

Figure 1 presents the main stages of our method for knowledge representation for *ISAS*. The first stage, as it is shown in Fig. 1 is intended to acquire the information from various and heterogeneous data sources which describe situations. This stage can be implemented in the form of various connectors to existing databases, sensor software, data exchange protocols and is a purely technical, and not a design problem for the development of *ISAS*. As an example, one can refer to the [12].

The second stage is the least formalizing stage relative to the others. We consider semantic procedures to form the ontologies, such as have been proposed in [13]. The essence of the proposed idea in [13] is that ontologies are based on "not specific concepts and relations between them, but the means of creating specific concepts." A separate concept is presented as a means of obtaining new knowledge, while the self-organizing concept process takes place.

The third stage performs dynamic forming of the collective ontology. From one side, that process is semi-formalized and requires significant efforts of the experts. But, from the other side, the efforts can be distributed among a group of the experts using shared resources to edit the collective ontology. On the technical side, the third stage of the proposed method is to synthesize a collective ontology from distributed ontologies created by individual domain experts. This process is explained in Fig. 2. Even though the process of collective ontologies development is as difficult to formalize as the second stage of the proposed method, the client-server implementation of the system presented in Fig. 2, does not cause difficulties.

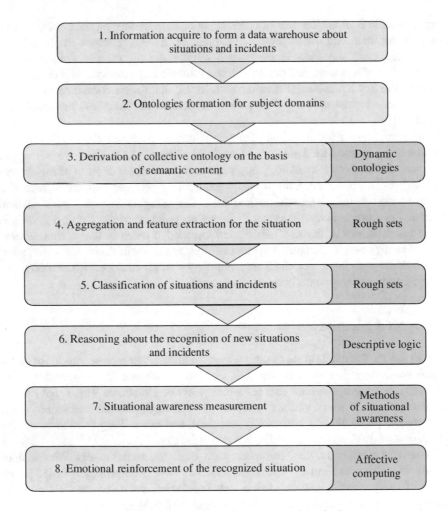

Fig. 1. The outline of the proposed method

Stages four and five are designed to extract the most significant features that accompany the situation or the incident occurrence, were considered in detail in [8, 10]. As a theoretical basis, methods of the rough sets theory are chosen. There are several reasons for this choice. First, using the methods of the rough sets theory, there is no need for additional a priori information about the statistical characteristics, or the probability distributions of the observed data. Secondly, only the facts hidden in the data are analyzed, and relationships are established without the use of statistical methods. And, thirdly, having a set of data with discrete attributes, one can always find a subset of the most informative ones, while reducing the dimension of the original data set with minimal information loss. This problem and the methods for solving it were considered in the works mentioned earlier [8, 10] as a problem of determining the irreducible attributes set of an information system. After building ontologies, as well as highlighting the most informative signs, the question about knowledge elicitation usually arises. What are the algorithms and procedures for extracting knowledge and whether automated and software tools will be used? Thus is the most important problem that arises in the design.

At the sixth stage, the description logic is proposed for reasoning about the new situations in *ISAS*. Of course, *ISAS* can be confidently attributed to a class of systems that have dynamic changes in the process of functioning. The reasoning mechanism used for *ISAS* should also provide the possibility of dynamic inference. In the next section, the possibilities of logical inference for *ISAS* will be presented, considering the dynamic change in the reasoning rules.

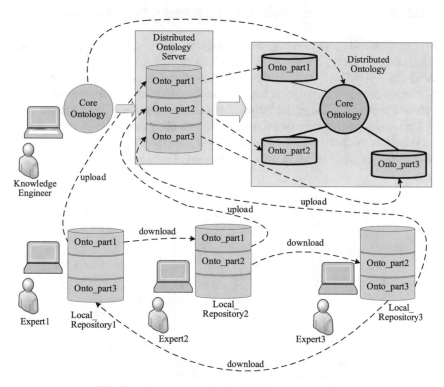

Fig. 2. Distributed collective ontology engineering

The seventh stage, where it is proposed to perform a *SA* measurement, is important for establishing numerical estimates that characterize the level of staff awareness in the current situation, and for comparing the estimates among themselves also. In this paper, the aim is not to examine this stage in detail. However, some measurement technologies can be found in [4]. However, it should be noted that for its successful solution it is required to develop *SA* measures, at least in three ways: (1) the possibility of making measurement for individual and collective use of *ISAS*; (2) the possibility of *SA* measurement when the sources of situations are concentrated in one place and for spatially-distributed situational processing; (3) the possibility of real-time *SA* measurement.

The final, eighth stage of the proposed method is the emotional reinforcement of the recognized situation. Research methods of this stage are in the field of affective computing, which is associated with the recognition and modeling of the emotional state of a person. In the proposed method, it is required not so much to recognize emotions, or to perform an analysis of the behavior of the *ISAS* operator, how many to perform the generation of reinforcing emotions, which correspond to the strengthening or weakening of the situation that has arisen. In this paper, such methods are not considered in detail, but one can refer to [14].

3   An Approach to Knowledge Representation for *ISAS*

The data acquired from heterogeneous sources and consequently, the distributed ontologies engineering, led to the development of this approach in the form of distributed descriptive logic [15]. It should be noted that knowledge representation systems that are based on description logic, as well as traditional first-order logic, used to represent a static knowledge, including mathematical theories. For dynamic environments, a variant of dynamic description logic has been proposed in [16], which has been enhanced further in [17]. On its base we present an approach to knowledge representation for *ISAS* using *Distributed Dynamic Description Logic* (D^3L) in Fig. 3.

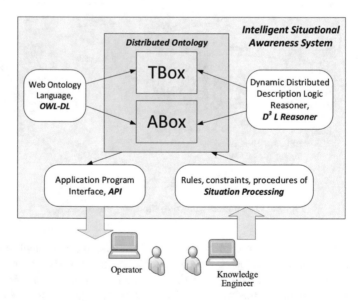

Fig. 3. Distributed collective ontology engineering

Core language for our representation is *OWL-DL*, which has correspondence with description logic. The distributed ontology has sets of terminology *TBox* and assertions *ABox*. We note that the addition of dynamics or, "temporality" in any logic, including description logic, can be performed in various ways. Among the others the following cases can be distinguished: (1) time is explicitly indicated; (2) the time is implicit and is indicated by discrete moments; (3) the time is indicated by the intervals; (4) time is linear flow; (5) time is branched or parallelized; (6) time is indicated through the internal representation of the logical structure; (7) time is indicated by a process external to the logical structure. In addition, the introduced temporal operators can be applied to different parts of the logical structure, and, most importantly, to the moments of concepts change, *TBox* axioms and *ABox* statements.

In terms of the description logic the temporal interpretation $\Im = \left(\Delta^{\Im}, \bullet^{\Im} \right)$ contains nonempty domain of interpretation $\Delta^{\Im}$, and interpretation function $\bullet^{\Im}$, which maps every atomic concept $A \in N_C$ to a subset $A^{\Im} \subseteq \mathbb{N} \times \Delta^{\Im}$, every atomic role $R \in N_R$ to a subset $R^{\Im} \subseteq \mathbb{N} \times \Delta^{\Im} \times \Delta^{\Im}$, and every individual $a \in N_R$ to $a^{\Im} \in \Delta^{\Im}$. The elements of the set of natural numbers $\mathbb{N}$ make discrete time points. Any temporal interpretation $\Im$ can be represented as infinite series of non-temporal interpretations $\Im(0), \Im(1), \ldots$, but it must be restricted by some time point for practical reasons.

In this paper we consider the changing the number of assertions *ABox* as a specific temporal event. To detail our definitions, we have chosen the *SHOIN* description logic than fully corresponds to the *OWL*-DL language. In Table 1 the constructors of our D^3L logic are given. In addition to *ABox* and *TBox* constructors and axioms, the *ACt* (*ACtions*) axioms are presented.

Further, we introduce the necessary notation. The local ontology, or local knowledge base KB_i we define as a pair $\langle T_i, A_i \rangle$, where T_i – a set of axioms *TBox*, A_i – a set of axioms *ABox* for i-th case, $i \in \mathbb{N} = \{1, 2, \ldots\}$. For every local knowledge base KB_i we denote signature $Sig(KB_i)$, which contains of the set of terms Σ, i.e. atomic concepts, roles and individuals and appearing in axioms *TBox* и *ABox*. The set of all possible axioms *TBox* and *ABox*, that can be constructed by the induction with rules in Table 1 we denote $Ax(\Sigma)$. The concepts in D^3L are defined by induction rules: (1) any atomic concept A is a concept; (2) if C, D are concepts, then $\neg C, C \sqcap D, C \sqcup D$ are concepts too; (3) if C is a concept, and R is a role, then $\forall R.C, \exists R.C$ are concepts too. The temporal interpretation $\Im$ of D^3L includes the atomic action $\alpha = ACt(x_1, \ldots, x_n) = \langle I_{ABox}, O_{ABox} \rangle$, where $\alpha \in N_A$ is a name of the action, $(x_1, \ldots, x_n)$ are variables that specify the objects which are undergo the actions, I_{ABox} is a finite set of assertions *ABox*, which describes the input actions parameters, and finally, O_{ABox} is a finite set of assertions *ABox*, which describes the output actions parameters and action effect. The formulas in D^3L are defined by inductive rules too, as following: (1) if C is a concept, R is a role, and (a, b) individuals, then assertions $C(a)$ and $R(a, b)$ are formulas too; (2) if φ and ψ are formulas, then $\neg\varphi, \varphi \wedge \psi, \varphi \rightarrow \psi$ are formulas too; (3) if φ is a formula and α is an action, then $[\alpha]\varphi$ is a formula too. The constructions of D^3L must allow to interpret complex actions, composed of atomic actions. Therefore, we denote complex actions in the following way. Let's α, β are atomic actions, φ is a formula, then complex actions are composed of these rules: (1) $(\varphi?; \alpha)$ test the φ, and performing the action α, if result is true; (2) $(\alpha \cup \beta \cup \ldots)$ choosing the case of the action α, β and so on; (3) $(\alpha; \beta; \ldots)$

sequential performing the action α, then β and so on; (4) $(\alpha^*(\beta^*)\ldots)$ cycle performing the action β, inside action α and so on; (5) $(\alpha\|\beta\|\ldots)$ parallel performing the actions α, β and so on. For example, the programming construction «if φ is true, then α, β, else cycle β» will be transformed into complex action $(\varphi?;\alpha;\beta)\cup((\neg\varphi)?;\beta^*)$. Now we are able to define the temporal interpretation $\Im$ for local knowledge base KB_i, with signature $Sig(KB_i)$ in D^3L.

Table 1. Syntax and semantics D^3L constructions

Construct name	Syntax	Semantics
Individual	a	$a^\Im \in \Delta^\Im$
Atomic concept	A	$A^\Im \subseteq \Delta^\Im$
Universal concept	$\top$	$\top^\Im = \Delta^\Im$
Empty concept	$\bot$	$\bot^\Im = \Delta^\Im$
Conjunction	$C \cap D$	$(C \cap D)^\Im = C^\Im \cap D^\Im$
Disjunction	$C \cup D$	$(C \cup D)^\Im = C^\Im \cup D^\Im$
Negation	$\neg C$	$\Delta^\Im / C^\Im$
Nominals	$\{a_1, \ldots, a_n\}$	$\{a_1^\Im, \ldots, a_n^\Im\}$
Atomic role	R	$R \subseteq \Delta^\Im \times \Delta^\Im$
Exists restriction	$\exists R.C$	$\{x \mid \exists y.\langle x, y\rangle \in R^\Im \wedge y \in C^\Im\}$
Value restriction	$\forall R.C$	$\{x \mid \forall y.\langle x, y\rangle \in R^\Im \Rightarrow y \in C^\Im\}$
Number restrictions	$\leq nR$	$\{x \mid \#\{y.\langle x, y\rangle \in R^\Im\} \leq n\}$
Number restrictions	$\geq nR$	$\{x \mid \#\{y.\langle x, y\rangle \in R^\Im\} \geq n\}$
	TBox axioms	Interpretation constraints
Subsumption	$C \sqsubseteq D$	$C^\Im \sqsubseteq D^\Im$
Role inclusion	$R \sqsubseteq S$	$R^\Im \sqsubseteq S^\Im$
Role transitivity	Trans (R)	$R^\Im = (R^+)^\Im$
	ABox axioms	Interpretation constraints
Class membership	$C(a)$	$a^\Im \in C^\Im$
Role membership	$R(a_1, a_2)$	$\langle a_1^\Im, a_2^\Im \rangle \in R^\Im$
Identity	$a_1 = a_2$	$a_1^\Im = a_2^\Im$
	ACt	Interpretation constraints
Actions	$ACt(x_1, \ldots, x_n)$	$ACt(x_1, \ldots, x_n) = \langle I_{ABox}, O_{ABox}\rangle$
Input actions *ABox*	I_{ABox}	$\{I_{ABox_i}\}, i \in \mathbb{N}$
Output actions *ABox*	O_{ABox}	$\{O_{ABox_i}\}, i \in \mathbb{N}$

Definition 1. (Temporal interpretation)

Temporal interpretation $\mathfrak{I}$ of all terms Σ for axioms construction $Ax(\Sigma)$ of local knowledge base KB_i, $i \in \mathbb{N}$, with signature $Sig(KB_i)$, is a pair $\langle \Delta^{\mathfrak{I}}, \bullet^{\mathfrak{I}} \rangle$, containing domain of interpretation $\Delta^{\mathfrak{I}}$, interpretation functions $\bullet^{\mathfrak{I}}$, mapping terms from Σ to terms in $\Delta^{\mathfrak{I}}$, concepts from Σ to subsets of $\Delta^{\mathfrak{I}}$, roles from Σ to subsets of $\Delta^{\mathfrak{I}} \times \Delta^{\mathfrak{I}}$, actions from Σ to $\Delta^{\mathfrak{I} \times n}$.

Let's turn to the distributed knowledge base, which we will denote **DKB**. This knowledge base **DKB** contains of 5 parts: (1) a set of local ontologies, local knowledge bases $KB_i, KB_j, KB_k, \ldots$; (2) a set of distributed terminology $TBox_i, TBox_j, TBox_k, \ldots$ according to $i, j, k, \ldots$ ontologies; (3) a set of distributed assertions $ABox_i, ABox_j, ABox_k, \ldots$, according to $i, j, k, \ldots$ ontologies; (4) a set of distributed actions $ACt_i, ACt_j, ACt_k, \ldots$, according to evolution of the assertions caused by actions in $i, j, k, \ldots$ ontologies; (5) a set of ontology evolution rules **Ev** caused by performed actions.

Finally, we define the ontology evolution rules in the following way.

Definition 2. (Ontologies evolution rules)

Let's KB_i и KB_j are knowledge bases. The evolution of the distributed knowledge base **DKB** acquiring ontologies from KB_i and KB_j can be caused by actions ACt_i and ACt_j, $i \neq j$ only by rules:

(1) $i : C \xrightarrow{\sqsubseteq} j : D$, as reducing the ontologies;

(2) $i : C \xrightarrow{\sqsupseteq} j : D$, as expanding the ontologies;

(3) $i : x = \{x_1, \ldots, x_n\} \xrightarrow{\cong} j : y = \{y_1, \ldots, y_n\}$, as fusion of the ontologies by partial identity;

(4) $i : x = \{x_1, \ldots, x_n\} \xrightarrow{\equiv} j : y = \{y_1, \ldots, y_n\}$ as fusion of the ontologies by full identity,

where C, D are the concepts or roles, x, y are individuals.

4 Conclusion

Among the scientific results that are presented in this article, it is necessary to distinguish several of the following results, which have scientific novelty in the field of theoretical foundations of computer science. The first result relates to the consideration of the *ISAS* as intelligent systems that have a human-machine interface. From this point of view, four groups of interconnected intellectual capabilities of the *ISAS* human-machine interface have been analyzed. These features made it possible to propose a second result. It consists of a new approach to the synthesis of *ISAS*, which can be named a fully non-metric approach. And, finally, our new result is a formal definition of description logic allows operating with distributed knowledge bases in dynamic environments.

References

1. Tretmans, J.: Introduction: situation awareness, systems of systems, and maritime safety and security. In: Tretmans, J., van de Laar, P., Borth, M. (eds.) Situation Awareness with Systems of Systems, pp. 3–20. Springer (2013). https://doi.org/10.1007/978-1-4614-6230-9
2. Mozzaquatro, B.A., Jardim-Goncalves, R., Agostinho, C.: Situation awareness in the Internet of Things. In: 2017 International Conference on Engineering, Technology and Innovation (ICE/ITMC), Madeira Island, Portugal, pp. 982–990 (2017). https://doi.org/10.1109/ice.2017.8279988
3. Endsley, M.R., Bolte, B., Jones, D.J.: Designing for Situation Awareness: An Approach to Human-Centered Design. CRC, Press, Taylor & Francis, London (2011)
4. Endsley, M.R.: Situation Awareness Analysis and Measurement. CRC Press, Atlanta (2000). Endsley, M.R., Garland, D.G. (eds.)
5. Chernov, A.V., Butakova, M.A., Karpenko, E.V.: Security incident detection technique for multilevel intelligent control systems on railway transport in Russia. In: 2015 23rd Telecommunications Forum Telfor (TELFOR), pp. 1–4 (2015). https://doi.org/10.1109/telfor.2015.7377381
6. Chernov, A.V., Bogachev, V.A., Karpenko, E.V., Butakova, M.A., Davidov, Y.V.: Rough and fuzzy sets approach for incident identification in railway infrastructure management system. In: 2016 XIX IEEE International Conference on Soft Computing and Measurements (SCM), St. Petersburg, pp. 228–230 (2016). https://doi.org/10.1109/scm.2016.7519736
7. Chernov, A.V., Butakova, M.A., Karpenko, E.V., Kartashov, O.O.: Improving security incidents detection for networked multilevel intelligent control systems in railway transport. Telfor J. **8**(1), 14–19 (2016). https://doi.org/10.5937/telfor1601014C
8. Chernov, A.V., Butakova, M.A., Vereskun, V.D., Kartashov, O.O.: Mobile smart objects for incidents analysis in railway intelligent control system. Adv. Intell. Syst. Comput. **680**, 128–137 (2017). https://doi.org/10.1007/978-3-319-68324-9_14
9. Chernov, A.V., Butakova, M.A., Vereskun, V.D., Kartashov, O.O.: Situation awareness service based on mobile platforms for multilevel intelligent control system in railway transport. In: 24th Telecommunications Forum, TELFOR, pp. 1–4 (2016). https://doi.org/10.1109/telfor.2016.7818714
10. Chernov, A.V., Kartashov, O.O., Butakova, M.A., Karpenko, E.V.: Incident data preprocessing in railway control systems using a rough-set-based approach. In: 2017 XX IEEE International Conference on Soft Computing and Measurements (SCM), St. Petersburg, pp. 248–251 (2017). https://doi.org/10.1109/scm.2017.7970551
11. Butakova, M.A., Chernov, A.V., Shevchuk, P.S., Vereskun, V.D.: Complex event processing for network anomaly detection in digital railway communication services. In: 25th Telecommunication Forum (TELFOR), Belgrade, pp. 1–4 (2017). https://doi.org/10.1109/telfor.2017.8249273
12. Yants, V.I., Chernov, A.V., Butakova, M.A., Klimanskaya, E.V.: Multilevel data storage model of fuzzy semi-structured data. In: 2015 XVIII International Conference on Soft Computing and Measurements (SCM), vol. 1, pp. 112–114 (2015). https://doi.org/10.1109/scm.2015.7190427
13. Rogozov, Y.: Approach to the construction of a systemic concept. In: Advances in Intelligent Systems and Computing, vol. 679, pp. 429–438 (2017). https://doi.org/10.1007/978-3-319-68321-8_44

14. Vlachostergiou, A., Caridakis, G., Kollias, S.: Investigating context awareness of affective computing systems: a critical approach. Proc. Comput. Sci. **39**, 91–98 (2014). https://doi.org/ 10.1016/j.procs.2014.11.014

15. Borgida, A.: Distributed description logics: assimilating information from peer sources. In: Spaccapietra, S., March, S., Aberer, K. (eds.) Journal on Data Semantics I. Lecture Notes in Computer Science, vol. 2800, pp. 153–184 (2003). https://doi.org/10.1007/978-3-540-39733-5_7

16. Shi, Z.: A logical foundation for the semantic Web. Sci. China Ser. F **48**, 161–178 (2005). Shi, Z., Dong, M., Jiang, Y., et al. https://doi.org/10.1360/03yf0506

17. Chang, L., Lin, F., Shi, Z.: A dynamic description logic for semantic web service. In: Third International Conference on Semantics, Knowledge and Grid (SKG 2007), pp. 74–79 (2007). https://doi.org/10.1109/skg.2007.59

Intelligent Support of Grain Harvester Technological Adjustment in the Field

Valery Dimitrov$^{(\boxtimes)}$, Lyudmila Borisova, and Inna Nurutdinova

Don State Technical University, Rostov-on-Don, Russian Federation
{dimitrovvalery,nurut.inna}@yandex.ru,
borisovalv09@mail.ru

Abstract. The problems of creating intelligent systems for information support in making decisions on preliminary technological adjustment of complex harvesting machines functioning in the field are considered. The solution of the problem for a combine harvester being a universal machine for harvesting grain, leguminous and other cultivated crops is presented. A combine harvester is considered as a complex mechatronic system that functions in a changing environment. Different types of uncertainty in the consideration of the semantic spaces of environmental factors and adjustable machine parameters cause the application of the logical-linguistic approach and the mathematical apparatus of fuzzy logic to find the optimal initial values of the adjustable parameters. The models of studied semantic spaces have been built. An expert knowledge base has been created, quantitative assessments of the consistency of expert information have been obtained. On the basis of the system of production rules, further fuzzy inference of solutions in the task of preliminary technological adjustment has been carried out. The proposed formal logical scheme of the decision-making process is applied to the selection of the values of the most important adjustable parameters of the combine, such as the speed, the rotational speed of the threshing drum, rotor speed of a separator fan.

Keywords: Intelligent system · Technological adjustment
Grain combine harvester · Expert knowledge · Membership function
Linguistic variable · Fuzzification · Defuzzification · Fuzzy inference

1 Introduction

Search for optimal values of adjustable parameters in the field is a sophisticated problem. The difficulty of solving this problem is stipulated not only by variability of external conditions but also by complexity of connections between the adjustable machine parameters and external factors [1]. Due to an operator's low professional skills, his decisions made on handling a harvester may be non-optimal, ant this results in direct grain losses, mechanical damage of grain, increase of a machine downtime and harvest time due to long-term search for reason of technological process breakage. All that leads to the increase of grain production costs and its biological losses.

Improving the quality level of harvesting can be done by simultaneous upgrading the harvester design and implementing intelligent automated systems [2]. Intelligent

© Springer Nature Switzerland AG 2019
A. Abraham et al. (Eds.): IITI 2018, AISC 875, pp. 236–245, 2019.
https://doi.org/10.1007/978-3-030-01821-4_25

information systems (IIS) accumulate different kinds of knowledge, expert and heuristic ones being among them, they use the gathered operational experience in different conditions, including extreme ones. The use of IIS helps to lower information load on an operator and quickly adapt to changing operational conditions. The IIS are used in agricultural industry. But they are mainly connected to the analysis of pictures, weather conditions, processing the products, identification of weeds, estimation of harvest and so on [3, 4 etc.]. Some problems of automatic selection of parameters of the harvester operation on the basis of quality indices with the use of production rules formed on the basis of expert knowledge were considered in [5, 6].

2 Problem Solution

2.1 Problem Statement

The paper considers the problem of designing decision-making processes about the values of the whole complex of the harvester technological adjustment main parameters on the basis of the initial fuzzy information. The main harvester parameters, determining the qualitative indices of harvesting, include driving speed, rotational speed of the threshing drum and rotor speed of the separator fan [7]. It is these three parameters that 80% of adjustments fall at. When selection of values of these parameters is incorrect, substantial deviations of harvesting quality indices occur.

We consider the problem solving procedure of selecting initial values of adjustable parameters and illustrate it by the example of selecting the main parameters' values mentioned above.

2.2 Methods

The problem of the harvester tools technological adjustment is the problem of decision making in a fuzzy environment which is characterized by the presence of relations between parameters with fuzzy boundaries and also statements with different truth degree. The problem solution on the basis of fuzzy control involves three stages: fuzzification, composition, and defuzzification [8].

The stage of fuzzification incorporates solution of a series of problems: definition of a carrier of the fuzzy set, selection of the basic term-set, checking the requirements for constructing a membership function (MF), selection of the MF construction method, estimation of consistency of fuzzy expert knowledge, construction of generalized MF. As a result of fuzzification of the studied object domain a linguistic description of the problem conditions was obtained, linguistic variables (LV) and MF of the adjustable parameters and external factors were determined. Let us consider normal fuzzy sets, for which the height equals to 1, i.e. upper bound of the MF equals to 1. Fuzzy sets can be both unimodal, i.e. $\mu_A(x) = 1$ only on one x of E, and also having tolerance domain.

On the basis of the methodology of linguistic approach to investigation of complex systems [9], we have developed the models of factors of external environment X and adjustable combine parameters Y in the form of semantic spaces and the corresponding them MF's:

$$\{X_i, T(X_i), U, G, M\}, \ \mu_R(x_1, x_2, \ldots, x_{i,}) \in (0; 1),$$
$$\{Y_j, T(Y_j), U, G, M\}, \ \mu_R\left(y_1, y_2, \ldots, y_{j,}\right) \in (0; 1),$$

where X and Y are the names of the LV's, T - set of its values, or terms, which are the names of the LV's defined over the set U, G - syntactic procedure describing the process of deriving the new values of LV's from the set T, M - semantic procedure, which allows one to map new value generated by procedure G into fuzzy variable, μ - MF's.

The generalized model of the object domain "preliminary adjustment" is adopted in the form of composition of fuzzy relations of the semantic spaces under consideration:

$$R = X \rightarrow Y,$$

where R is a fuzzy relation between environmental factors and adjustable parameters:

$$R\{X_i, T(X_i), U, G, M\} \times \{Y_j, T(Y_j), U, G, M\}; \forall(x, y) \in X \times Y.$$

Description of the LV's, characterizing a set of the external environment factors and adjustable parameters, incorporates a definition of the basic term-sets for all LVs. In general case the basic term-set of LV has the form [10]:

$$T_i = \left\{T_1^i, T_2^i, \ldots T_m^i\right\}, (i \in K = \{1, 2, \ldots, m\}).$$

Here $\langle T_i, X; \tilde{C}_i \rangle$ is a fuzzy variable corresponding the term $T_i \in T$; $\tilde{C}_i = \left\{\langle \mu_{C_i}(x)/x \rangle\right\}$, $x \in X$; C_i is a carrier of the fuzzy set $\tilde{C}_i$, $\mu_{C_i}(x)$ - MF. As a carrier the subset of LV values is used. The basic term-set is stated on the basis of expert judgments.

Representation of MF with the help of standard functions, defined parametrically, is most preferable. The function form is defined axiomatically, and its parameters are estimated by the experts, this provides convenience and simplicity of construction. For example, in case of a triangle form of the MF, x_1, x_2, x_3 parameters are assigned, wherein it takes on unit and zero values, i.e. $\mu_A(x_2) = 1$ and for all $x \le x_1, x \ge x_3$ one has $\mu_A(x) = 0$. However, the use of this representation involves for examination of adequacy of typical forms (triangle, trapezoidal, etc.). The reason for choosing the specific kind of MF is different assumptions on the properties of these functions, such as symmetry, monotonicity and so on, fuzziness feature and also its physical meaning are taken into account.

Consistency of expert information is determined by several criteria. The characteristics of paired consistency include d_{ij}^l index of models difference between i-th and j-th experts in terms of l-th term (Hamming linear distance [11] between fuzzy sets with MF's $\mu_{il}(x)$ and $\mu_{jl}(x)$) and k_{ij}^l consistency index [12]:

$$d_{ij}^l = \int\limits_0^1 \left|\mu_{il}(x) - \mu_{jl}(x)\right| dx, \; k_{ij}^l = \frac{\int\limits_0^1 \min\left[\mu_{il}(x), \mu_{jl}(x)\right]}{\int\limits_0^1 \max\left[\mu_{il}(x), \mu_{jl}(x)\right]} \qquad (1)$$

Elements d_{ij}^l and k_{ij}^l form matrices of fuzziness D^l and paired consistency K^l for l-th term. On the basis of these matrices, obtained for each of the terms, there are matrices of fuzziness D and paired consistency K of the models for all terms. Their elements are determined as arithmetic mean of the corresponding elements of the matrices D^l and K^l.

To characterize general consistency of multiple models of an attribute assessment by an expert, additive k_a and multiplicative k_m indices are calculated by formulae [12]:

$$k_a = \frac{1}{m}\sum_{i=1}^m \frac{\int\limits_0^1 \min_{\forall i=1,2,\dots,n} \mu_{il}(x)\, dx}{\int\limits_0^1 \max_{\forall i=1,2,\dots,n} \mu_{il}(x)\, dx}; \quad k_m = \sqrt[m]{\prod_{i=1}^m \frac{\int\limits_0^1 \min_{\forall i=1,2,\dots,n} \mu_{il}(x)\, dx}{\int\limits_0^1 \max_{\forall i=1,2,\dots,n} \mu_{il}(x)\, dx}}, \qquad (2)$$

where $l = 1, 2, \dots, m$ - term number, $i = 1, 2, \dots, k$ - expert number, $\mu_{il}(x)$ - MF, which was given by i-th expert for l-th term.

Expert information consistency assessment is a necessary step at the stage of fuzzification since it determines the quality of information and depicts the degree of adequacy of formal description of a real situation, and that makes it possible to use that description in the system of fuzzy logic inference. The procedure for expert information assessment is presented in [12]. To form the block of expert information, satisfying all the requirements, it is expedient to use algorithm suggested in [13].

As a result of the analysis of the considered object domain, a base of knowledge was created, and a solution inference is based on it. Production rules of the fuzzy inference system are intended for formal presentation of empirical knowledge, they represent a finite set of rules of fuzzy products, which are consistent in respect to linguistic variables used in them. Usually a base of production rules takes the form of a structured text of the kind: IF "Condition_1" THEN "Conclusion_1".

At the heart of the inference mechanism of the IIS there is a model of the object domain, representing a composition of fuzzy relations of semantic spaces of external environment factors and a harvester adjustable parameters [9]. An expanded form of fuzzy inference for the system of knowledge can be presented in the form [14]:

$$\mu_{B'} = \bigvee_{x \in X}(\mu_{A'}(x) \wedge \mu_R(x, y)).$$

At the stage of defuzzification, exact meanings of the resulting LV are calculated. The most commonly encountered method is a "barycenter" method [11, 14], which possesses sufficiently high accuracy. This method is implemented in MatLab environment with the help of application software package Fuzzy Logic Toolbox.

2.3   The Result of Investigation

All the mentioned above problem solving stages of optimal technological harvester adjustment have been applied for inference of values of the main adjustable harvester parameters: driving speed, rotational speed of the threshing drum, and rotational speed of the separator fan rotor.

At the first stage, let us determine a linguistic scale. Practical experience and the object domain analysis have shown that the ranges of parameters changing will be different for different crops. For example, changing rotor speed of the separator fan for wheat is in the range of [600–920] rev/min, while for barley it is in the range of [550–800].

To describe the terms in this paper we applied typical functions of triangle and trapezoidal form [11], as the most convenient ones for estimation by experts and further application at the stages of solution inference [15].

A linguistic description of input factors will be stated for a certain crop – wheat. It is defined, that in this case the output parameters under consideration are substantially influenced by such external environment factors as: crop yield, stand of grain humidity, rough straw, stand of grain dockage [9]. A variety of the harvester operation conditions requires differentiation of such indices as crop yield, therefore, the parameter of crop yield is expedient to be considered for different values, in particular, crop yield of about 50 q/ha, 40 q/ha etc. In the following example let us consider two cases: crop yield of about 40 q/ha and crop yield of about 50 q/ha.

We have obtained the following linguistic description of the external factors:

<CROP YIELD-40, q/ha {Less 40, Approximately 40, More 40}, [34–46]>
<CROP YIELD-50, q/ha {Less 50, Approximately 50, More 50} [44–56] > (Fig. 1a).
<ROUGH STRAW, % {Small, Normal}, [40–70] > (Fig. 1b).
<STAND OF GRAIN HUMIDITY, % {Dry, Normal, Humid}, [0–30] > (Fig. 1c).
<STAND OF GRAIN DOCKAGE, % {Low, Large}, [0–40] > (Fig. 1d).
The tuples of the output LV's are stated for wheat - 50 i.e. for crop yield approximately 50 q/ha:
<SPEED OF HARVESTER, km/h {Very low, Low, Lower than nominal, Nominal, Higher than nominal, High, Very high}, [2, 5]>
SH = {VL, L, LN, N HN, H, VH, km/h}(Fig. 1e).
<ROTATIONAL SPEED OF THRESHING DRUM, rev/min {Very low, Low, Lower than nominal, Nominal, Higher than nominal, High, Very high}, [620–940]>
RSTD = {VL, L, LN, N, HN, H, VH, rev/min} (Fig. 1f).
<ROTOR SPEED OF SEPARATOR FAN, rev/min {Very low, Low, Lower than nominal, Nominal, Higher than nominal, High, Very high}, [600–920]>
RSSF = {VL, L, LN, N, HN, H, VH, rev/min} (Fig. 1g).

A substantive issue is a question of choosing the number of LV terms. On the one hand, the number of terms is limited by virtue of measurement accuracy of the parameter under consideration, and from the other hand, the number of terms should be enough to identify and describe interactions of this factor with performance indices. Therefore, before choosing the optimal number of terms a priori analysis of the object domain was carried out, and it was recognized that the indicated number of terms for all

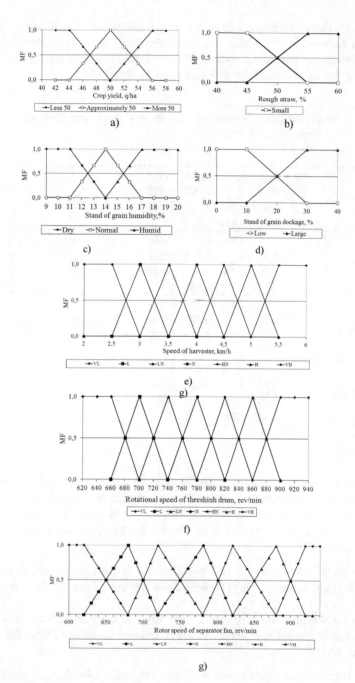

Fig. 1. Membership functions of LV terms (a) crop yield – 50, (b) rough straw, (c) stand of grain humidity, (d) stand of grain dockage, (e) speed of harvester, (f) rotational speed of threshing drum, (g) rotor speed of separator fan

LVs satisfy all the mentioned above requirements. Moreover, a posteriori analysis was carried out, the criterion of which was the consistency of expert information. Let us show, as an example, part of the results of the consistency assessment (formulae (1) and (2)) obtained for «RSSF» LV for different crops (Table 1).

Table 1. The matrix of paired consistency for all the terms and the indices of general consistency for «RSSF» LV for different crops

Crop	Matrix K				k_a	k_m
Wheat	1	0,992	0,735	0,907	0,734	0,727
	0,992	1	0,795	0,832		
	0,735	0,795	1	0,813		
	0,907	0,832	0,813	1		
Barley	1	0,927	0,853	0,888	0,783	0,774
	0,927	1	0,796	0,873		
	0,853	0,796	1	0,884		
	0,888	0,873	0,884	1		
Rye	1	0,873	1	0,911	0,873	0,865
	0,873	1	0,873	0,955		
	1	0,873	1	0,911		
	0,911	0,955	0,911	1		
Oats	1	0,807	0,875	0,92	0,706	0,674
	0,807	1	0,829	0,743		
	0,875	0,829	1	0,795		
	0,92	0,743	0,795	1		

These values show us that the expert knowledge consistency level is high enough, and the considered description of the basic term-set can be applied further on.

As a result of the object domain analysis, a base of knowledge was created, and solution inference is based on it. Let us show a fragment of knowledge base for the output «RSSF» LV:

1. If (crop yield is less than 50) and (rough straw is low) and (dockage is low) and (stand of grain humidity is dry) then (rotational speed of the threshing drum is very low);
2. If (crop yield is less than 50) and (rough straw is low) and (dockage is low) and (stand of grain humidity is normal) then (rotational speed of the threshing drum is less than nominal).

Fuzzy inference was performed with the help of application software package Fuzzy Logic Toolbox (MatLab). The parameters of the generalized MFs, as a rule, mean values of the parameters presented by experts. The method of averaging may be different [16]; in this paper the parameters' arithmetical average values were used.

Fuzzy inference is an application of a maximin composition as a composition rule of fuzzy inference and an operation of taking minimum as a fuzzy implication [9]. Figure 2 gives examples of surfaces "inputs-output" for wheat – 40 and wheat – 50, corresponding to the synthesized fuzzy system of production rules.

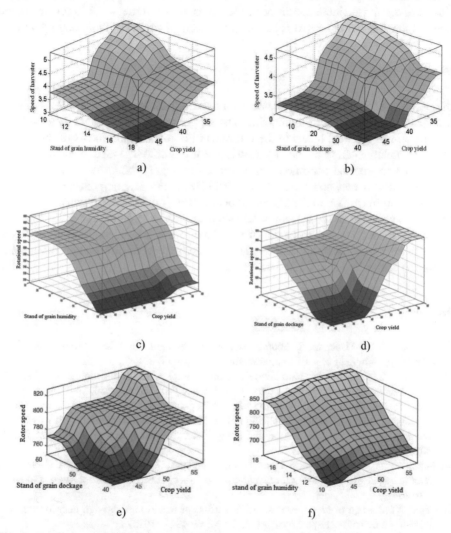

Fig. 2. Response surfaces of (a) speed of the harvester vs stand of grain humidity and crop yield; (b) speed of the harvester vs stand of grain dockage and crop yield; (c) rotational speed of the threshing drum vs stand of grain humidity and crop yield; (d) rotational speed of the threshing drum vs stand of grain dockage and crop yield; (e) rotor speed of separator fan vs rough straw and crop yield; (f) rotor speed of separator fan vs stand of grain humidity and crop yield

To obtain certain values of the adjustable parameters it is necessary to assign input factor values. In MatLab environment an inference is produced by the "barycenter" method of each output parameter separately according to the system of production rules for this parameter. As an example, let us choose the values of input factors: crop yield – 50 q/ha, stand of grain dockage – 20%, rough straw – 50%, stand of grain humidity – 14%. The results of inference in MatLab: speed of harvester - 3,2 km/h, rotational speed of threshing drum - 810 rev/min, rotor speed of separator fan - 800 rev/min.

3 Conclusion

The problem of preliminary adjustment of a harvester tools refers to the class of non-formalized problems of decision making. The suggested linguistic approach for solving this problem in the field meets to the full extent the main requirements of the system analysis. Implementation of the approach made it possible to construct models which are relevant to external conditions, to consider thoroughly the system main elements and interrelations between them. The possibility of the proper presentation on the ground of a uniform formal description of quantitative, qualitative (linguistic), and also heuristic knowledge of the system under consideration is important, and this makes it possible to develop a unified model of the object domain corresponding to real conditions of harvesters operation. As a result, we have prepared information for creating IIS for the harvester technological adjustment.

References

1. Rybalko, A.G.: Osobennosti uborki visokourojainih zernovih kultur (Some features of harvesting high-yield crops). Agropromizdat, Moscow (1988). (in Russian)
2. Borisova, L., Dimitrov, V., Nurutdinova, I.: Intelligent system for technological adjustment of the harvesting machines parameters. Adv. Intell. Syst. Comput. **680**, 95–105 (2018)
3. Zareiforoush, H., Minaei, S., Alizadeh, M.R., Banakar, A., Samani, B.H.: Design, development and performance evaluation of an automatic control system for rice whitening machine based on computer vision and fuzzy logic. Comput. Electron. Agric. **124**, 14–22 (2016)
4. Sujaritha, M., Annadura, S., Satheeshkuma, J., Sharan, S.K., Mahesh, L.: Weed detecting robot in sugarcane fields using fuzzy real time classifier. Comput. Electron. Agric. **134**, 160–171 (2017)
5. Omid, M.: Design of an expert system for sorting pistachio nuts through decision tree and fuzzy logic classifier. Expert Syst. Appl. **38**, 4339–4347 (2011)
6. Craessaerts, G., de Baerdemaeker, J., Missotten, B., Saeys, O.: Fuzzy control of the cleaning process on a combine harvester. Biosys. Eng. **106**, 103–111 (2010)
7. Yerokhin, S.N., Reshetov, A.S.: Vliyanie tekhnologicheskih regulirovok na poteri zerna za molotilkoj kombajna Don-1500 (Influence of technological adjustments on grain loss behind the thresher of the combine Don-1500). Mech. Electrification Agric. **6**, 18–19 (2003). (in Russian)
8. Zadeh, L.: Knowledge representation in fuzzy logic. In: Yager, R.R., Zadeh, L.A. (eds.) An Introduction to Fuzzy Logic Applications in Intelligent Systems, The Springer International Series in Engineering and Computer Science, vol. 165, pp. 1–27. Springer, New York (1992)

9. Borisova, L.V. Nurutdinova, I.N., Dimitrov, V.P.: Approach to the problem of choice of values of the adjustable parameters harvester based on fuzzy modeling. Don State Tech. Univ. Bull. **5-2**(81), 100–107 (2015)
10. Averkin, A.N., Batyrshin, I.Z., Blishun, A.F., Silov, V.B., Tarasov, V.B.: Nechetkie mnojestva v modelyah upravleniya i iskusstvennogo intellekta (Fuzzy sets in the models of management and artificial intelligence). Nauka, Moscow (1986). (in Russian)
11. Kofman, A.: Vvedenie v teoriyu nechyotkih mnozhestv (Introduction in the theory of fuzzy sets). Radio i svyaz', Moscow (1982). (in Russian)
12. Dimitrov, V., Borisova, L., Nurutdinova, I.: Modelling of fuzzy expert information in the problem of a machine technological adjustment. In: MATEC Web of Conference 13. Ser. 13th International Scientific-Technical Conference "Dynamic of Technical Systems", DTS-2017. P. 04009 (2017)
13. Borisova, L., Dimitrov, V., Nurutdinova, I.: Algorithm for assessing quality of fuzzy expert information. In: Proceedings of IEEE East-West Design & Test Symposium (EWDTS 2017), Serbia, pp. 319–322 (2017)
14. Asai, K., Vatada, D., Sugeno, S.: Prikladnie nechetkie sistemi (Applied fuzzy systems). Mir, Moscow (1993). (in Russian)
15. Dimitrov, V.P., Borisova, L.V., Nurutdinova, I.N.: O metodike defazzifikacii nechyotkoj ehkspertnoj informacii (On defuzzification method in fuzzy expert information processing). Don State Tech. Univ. Bull. **10-6**(49), 868–878 (2010). (in Russian)
16. Nurutdinova, I.N., Shumskaya, N.N., Dimitrova, L.A.: Ob ispol'zovanii vesovyh koehffi-cientov pri formirovanii ehkspertnoj informacii (On the use of weight coefficients in the formation of expert information). In: sbornik statej 10-j Mezhdunarodnoj yubilejnoj nauchno-prakticheskoj konferencii v ramkah 20-j Mezhdunarodnoj agropromyshlennoj vystavki "Interargomash-2017" Sostoyanie i perspektivy razvitiya sel'skohozyajstvennogo mashinostroeniya, pp. 332–334. Don State Technical University, Rostov-on-Don (2017). (in Russian)

Development and Research of the Hybrid Approach to the Solution of Optimization Design Problems

Leonid A. Gladkov[✉], Nadezhda V. Gladkova, Sergey N. Leiba, and Nikolay E. Strakhov

Southern Federal University, Taganrog, Russia
leo_gladkov@mail.ru, nadyusha.gladkova77@mail.ru,
lejba.sergej@mail.ru, kolanfantagan@gmail.com

Abstract. The article suggests a hybrid approach to solving optimization problems of computer-aided design. As an example, to illustrate the proposed approach, the problems of location and tracing of fragments of circuits of digital electronic computing equipment are chosen. The statement of the problem is given, limitations of the domain of admissible solutions are chosen and a criterion for estimating the quality of the solutions is formulated. A hybrid approach is described on the basis of a combination of evolutionary search methods, the mathematical apparatus of fuzzy logic and the possibilities of parallel organization of the computational process. A modified migration operator is proposed to exchange information between solution populations in the process of performing parallel computations. The structure of the parallel search algorithm is developed. Features of software implementation of the proposed hybrid algorithm are considered. A brief description of the computational experiments that confirm the effectiveness of the proposed method is presented.

Keywords: Design tasks · Bioinspired algorithms · Neural networks
Hybrid methods · Parallel computing

1 Introduction

Many optimization tasks that are solved in the process of the design stage of computer-aided design of elements of electronic computing equipment (EVA) require huge time and computational resources, which is explained by the need to search through a huge number of different solutions. Therefore, in practice, various meta-heuristic algorithms are developed to solve such problems, which make it possible to find solutions close to optimal (quasi-optimal) [1–3].

One of the approaches that make it possible to successfully solve the problem of increasing the efficiency and quality of the solutions obtained is the integration of various scientific methods of computing intelligence [4–7].

The scientific literature describes a large number of modifications of evolutionary algorithms for solving various problems of the design stage of design [4–8]. As a rule, such algorithms quite successfully cope with the solution of the problem of finding a quasi-optimal solution for polynomial time. However, most of the algorithms

© Springer Nature Switzerland AG 2019
A. Abraham et al. (Eds.): IITI 2018, AISC 875, pp. 246–257, 2019.
https://doi.org/10.1007/978-3-030-01821-4_26

mentioned above have problems with premature convergence of the search process. Also, when organizing genetic algorithms, it is important to consider the need to expand the search area, the ability to organize search in various areas of the space of permissible solutions that are remote from each other.

In this paper, we propose a hybrid approach to the solution of the optimization problems under consideration based on the integration of various approaches, such as evolutionary search algorithms, artificial neural networks, fuzzy models for controlling algorithm parameters and parallel computations [9–11].

2 Problem Definition

The most complex and responsible tasks of the design phase of designing electronic equipment are placement and routing tasks. The most complex and responsible in terms of the quality of future products are the placement and tracing tasks. These tasks are closely interrelated with each other, since the result of solving the task of placing the elements is the initial information for the routing task, and the quality of the solution of the allocation problem directly affects the complexity and quality of the routing task. Therefore, it is of practical interest to develop integrated methods that make it possible to solve these problems in a single cycle, taking into account mutual limitations and current results [12, 13].

Let there be given a set of elements E:

$$E = \{e_i \,|\, i = 1,\ldots,N\},$$

where e_i - is the placed element, N – is the number of placed elements. And

$$e_i = (l_i, h_i, T_i),$$

where l_i – is the length of the element, h_i – is the height of the element, T_i – is the list of contacts of the placement element

$$T_i = \{t_j \,|\, j = 1,\ldots,K\},$$

where t_j – contact, K – number of contacts of the element;

$$t_j = (x_j, y_j),$$

where x_j, y_j – are the coordinates of the contact relative to the base point of the element.
A set of circuits connecting elements:

$$U = \{u_h \,|\, h = 1,\ldots,L\},$$

where u_h – circuit, L – number of electrical circuits;

$$u_h = \{(Ne_k, Nc_k) \,|\, k = 1,\ldots,M\},$$

where Ne_k – is the element number, Nc_k – is the contact number, M – is the number of contacts connected by the current circuit.

It is required to find an option for placing elements on the installation space

$$V = \{(x_i, y_i) \mid i = 1, \ldots, N\},$$

where (x_i, y_i) - are the coordinates of the upper-left corner of the mounting area of the placement element i, such that the total overlapping area of the placed elements is zero, and the sum of the remaining criteria is minimal.

For each chain, it is necessary to find a list of the positions of the switching field through which it passes:

$$W_h = \{(x_q, y_q) \mid i = 1, \ldots, Q\},$$

where Q – is the number of positions through which the h-th circuit passes.

3 Description of the Algorithm

When coding solutions, a set of positions is represented as a regular structure (grid). Each position p_i has coordinates x_i, y_i. The positions are numbered in ascending order of the x_i coordinate within the string from left to right, and the rows are in turn ordered in ascending order of the y_i coordinate from top to bottom.

Each element has a base point O_i^δ and basic axes of coordinates $O_i^\delta X_i^\delta, O_i^\delta Y_i^\delta$, with respect to which a contour description of the element e_i is given. The base point is the bottom-left corner of the element. We assume that the element e_i is assigned to the position p_j if its base point O_i^δ is aligned with the point of the commutation field having the coordinates x_j y_j. The intersection points of the dashed lines correspond to the seating positions on the commutation field.

Each solution is represented as a chromosome H_i. The sequence number of the gene in the chromosome corresponds to the ordinal number of the placed element. The gene value corresponds to the position number on the switching field. The number of genes in the chromosome is equal to the number of placed elements.

When calculating the value of the fitness function (FF) (Fig. 1), a normalized estimate of the amount of the penalty for overlapping the areas of the placed elements, estimating the interconnection lengths, the traceability index, and also estimating the thermal and electromagnetic compatibility of the elements is calculated

$$FF = k_1 * S + k_2 * L + k_3 * T + k_4 * J + k_5 * Q,$$

where S – is the total overlapping area of elements, L – is the estimate of the interconnection lengths, T – evaluation of routing, J – is the total electromagnetic effect of the elements on each other, Q – is the total thermal effect of the elements on each other, k_1, k_2, k_3, k_4, k_5 - weight coefficients that determine the impact of each component of the FF on the overall estimate.

The size of the penalty for overlapping the areas of the placed elements depends on the total area of intersection of all elements. This takes into account the minimum allowable distance between the elements.

The intersection area of two elements is calculated as follows:

$$x_{11} = max(min(x_1, x_2 + d), min(x_3, x_4 + d));$$
$$x_{12} = min(max(x_1, x_2 + d), max(x_3, x_4 + d));$$
$$y_{11} = max(min(y_1, y_2 + d), min(y_3, y_4 + d));$$
$$y_{12} = min(max(y_1, y_2 + d), max(y_3, y_4 + d));$$

$$\text{IF}((x_{12}-x_{11} > \ = 0)\,\text{AND}\,(y_{12}-y_{11} > \ = 0))$$

$$\text{THEN}\,R_{ij} = (x_{12}-x_{11}) * (y_{12}-y_{11})\text{ELSE}\,R_{ij} = 0,$$

where

x_1, y_1 – are the coordinates of the upper left corner of the first rectangle,
x_2, y_2 – are the coordinates of the lower right corner of the first rectangle,
x_3, y_3 – are the coordinates of the upper left corner of the second rectangle,
x_4, y_4 – are the coordinates of the lower right corner of the second rectangle,
d – is the minimum allowable distance between elements.

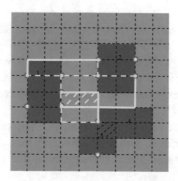

Fig. 1. Example of determining placement quality

To estimate the lengths of interconnections, it is advisable to use the semi -perimeter of the describing rectangle of the circuit. When solving the task of routing, the indicator of the quality of the solution is the percentage of non-routed connections. An additional criterion for the quality of the routing is the total area of intersection of the regions describing the rectangles of all circuits.

A parallel multipopulation genetic algorithm is used to jointly solve the placement and routing problems [9–11]. It assumes the parallel fulfillment of evolutionary processes on several populations. For the exchange of individuals, the island and buffer models of the parallel genetic algorithm are used.

In the island model, asynchronous processes are synchronized at migration points. The migration operator is used to exchange individuals between populations. The selection of individuals for migration is performed from a certain number of chromosomes of the population that have the best FF value. The selection is based on an estimate of the number of non-routed connections (Fig. 2).

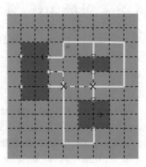

Fig. 2. Example routing quality evaluation

For each placement variant described by the chromosome, a routing is performed. Then, from one population to another, a certain number of chromosomes with the best value of this index are copied. In this case, the same number of chromosomes with the worst value of the indicator is removed from the populations. Figure 3 shows a diagram of a model of a parallel genetic algorithm performed on two populations. In practice, the number of populations can be much larger.

In the buffer model, the exchange of individuals between populations is carried out through a common intermediate buffer of chromosomes. Exchange is performed at migration points. The migration operator is applied to the population that has reached the migration point.

A wave algorithm is used for routing. The work of the algorithm includes three stages: initialization, propagation of the wave and restoration of the path. During the initialization, an image of the set of cells of the printed circuit board is built, each cell is attributed with the patency/obstruction attributes, the starting and finishing cells are memorized.

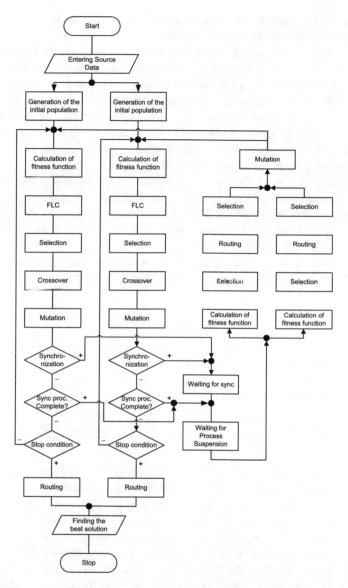

Fig. 3. The scheme of the island model of the placement algorithm

Further, from the starting cell a step is generated in the neighboring cell, it is checked whether it is passable and whether the cell previously marked in the path does not belong. The adjacent cells are 4 cells along the vertical and horizontal lines. When the conditions of patency are fulfilled and are not attached to the already marked cells, a number equal to the number of steps from the starting cell is written to the cell attribute, in the first step it will be 1. Each marked cell becomes the starting cell and the next steps to neighboring cells are generated from it. Obviously, with such a search, the path from the initial cell to the final cell will be found, or the next step from any cell generated in the path will be impossible.

The search for optimal solutions is carried out with the help of a bioinspired algorithm with modified evolutionary operators.

The development of hybrid approaches and systems based on the integration of various scientific directions gives reason to believe that using the parallel structure of calculations and constructing a scheme of the evolution process, it is very promising to use the principles of constructing multi-agent systems [14]. Methods and models of organization of multi-agent systems and technologies are one of the most rapidly developing areas of science in recent years. The issues related to the use of intelligent organizations of agents in various fields of science and economy, including design, robotics, and military purposes are already being actively studied.

Evolutionary modeling and multi-agent methods are closely interrelated. On the one hand, the application of the principles of evolutionary development allows solving the problems of adaptation of multi-agent systems to changes in the external environment. On the other hand, evolution can be used as a driving force, as a mechanism and as a motivation for achieving the set goals.

We can identify some parallels (correlations) between the basic concepts of evolutionary modeling and the theory of multi-agent systems (Table 1).

Table 1.

Evolutionary modeling	Theory of multi-agent systems
Gene	Agent property
Chromosome	Set of properties
Individual (solution)	Agent
Family (2 parents and 1 offspring)	Society of agents
Population	Evolving multi-agent system

We can also assume that the principles of organizing multi-agent systems can be successfully used to solve the problem of parallelizing computations.

To improve the quality of solutions obtained by the bioinspired algorithm, a fuzzy logic controller that regulates the values of the parameters of the main evolutionary operators is included in the search scheme. Various values can be used to assess the current state of the population, for example, the variety of the population's genotype, the growth rate of the average fitness function, etc. For the developed algorithm, the following parameters were chosen to evaluate the evolution efficiency [12, 13]:

$$e_1(t) = \frac{f_{ave}(t) - f_{best}(t)}{f_{ave}(t)}; \; e_2(t) = \frac{f_{ave}(t) - f_{best}(t)}{f_{worst}(t) - f_{best}(t)};$$

$$e_3(t) = \frac{f_{best}(t) - f_{best}(t-1)}{f_{best}(t)}; \; e_4(t) = \frac{f_{ave}(t) - f_{ave}(t-1)}{f_{ave}(t)},$$

where t – is the time step, $f_{best}(t)$ - is the best value of the FF at iteration t, $f_{best}(t-1)$ - is the best value of the FF at iteration $(t-1)$, $f_{worst}(t)$ – is the worst value of the FF at iteration t, $f_{ave}(t)$ – is the average value of the FF at iteration t, $f_{ave}(t-1)$ – is the average value of the FF at iteration $(t-1)$.

The variables e_1, e_2, e_3, e_4 are defined on the following intervals:

$$e_1 \in [0; 1]; e_2 \in [0; 1]; e_3 \in [-1; 1]; e_4 \in [-1; 1].$$

The output parameters are the probabilities of crossover, mutation and migration, respectively - $Pc(t), \Delta Pm(t), \Delta Ps(t)$:

$$Pc(t) \in [0; 1]; Pm(t) \in [0; 1]; Ps(t) \in [0; 1].$$

We use the parameters e_i as the input variables $\bar{x}_i$. The obtained values of $\bar{y}$ will be equivalent to the parameters ΔPc, ΔPm, $Ps(t)$. To calculate each of these parameters, a separate fuzzy control module will be used.

The final stage in the process of designing the fuzzy control module is the definition of the form of representation of fuzzy sets A_i^k, $1, \ldots, n;$ $k = 1, \ldots, N$. For example, this can be a Gaussian function

$$\mu_{A_i^k}(x) = \exp\left(-\left(\frac{x_i - \bar{x}_i^k}{\sigma_i^k}\right)^2\right),$$

where the parameters $\bar{x}_i^k$ and σ_i^k have a physical interpretation: $\bar{x}_i^k$ is the center, and σ_i^k is the width of the Gaussian curve.

As will be shown below, these parameters can be modified during the learning process, which allows changing the position and structure of fuzzy sets [12, 13]. After combining all the elements, the function for the fuzzy control module acquires the final form:

$$\bar{y} = \frac{\sum_{k=1}^{N} \bar{y}^k \left(\prod_{i=1}^{n} \exp\left(-\left(\frac{\bar{x}_i - \bar{x}_i^k}{\sigma_i^k}\right)^2\right)\right)}{\sum_{k=1}^{N} \left(\prod_{i=1}^{n} \exp\left(-\left(\frac{\bar{x}_i - \bar{x}_i^k}{\sigma_i^k}\right)^2\right)\right)}.$$

Each element of this formula can be specified in the form of a functional block (sum, product, Gauss function), which, after the appropriate combination, allows creating a multilayer neural network. In our case, the neural network will contain 4 layers. Each element of the first layer implements the fuzzy set membership function A_i^k, $1, \ldots, n;$ $k = 1, \ldots, N$. In this layer the input signals $\bar{x}_i$ come, and at its output the values of the membership function for these signals are formed. At the output of the first layer, the values of the membership function of fuzzy sets are formed. The configuration of the links of layer 2 corresponds to the rule base, and the multipliers to the output block.

The use of multipliers as nodes of layer 2 is due to the fact that in fuzzy operations the multiplication operation is used. The number of elements in this layer is equal to the number of rules stored in the database. Layers 3 and 4 realize the functions of the defuzzification block [12, 13].

Obviously, the described structure is a multi-layer network based on the idea of fuzzy inference. Unlike "pure" neural networks, each layer as a whole and its constituent elements, as well as the configuration of connections, all parameters and weights have a physical interpretation. This property is extremely important, because knowledge is not distributed over the network and can be easily localized and, if necessary, adjusted by an expert-observer.

The fuzzy control block uses 6 membership functions for the sets A^k and 2 functions for the sets B^k. In this case, $k_j[i]$ - are equivalent to the parameters y^k and are interpreted as centers of the fuzzy set membership functions B^k; $f_p[i]x$ and $f_p[i]y$ are interpreted respectively as the center and width of the Gaussian function to estimate the degree of belonging of the input data $\bar{x}_i$, to the corresponding fuzzy sets A_i^k and correspond to the parameters x_i^k and σ_i^k in the fuzzy control module; value "a" - corresponds to the output value of the control module. $\bar{y}$.

During the operation of the block, 24 membership functions are calculated for the sets A_i^k and 6 functions for the sets B^k.

4   Results of Experiments

Algorithms for placement and routing are implemented in the form of generalized algorithms that can handle various input data. The algorithm for the joint solution of the placement and routing problems is implemented on the basis of the generalized allocation algorithm, for which the method of calculation of the target allocation function is defined. The input of the algorithm is given data on the topology of the printed circuit board. The routing algorithm is auxiliary and is used to evaluate the resulting layouts. The fuzzy control module is used to dynamically adjust the genetic algorithm. The base of the rules of the fuzzy control module is read from the file. The parser is used to read data about the topology of the printed circuit board from a file, and also to write data to a file. The developed architecture allows supplementing the system with new properties and models of behavior.

To analyze the effectiveness of the algorithms being developed, graphs are used to change the mean and minimum value of the target allocation function. At each iteration, the mean values of the objective function of all populations in which the evolutionary process is launched are calculated. Also used are graphs of the mean and minimum value of the target routing function, the values of which are calculated at the migration points. To analyze the operation of the fuzzy logic controller, graphs are used to change the input and output parameters of the controller (Fig. 4).

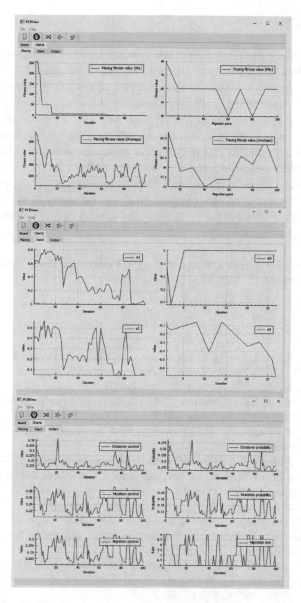

Fig. 4. Graphs of output parameters change of fuzzy controller

5   Conclusions

The development of hybrid approaches and systems based on the integration of various scientific directions gives reason to believe that using the parallel structure of calculations and constructing a scheme of the evolution process, it is very promising to use the principles of constructing multi-agent systems [14].

The efficiency of the controller can be improved by the introduction of a training unit based on the neural network. As a result of using the training block, the optimal parameters were determined. The training was based on statistical information on the dependence of NLC parameters and the efficiency of the allocation algorithm.

Analyzing the obtained data, it can be concluded that in order to obtain the best solution for the placement and tracing problems, it is necessary to use 4 streams of the parallel genetic algorithm. Further increase in the number of flows adversely affects the result of solving problems.

Acknowledgment. This research is supported by the grant from the Russian Foundation for Basic Research (projects 17-01-00627).

References

1. Cohoon, J.P., Karro, J., Lienig, J.: Evolutionary algorithms for the physical design of VLSI circuits. In: Ghosh, A., Tsutsui, S. (eds.) Advances in Evolutionary Computing: Theory and Applications, pp. 683–712. Springer, London (2003)
2. Alpert, C.J., Mehta, D.P., Sapatnekar, S.S.: Handbook of Algorithms for Physical Design Automation. CRC Press, New York (2009)
3. Shervani, N.: Algorithms for VLSI Physical Design Automation, 538 p. Kluwer Academy Publisher, Norwell (1995)
4. Michael, A., Takagi, H.: Dynamic control of genetic algorithms using fuzzy logic techniques. In: Proceedings of the 5th International Conference on Genetic Algorithms. Morgan Kaufmann, pp. 76–83 (1993)
5. Im, S.-M., Lee, J.-J.: Adaptive crossover, mutation and selection using fuzzy system for genetic algorithms. Artif. Life Robot. **13**(1), 129–133 (2008)
6. Herrera, F., Lozano, M.: Fuzzy Adaptive Genetic Algorithms: design, taxonomy, and future directions. Soft Comput. **7**, 545–562 (2003)
7. Herrera, F., Lozano, M.: Adaptation of genetic algorithm parameters based on fuzzy logic controllers. In: Herrera, F., Verdegay, J.L. (eds.) Genetic Algorithms and Soft Computing, pp. 95–124. Physica-Verlag, Heidelberg (1996)
8. King, R.T.F.A., Radha, B., Rughooputh, H.C.S.: A fuzzy logic controlled genetic algorithm for optimal electrical distribution network reconfiguration. In: Proceedings of 2004 IEEE International Conference on Networking, Sensing and Control, Taipei, Taiwan, pp. 577–582 (2004)
9. Rodriguez, M.A., Escalante, D.M., Peregrin, A.: Efficient distributed genetic algorithm for rule extraction. Appl. Soft Comput. **11**, 733–743 (2011)
10. Alba, E., Tomassini, M.: Parallelism and evolutionary algorithms. IEEE T. Evolut. Comput. **6**, 443–461 (2002)

11. Zhongyang, X., Zhang, Y., Zhang, L., Niu, S.: A parallel classification algorithm based on hybrid genetic algorithm. In: Proceedings of the 6th World Congress on Intelligent Control and Automation, Dalian, China, pp. 3237–3240 (2006)
12. Gladkov, L.A., Gladkova, N.V., Leiba, S.N.: Manufacturing scheduling problem based on fuzzy genetic algorithm. In: Proceedings of IEEE East-West Design & Test Symposium (EWDTS 2014), Kiev, Ukraine, 26–29 September 2014, pp. 209–213 (2014)
13. Gladkov, L.A., Gladkova, N.V., Legebokov, A.A.: Organization of knowledge management based on hybrid intelligent methods. In: Software Engineering in Intelligent Systems. Proceedings of the 4th Computer Science On-line Conference 2015 (CSOC 2015). Software Engineering in Intelligent Systems, vol. 3, pp. 107–113. Springer, Cham (2015)
14. Tarasov, V.B.: Ot mnogoagentnykh sistem k intellektual'nym organizatsiyam. Editorial URSS (2002)

The Concept of Methodological Framework for the Design of Information Systems

Yury Rogozov and Sergei Kucherov(✉)

Southern Federal University,
44 Nekrasovsky Street, Taganrog, Russian Federation
{yrogozov, skucherov}@sfedu.ru

Abstract. The problem of agreeing points of view on the target system being developed, often inseparable from the problem of mutual understanding of stakeholders in the process of developing information systems, as a special case of complex technical systems, is one of the fundamental for today. According to researchers, the reason for this situation is the inconsistency of the image of the target system, which arises from a misunderstanding between the customer and the developer. The overwhelming majority of modern research and practical developments are aimed at solving the problem of mutual understanding between two participants in the life cycle of information systems, located at its neighboring stages. The analyst, using the most advanced methods and techniques, can get a 100% reliable description of the customer's wishes (requirements), but he is still neither a domain specialist nor a designer (programmer). To solve this problem, we propose the methodology for the creation of information systems on the basis of a single consistent representation in the form of a means of creating concepts and its implementation in the development process.

Keywords: Information system · Design · Life cycle · Structure
Coordination · Process · Graph

1 Introduction

Despite the development of technology and methodological support in the field of IT, one of the key problems remains to be solved - the problem of mutual understanding between the customer and the developer (stakeholders). The essence of the problem lies in the complexity of the design process, which is characterized by conflicts between the results of the design of development teams that are jointly working on the design of one system [1], deliberate distortion or limited exchange of information between developers [2], semantic barriers between development teams [3], the protection of experts by sensitive design data [4]. By itself, the process of developing complex technical systems requires interaction between several technical disciplines, the authority over which is distributed among a multitude of experts. In this case, experts can be distributed within the organization and carry out activities at various stages of the life cycle of the product and at various stages of its production chain. The complex nature of such projects requires the interaction of many disciplines in the context of decentralized design authorizations and imperfect knowledge of development teams about the system.

© Springer Nature Switzerland AG 2019
A. Abraham et al. (Eds.): IITI 2018, AISC 875, pp. 258–264, 2019.
https://doi.org/10.1007/978-3-030-01821-4_27

Researchers around the world link the reasons for the emergence of the problem of mutual understanding of stakeholders with the lack of knowledge about the specifics of different subject areas related to different stages of the system design. For this reason, methods, approaches and tools are being developed to extract and transmit meaning between subjects from two different subject areas:

- Requirements extraction techniques [5–7] (for the customer-analyst/project manager bundle)
- Model-oriented approaches [8–10] (for the liaison of the analyst/project manager - the designer)
- Interpreted models and code generators [11] (for the link designer-programmer).

Also in a separate class can be selected object-oriented languages (DSL), ideally aimed at a direct transition between the customer and the system. However, such languages require preliminary work of analysts, designers and programmers with a deep immersion in the subject area, and at the output of the development process, they give a highly specialized tool that is not suitable for mass use.

Despite some successful attempts to solve the problem of mutual understanding between two neighboring participants of the software life cycle, all the above approaches have one significant drawback: the tools created do not agree with each other, and, most importantly, they do not allow one time to fix the meaning (semantics) system and transfer it through all stages of the life cycle.

This causes the urgency of developing new approaches to formulating, documenting and using a single evolutionary semantic model of the target system at all stages of the life cycle.

This article presents a methodology for creating information systems on the basis of a single consistent representation in the form of a methodological framework.

2 Related Work

Fundamental scientific research devoted to the problem of the synthesis of technical systems in terms of the completeness and consistency of the original description of the target object, for the most part, refers to the second half of the 20th century. Among them, in particular, are:

1. Work in the field of systemology [12, 13], in which, from the point of view of project objectives, the subject domain is viewed through the prism of the generating systems. These works are the closest to the project in its ideology.
2. Works by Pospelov DA, devoted to the theory of situational management [14]. From the standpoint of the tasks of the project, the author proposed the basic principles of formalizing the description of processes through the concept of "situation".
3. DSM-method of Finn [15]. The work of this researcher from the position of the project contains attempts to automate the formalization of knowledge about the subject area by means of so-called quasi-axiomatic theories (CAT). At the moment, the DSM-method is used in the synthesis of anonymous procedures: induction, analogy and abduction.

4. The theory of solving inventive problems Altshuller GS [16]. From the project's point of view, attempts have been made to methodologically obtain new knowledge in this work.

Among the most recent works in the field of fundamental research are the works of S.P. Nikanorova, dedicated to the variability in the creation of new design solutions by formalizing the conceptual concepts of subject areas using the apparatus of steps, based on the theory of the genera of Bourbaki structures [17–21].

The most relevant for the time of appearance are applied scientific research aimed at solving current problems in the development of complex technical solutions:

- Parallel design [22], consisting in the decentralization of the process of designing complex technical systems and based on joint working groups, which include specialists from different fields whose purpose is to harmonize the different concepts during the process of creating the system.
- Applied methods of game theory [23], solving problems of interaction between experts in decision-making tasks. Directly these methods in the design of complex technical systems are not used, but they have a developed formal apparatus and can be applied to achieve consistency between participants in the life cycle;
- System architecture [24] and Solution space research method [25], aimed at studying key decisions made in the early stages of system design and having a great impact on the total cost and success of the project. The method of investigating the solution space in this context allows us to evaluate possible design solutions in order to determine the most effective variant.
- The work of the international community on system engineering INCOSE [26], namely the development of methods, tools and technologies within the framework of the model-oriented approach (MBSE) [27], the main essence of which is the transition from the documentary presentation at the development stages to the modeling one. In this case, all the models are consistent with each other and thus form a general model of the system. The disadvantage of this approach is the formation of a unified model of the system in the late stages of development (when all types of models are ready), as well as the consideration of the system from the standpoint of its components and functions;
- Requirements extraction techniques. This includes methods for analyzing and tracing requirements [28], various language and CASE-tools, customer-oriented [7], techniques involving the customer or its representatives in the development process [6]. Disadvantages of this scientific direction are limited use possibilities (only at the level of the customer-analyst) and high labor input;
- Approaches based on scenario descriptions [29]. The idea of these approaches is contained in a single view of the system as a certain scenario of its use, which is consistent in the early stages of development, and is used as a reference model (and sometimes the initial one for the generation of system components).

In the approaches considered, a number of fundamental shortcomings can be singled out low average expressiveness of the scenario for all participants in the life cycle (scenarios are more understandable for analysts), as well as the development of a target system in isolation from the scenario where compliance is tested empirically.

3 The Concept of a Methodological Framework for the Design of Information Systems

Regardless of the scope of application, all existing approaches to automating the synthesis of complex systems are built around the formalization and conceptualization of the individual concepts from which it consists. In the case of successful completion of research, this leads to the emergence of highly formalized libraries of ready-made components (concepts), automated search tools and the synthesis of private systems based on them. Thus, the variability of the system is limited to a set of available modules and the development of search tools that can assess the relevance of the module in relation to the task.

The concept of the methodological framework is based on a fundamentally different view of the synthesis of complex technical systems. The main idea is to consider the component of the system (concept) as a procedure for obtaining it. Thus, the concept acquires two aspects - structural in the form of an insoluble structure of the medium and meaningful. Structural determines the purpose of the concept in the context of the information system (data collection, calculation of values, processing of interaction with the user, etc.) and is expressed by the structure of the tool consisting of characteristics: elements, functions, tools, result. The content aspect determines the embodiment of the concept in a specific environment (data model, user interface technology, way of performing computational operations) and is expressed in specific values of the characteristics of the facility.

The components of the system (the means of creating the concept) must form an organic unity. This means that in the target system, as well as in the initial semantic model, the coordination of components must be achieved. The coordination of components means the absence of connections in the form of inter-feysov (transducers), and the establishment of interpenetration of means of co-building concepts. The only form of interpenetration of the means of creating concepts is the intersection of the characteristics of various means.

In the methodological framework, the basis for a uniform representation and understanding of the meaning of the target system is a certain generality of not concepts and relations between them, but the totality of the structures of means for creating concepts of different levels of design and the rules for organizing their interpenetration. This commonality of tools and rules is a space of possibilities for creating object design processes. Under the meaning of a particular concept means means for its creation. Using this community, it is possible to define the structure of the means (sense) of the target information system being created in the form of a certain sequence of inter-penetrating means of creating separate concepts (to create a sense of the system from the meanings of concepts). At the early stages of the life cycle, the given community, through its structure of interpenetrating resources, gives a presentation on the process of creating the target properties of systems of this type. At the further stages of the life cycle, the responsible specialists (experts of their domain) fill the formed structure with content (specific entities, attributes, functions, etc.), and expand the structure with new means if required by technology.

The semantic model of the target system will thus be based on the basic abstraction of the structure of the means (meaning) of the concept creation. Using the basic abstraction of the structure of the means to create the concept, it is proposed to develop the structure of the process of creating a system in the form of a structure of organized interpenetration of means for creating separate concepts of different stages of the life cycle. Organized interpenetration creates an organic unity of means for creating separate concepts of different stages of the life cycle, which will ensure their mutual coherence. Such coherent unity of interpenetrating structures of means of creating concepts from different stages of the life cycle is the structure of the process of designing systems of a given kind, which can be considered a methodological framework. The methodological framework is due to the fact that methodological tools can also be used as tools.

The concept of a methodological framework is to divide the process of creating an information system into three phases:

1. Creating a space of opportunities to create a specific information system. The space of possibilities is a graph in which the vertices correspond to the means of creating concepts (components of the system) expressed, for example, by the structure of the action, and the edges - possible connections between the means. Thus, the admissible framework for the composition and structure of information systems of this type is marked.
2. Development of the structure of the process of creating a system in the form of a structure of organized interpenetration of the means for creating individual concepts from different stages of the life cycle. At this stage, experts from different technical disciplines and stages of the life cycle form an agreed view of the future system. The coordinated representation is a graph of a special form, in which the edges are degenerate, and the existence of a connection between two vertices, which are the means of creating individual concepts, is represented by the intersection of vertices by individual characteristics of the action (mutual penetration). It is important to note that the structure does not have content (specifics), it indicates only the typing of the characteristics of the action to create concepts.
3. Coherent filling of the structure of the system creation process. At this stage, experts from various technical disciplines and stages of the life cycle fill the structure of the process of creating a system with concrete values of action characteristics, thereby developing a specific information system.

4 Conclusion

Thus, a methodological framework that represents the structure of the information system of a given type as a coherent structure of mutually reinforcing means of creating concepts, and the process of its creation as filling the structure with content, will solve the problem of conflict between the results of designing the development teams and agreeing on the points of view of the stakeholders. parties. And in the methodological framework itself, a whole set of new approaches and methods to software development will be concentrated.

The main differences between the proposed concepts of the methodological framework from the known approaches are:

(1) Representation of a single process of creating systems of this type as an organic unity of the structures of interpenetrating facilities (representing the meaning) of creating concepts of different design stages, and the subsequent use of the structure of the process of creating the system by all participants of all stages of the life cycle to create specific systems of this type;

(2) Consideration of the meaning of the target system as a structure of the process of its creation, and not a description of empirically created or, more often, necessary to create parts of the system and their interrelations, as is done today.

Such an evolving semantic structure of the system can be represented by a graph of a special kind, in which vertices correspond to the means for the formation of concepts from which the process of creating the target information system can consist, the edges in their classical representation correspond to the uncertainty (variability, possibility) of establishing interpenetration between means, and the established interpenetration of the means (vertices) of the creation of concepts will be reflected in the partial or total interpenetration of the vertices (cf. meals) into each other. Documenting such models can be provided with the use of modern graph database management systems such as Neo4 J.

Acknowledgment. The reported study was funded by RFBR according to the research project № 18-07-00908.

References

1. Chanron, V., Lewis, K.: A study of convergence in decentralized design processes. Res. Eng. Des. **16**(3), 133–145 (2005)
2. Austin-Breneman, J., Yu, B.Y., Yang, M.C.: Biased information through subsystems over time in complex system design. 17 August 2014. V007T07A023
3. Yassine, A., Braha, D.: Complex concurrent engineering and the design structure matrix method. Concurr Eng. **11**(3), 165–176 (2003)
4. Li, F., Wu, T., Hu, M.: Design of a decentralized framework for collaborative product design using memetic algorithms. Optim. Eng. **15**(3), 657–676 (2014)
5. Athula, G.: Meta-design paradigm based approach for iterative rapid development of enterprise WEB applications. In: Proceedings of the Fifth International Conference on Software and Data Technologies, ICSOFT, pp. 337–343 (2010)
6. Fischer, G., Giaccardi, E.: Meta design: a framework for the future of end user development. In: Lieberman, H., Paterno, F., Wulf, V. (eds.) End User Development: Empowering People to flexibly Employ Advanced Information and Communication Technology, vol. 9, 427–457. Springer
7. Rossi, C., Guevara, A., Enciso, M., Caro, J.L., Mora, A.: A tool for user-guided database application development - automatic design of XML models using CBD. In: Proceedings of the Fifth International Conference on Software and Data Technologies, ICSOFT 2010, vol. 2, pp. 195–201 (2010)

8. Liang, X.D., Kop, C., Ginige, A., Mayr, H.C.: Turning concepts into reality - bridging requirement engineering and model-driven generation of web applications. In: Proceedings of the Second International Conference on Software and Data Technologies, ICSOFT 2007, Volume ISDM/ EHST/ DC, pp. 109–116 (2007)
9. Stahl, T., Völter, M.: Model-Driven Software Development: Technology, Engineering, Management, 1st edn. Wiley (2006)
10. Frankel, D.: Model Driven Architecture - Applying MDA to Enterprise Computing. OMG Press, Indianapolis (2003)
11. Lyadova, L.N.: The technology of creation of dynamically adapted information systems. In: Proceedings Scientific-Technology Conference "Intellectual Systems" (AIS 2007). Fizmatlit, Moscow T. 2 (2007)
12. Clear, J.: Systemology Automation of the solution of system problems. Radio Commun. (1990)534 pp. Translated from English by Zuev, M.A., edited by Gorlin, A.I
13. Egenev, G.B.: Systemology of engineering knowledge: a manual for universities on the specialty "Computer-aided design systems" in the direction "Computer science and computer technology"/GB Egenev. MSTU. N.E. Bauman (2001. - 376 sec. - (Informatics in Technical University). - ISBN 5-7038-1524-X
14. Pospelov, D.A.: Situational management: theory and practice. Science, Moscow. Ch. Ed. fiz.-mat. lit. (1986). 288 p
15. Finn, V.K.: On the possibility of formalizing plausible reasoning by the means of many-valued logics. In: Vsesoyuzn. Simp. by the Logic and Methodology of Science, pp. 82–83. Naukova Dumka, Kiev (1976)
16. Algorithm for solving inventive problems (ARIZ) 1959 (together with R. Shapiro) (1985)
17. Nikanorov, S.P., Nikitina, N.K., Teslinov, A.G.: Introduction to the Conceptual Design of Automated Control Systems: Analysis and Synthesis of Structures. Izd, Moscow. Strategic Missile Forces (1995). 185c
18. Nikanorov, S.P.: Theoretical and system constructs for conceptual analysis and design. Concept (2006). 312 p
19. Nikanorov, S.P.: Postulate of conceptual models of subject domains containing hundreds of thousands of concepts. Subset, Concept, Moscow, issue 21, pp. 42–61 (2006)
20. Nikanorov, S.P.: Conceptualization of Subject Areas. Methodology and Technology, Moscow, Concept (2009). 268 p
21. Nikanorov, S.P.: Introduction to the apparatus stages and its application. Moscow, Concept (2010). 179 p
22. Addo-Tenkorang, R.: Concurrent Engineering (CE): a review literature report. In: Proceedings of the World Congress on Engineering and Computer Science 2011 Vol II. USA, San Francisco, IAENG, International Association of Engineers (2011)
23. Myerson, R.B.: Game Theory. Harvard University Press (1997). 588 p
24. Crawley, E., Cameron, B., Selva, D.: System architecture: strategy and product development for complex systems, p. 448. Prentice Hall, Pearson (2015)
25. Ross, A.M., Hastings, D.E.: The tradespace exploration paradigm. INCOSE Int Symp. 15(1), 1706–1718 (2005)
26. Systems Engineering Handbook. A guide for system life cycle process and activities. International Council on System Engineering. INCOSE-TP-2003-002-03, June 2006
27. Gianni, D., D'Ambrogio, A., Tolk, A., eds.: Modeling and Simulation-Based Systems Engineering Handbook, 1 edn. CRC Press, December 2014. ISBN 9781466571457
28. Leffingwell, D., Widrig, D.: Software Requirements Management, Addison Wesley (1999)
29. Go, K., Carroll, J.M.: The Blind Men and the Elephant: Views of the Scenario-Based System Design. Interactions 11(6), 44–53 (2004)

Intelligent and Fuzzy Railway Systems

Detection of Point Anomalies in Railway Intelligent Control System Using Fast Clustering Techniques

Andrey V. Chernov[1(✉)], Ilias K. Savvas[2], and Maria A. Butakova[1]

[1] Department of Computers and Automated Control Systems,
Rostov State Transport University, Rostov-on-Don, Russia
{avcher, butakova}@rgups.ru
[2] Department of Computer Science and Engineering,
T.E.I. of Thessaly, Larissa, Greece
savvas@teilar.gr

Abstract. This work presents a new area of application for clustering techniques in industrial and transport applications. The main aim of the research is to propose the technique for detection of point anomalies in telecommunication traffic produced by network subsystems of railway intelligent control system. The central idea behind is to apply enhanced *DBSCAN* algorithms for finding the outliers in traffic which are associated with unintended erroneous events or deliberated attacks targeted to infrastructure malfunction. The traffic flows in a part of the railway intelligent control system has been described in detail. Point anomaly detection in *IP*-networks data using distributed *DBSCAN* has been proposed. Series of computation experiments for outlier detection in network traffic has been implemented. The experiments showed the applicability of distributed *DBSCAN* technique to the robust detection of point anomalies caused by various incidents in the network infrastructure of railway intelligent control system.

Keywords: Network anomaly detection · Point anomalies · Clustering
DBSCAN · Intelligent transportation system

1 Introduction

Anomaly detection [1] in telecommunications traffic has been a promising research area for computer networks applications. Traffic anomalies are one of the main reasons affecting the telecommunications systems stable operation. They could be caused by various incidents, such as network interfaces failures, errors in the network software, deliberate or not intentional users actions leading to traffic overloads. In some cases, the telecommunication infrastructure undergoes targeted computer cyber-attack in which ongoing traffic volumes increase suddenly. The survey [2] is available to discuss the

The work was financially supported by Russian Foundation for Basic Research (projects 16-01-00597-a, 16-07-00888-a, 17-07-00620-a, 18-01-00402-a).

A. Abraham et al. (Eds.): IITI 2018, AISC 875, pp. 267–276, 2019.
https://doi.org/10.1007/978-3-030-01821-4_28

methods and tools for network anomalies detection which are classified as computer attacks.

Despite the considerable research efforts and practical experience, the broadly used term "network anomaly" is not strictly defined up to nowadays. Perhaps, due to their complexity and diversity network anomalies cannot be unambiguously classified and there is no general approach to define the term mentioned above in the formal language. However, the vast majority of researchers agree with the definition that anomaly in *IP* networks [3] arises the deviation of traffic behavior from the normal one. Certainly, the typical traffic behavior should be considered as some pattern, and the outliers are recognized as anomalies. In any case, the consequences of network anomalies in telecommunication traffic lead to the decreasing the quality of service until the denial of service.

Safety-critical network infrastructures, for example, intelligent transportation systems require nonstop service both for cargo and passenger transportation. Intelligent transport systems are integrated systems for automating and controlling the management of vehicles, transport infrastructure and technological processes. In this area, the *JSC "Russian Railways"* has developed a large-scale project – a railway intelligent control system (*RICS*). The *RICS* has a distributed hardware and software computing facilities that generate a huge amount of telecommunication traffic to interchange data in the real time. Despite being separated from global telecommunications, the subsystems of the *RICS* are vulnerable to various malfunctions and failures. Due to the critical *RICS* infrastructure, each erroneous event must be registered, stored and processed as infrastructure incident, especially if it concerns the information security violation. Security incident detection methods which are adapted for use in *RICS* was discussed in [4–7] previously. It is worth to note that approaches to network anomaly detection in transport applications such as *RICS* have been attracted much attention in last years [8]. However, unfortunately, we cannot highlight the most appropriate general network anomaly detection technique for *RICS* because of JSC *"Russian Railways"* telecommunications complexity and traffic sources diversity. In this paper, we describe our progress in point anomaly techniques and concentrate our efforts on fresh new applications of clustering techniques in industrial and transportation area.

Point anomaly techniques are mainly classified to statistical and distance-based approaches. While the statistical techniques try to find out the point that does not fit into the pattern, the distance based approaches form clusters of data according to a similarity measure and point which do not belong to any clusters are characterized as noise points or outliers. Many clustering algorithms have been proposed in the literature like *k-means*, *DBSCAN*, *OPTICS* etc. [9, 10]. The main disadvantage of the clustering algorithms is their computational complexity which makes them inefficient to produce results in real time. In this work, *DBSCAN* employed to find out outliers of data, and since its computational complexity is significant, a distributed version of it was applied and tested [13]. *DBSCAN* applied on a computational cluster using the *Message Passing Interface* (*MPI*) [12].

MPI offers the necessary tools for the data transmission among the worker nodes of a computational cluster (send and receive operations), synchronization (barrier operation) and consolidation of the results received by the nodes (reduce operation). Also, *MPI* provides many functions to mine network's information such as the number of participating nodes, can establish several virtual network topologies while peer-to-peer and broadcast operations can work efficiently. It is worth to underline that the possible number of the participating working nodes is unlimited while different and geographically distributed computational cluster can co-operate. Hence, *MPI* can work efficiently on big data scaling the number of workers (nodes) when it is required.

The results obtained were very promising on speed making the technique very efficient while their quality was very similar to the original sequential *DBSCAN*.

The rest of the paper is organized as follows. In Sect. 2 clustering techniques are presented while the traffic flows in the part of the *RICS* is given in Sect. 3. In Sect. 4, the experimental results are presented, and Sect. 5 concludes the paper and highlights the future research directions.

2   Clustering Techniques

2.1   *K*-means Algorithm

K-means algorithm is a very popular and straightforward clustering technique (Algorithm 1) [9]. Besides the data set, *k*-means requires as input the number of the desired cluster (*k*) and a minimum distance in order to considered as neighbors two points. The measure of distance used in this work is the Euclidean distance.

Algorithm 1. Original *k*-means

Input: number of clusters *k*

Choose *k* points from the data set *D*. // these represent the initial centroids.

REPEAT

      Assign each point to its closest centroid

      Recalculate the centroids

UNTIL (no centroid changes)

Output: List of centroids

The main disadvantages of *k*-means are twofold. Firstly, it requires the number of clusters beforehand and secondly its computational complexity which makes the technique not appropriate to be applied to big data sets on time. The time complexity is $O(kNM)$, where k is the number of clusters, N stands for the size of data set, and M is the number of iterations needed to finalize the algorithm. *K*-means produces the k predefined clusters where all data points are assigned to one of them no matter their distance from the centroid and does not points any outliers. Therefore, *k*-means is not the right technique to produce noise (outliers) points. On the other hand, *DBSCAN* (Density-Based Spatial Clustering of Applications with Noise) produces high-quality clusters and additionally characterizes the points of the data set as the core, border, or noise (outliers). So, *DBSCAN* seems a suitable algorithm to find out the outliers.

2.2   DBSCAN Algorithm

The original *DBSCAN* algorithm [9] is presented in Algorithm 2. On the contrary to *k*-means, *DBSCAN* requires as input two parameters: (1) the minimum distance (*eps*) in order two points to be considered as neighbors, and (2) the minimum number of points which are accepted to form a cluster (*minPts*).

Algorithm 2. Original *DBSCAN*

Input: *minPts, eps*
for each data point d_i of D
    if d_i is unvisited then
        Mark d_i as visited
        NumPts ← explore the rest of unvisited data points and examine how many and which of them are neighbors to d_i
        if *NumPts* < *MinPts*
            Mark d_i as Noise
        else
            Generate new cluster containing d_i and its neighbors
            Mark d_i as Core
            Determine all the rest data points that belong to the cluster if they are either core or border points
        endif
    endif
endfor

for each data point d_i characterized as noise
    Examine if d_i belongs to any cluster
    if d_i belongs to a cluster
        Unmark d_i as Noise
    endif
endfor
Output: List of clusters

The main advantages of *DBSCAN* are that understands the outliers, and that can find clusters of arbitrary shape. On the other hand, it is complicated to define *eps*, and the combination of *eps* and *minPts* is even more difficult especially when data is unknown or of high dimensionality (the curse of dimensionality). Finally, its computational complexity varies from $O(N\log N)$ to $O(N^2)$ depending on the meaningful or not choice of *eps*. This computational complexity makes the technique not acceptable when applies to big data sets and the results are need on the fly. To overcome this and speeds up the algorithm a fully distributed version of it is presented.

2.3   Parallel *DBSCAN*

The distributed version of *DBSCAN* is presented in Algorithm 3 [11]. The main idea is to split the data set into $P-1$ subsets (P represents the number of participating nodes in the computational cluster) and then to apply *DBSCAN* independently to these nodes. In addition to the original *DBSCAN*, each node calculates the centroids of the local clusters and their radiuses. This information is sent to its neighbor node which in turn produces the new clusters (and centroids – radiuses). This procedure is repeated until all required information has been sent to the final node which produces the global clusters. More detailed information about the technique can be found in [11].

Algorithm 3. Distributed *DBSCAN*

Input: *minPts, eps*

Master node, v_0 collects the number of available worker nodes, $(P\text{-}1)$

v_0: splits data set D into $D/(P\text{-}1)$ subsets

v_0: transfers data, *eps* and *minPts* to worker nodes

for all worker nodes in parallel

    Receive data set

    Apply *DBSCAN* algorithm

endfor

$i \leftarrow myid$ //node identification number

while $(i \leq \log(P\text{-}1))$

    forall worker nodes in parallel

      if $(i \bmod 2 = 1)$

        Send the local centroids, and the corresponding radius's to v_{i+1} //sender

      else

        Receive from v_{i-1} //node is a receiver

        Reform clusters

      endif

    endfor

    $i \leftarrow 2^i - 1$

endwhile

Last worker node transfers the clusters to the master node

Output (from master node): List of cluster

The computational complexity (worst case) of the proposed technique is $O(N^2/(P{-}1) + C)$ where C represents the communication overhead, and P is the number of participating nodes [11]. This proves that the technique could apply to large data sets and to produce results very fast. Also, the results obtained are similar to what the original sequential algorithm produces proving the quality of the technique.

3   Traffic Flows in *RICS*

In our study, we consider a part of corporate *JSC "Russian Railways"* telecommunication and computing facilities of the *RICS* that provide the primary ticket-selling service – the *ITS "Express"* server. Figure 2. shows traffic flows, hardware and software of the *ITS "Express"* server, virtual machines and configured internal names of the network interfaces:

(1)   input (800) and output (A000) connections with high-level user's applications through *HTTP* port 80;

(2)   input (B000) and output (D000) connections with middle-level users applications through *HTTP* port 81;

(3) output (OSA_00) connection with the Email server by *SMTP* port 25;
(4) input (CPU) connection by *SSH* port 22 and output (CISCO_5000) and (CISCO_6000) connections through ports 3306, 3307 with network and database administrator's workplaces.

Traffic tracing logs from network ports mentioned above were collected by `tcpdump 4.9.2` [12] utility of the `Linux CentOS 7.2` operating system.

4 Point Anomaly Detection in *IP*-Networks Data Using Distributed *DBSCAN*

4.1 Experimental Results

For the experiments of this work, 20 computational nodes were used (`AMD@ FX(tm) – 4100 Quad-Core Processorx4`, with `UBUNTU 14.04 LTS` operating system, `MPICH 3.04`, `gcc 4.84`, and Ethernet 100 Mbit/s). Traffic tracing logs from network internal ports 800, A000, B000 and D000 were preprocessed and we have obtained traffic volumes registered every 10 min on each port. The traffic measurements were performed during a day of the month (from 0:00 am until 23:50 pm, 144 time points) and from *January* 2017 until *November* 2017. In the next step the traffic measurements were refined to the raw datasets suited to the clustering techniques. One raw dataset consists of 4 numerical attributes meaning input/output data volumes from selected network ports (Fig. 1, 800, A000, B000, D000) of *ITS* "*Express*" server.

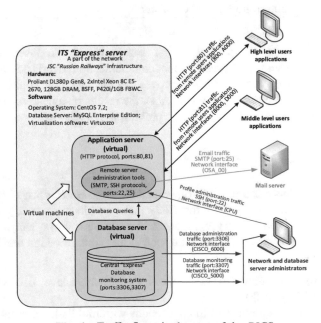

Fig. 1. Traffic flows in the part of the *RICS*

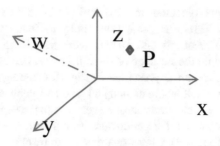

Fig. 2. 4d space

So, the data space is consisting of 4-dimensional points. Since we have a 4d space in order to clarify the following figures, we present for each case 4 graphs where each one of them represents the orthogonal projection of the 4th dimension to a 3d space. In all graphs, we assume that the representation of any data point according its coordinates (or characteristics) is of the form $P(x_P, y_P, z_P, w_P)$ (Fig. 2). The distance measure used in this work is the Euclidean distance.

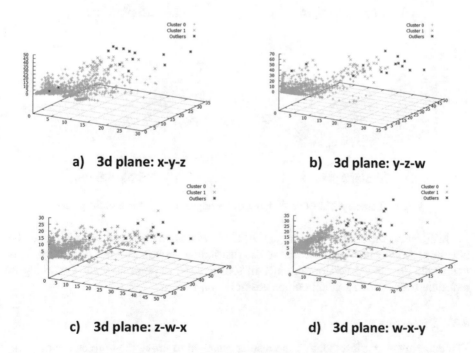

a) **3d plane: x-y-z**   b) **3d plane: y-z-w**

c) **3d plane: z-w-x**   d) **3d plane: w-x-y**

Fig. 3. November 3rd, Orthogonal projections to 3d planes

In Fig. 3, the results obtained for November the 3rd 2017 are presented in orthogonal projections. The four projections help to identify precisely the point anomalies (outliers). Different *eps* and *MinPts* were tested and in all cases only one cluster was produced while the number of the outliers was small.

In addition, a whole month is presented in Fig. 4. Once again, the set of outliers (point anomalies) is very small where actually 11 point anomalies detected in the whole month. Again, the number of clusters was one even we run the experiments testing rather large set of different eps and *MinPts* parameters. In only one case (February 2017) two clusters were produced (Fig. 4) where again similar results related with the small number of outliers were detected.

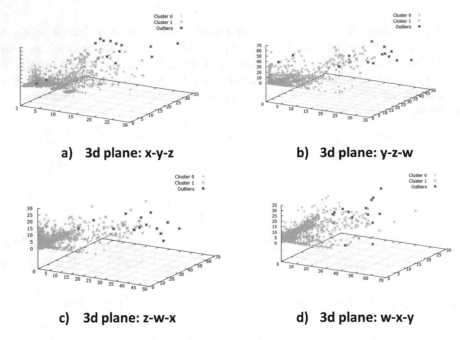

Fig. 4. February 2017 (whole month), Orthogonal projections to 3d planes

In all cases, the original sequential *DBSCAN* was tested also in order to compare the obtained results which were similar to the distributed version proving its quality. This makes the distributed version an efficient tool to explore point anomalies working in real time because of its small computational complexity.

4.2   Discussion

The experiments performed, and the results obtained to prove the correctness of using *DBSCAN* for detecting point anomalies in *IP*-networks data. The outliers checked and represent real anomalies. In addition, the clusters show that the data is smooth and the fact that in almost all cases one cluster and in very few one-two clusters produced even when different input parameters were used prove that a density-based clustering technique is the right tool of detecting point anomalies on this type of data.

From the technical point of view, the obtained clustering results are agreed with data that have been acquired from open-source software *Apache ZooKeeper* [14] installed on *ITS "Express"* server. The *Apache Zookeeper* provides coordination services for a group of distributed applications in *RICS* and performs load-balancing of queries from distributed network clients by service discovery and queries blocking features. In the time periods have been examined in the previous section if the number of outliers trend to rise then it leads to performance decrease of *ITS "Express"* virtual server applications and the number of blocked queries from network clients is increasing.

5 Conclusions

In this work, a distributed version of *DBSCAN* was used to determine point anomalies in traffic of *IP*-networks. The traffic flows in ticket-selling service of *RICS* was examined. It is essential, but not so complicated part of the huge corporate *JSC "Russian Railways"* telecommunication and computing network. It should be noted despite the fact that this part of the *RICS* is operating in the real-time, it is processing the traffic amounts which is not so huge as the subsystems involved into primary railway transportation processes. Thus we intend to extend our study in the future work.

Future work will focus on the following aspects:

- Firstly, we will try to include all the characteristics of the *IP*-networks producing a multi-dimensional space which in turn increases the computational complexity of the techniques. To overcome the "curse of dimensionality" several techniques will be explored like *Principal Component Analysis* (*PCA*), *Non-negative Matrix Factorization* (*NMF*), etc., to decrease both dimensionality and consequently the complexity while keeping high the quality of the network attributes.
- Nowadays, *Graphical Processing Units* (*GPU*) became *General Purpose GPUs* containing large numbers of cores at reasonable prices. The development of a variation of the distributed *DBSCAN* applying on *GPUs* will be examined to further increase its efficiency and the possibility to embed it on the hardware examined also.
- Moreover, at the third, we will consider more complex nature of the *IP* telecommunication traffic, trying to predict outliers rising under various system operating conditions. Traffic prediction features in complex corporate control systems, in the *RIMS* as an example, allows implementing necessary functions in the controlling of high loaded distributed network systems, such as performance balancing and service priority tasks.

References

1. Chandola, V., Banerjee, A., Kumar, V.: Anomaly detection: a survey. ACM Comput. Surv. **41**(3), 15:1–15:58 (2009). https://doi.org/10.1145/1541880.1541882
2. Bhuyan, M.H., Bhattacharyya, D.K., Kalita, J.K.: Network anomaly detection: methods, systems and tools. IEEE Commun. Surv. Tutor. **16**(1), 303–336 (2014). https://doi.org/10.1109/SURV.2013.052213.00046
3. Thottan, M., Ji, C.: Anomaly detection in IP networks. IEEE Trans. Signal Process. **51**(8), 2191–2204 (2003). https://doi.org/10.1109/TSP.2003.814797
4. Chernov, A.V., Butakova, M.A., Vereskun, V.D., Kartashov, O.O.: Mobile smart objects for incidents analysis in railway intelligent control system. Adv. Intell. Syst. Comput. **680**, 128–137 (2017). https://doi.org/10.1007/978-3-319-68324-9_14
5. Chernov, A.V., Butakova, M.A., Karpenko, E.V.: Security incident detection technique for multilevel intelligent control systems on railway transport in Russia. In: 2015 23rd Telecommunications Forum Telfor (TELFOR), pp. 1–4 (2015). https://doi.org/10.1109/telfor.2015.7377381
6. Chernov, A.V., Bogachev, V.A., Karpenko, E.V., Butakova, M.A., Davidov, Y.V.: Rough and fuzzy sets approach for incident identification in railway infrastructure management system. In: 2016 XIX IEEE International Conference on Soft Computing and Measurements (SCM), St. Petersburg, pp. 228–230 (2016). https://doi.org/10.1109/scm.2016.7519736
7. Chernov, A.V., Kartashov, O.O., Butakova, M.A., Karpenko, E.V.: Incident data preprocessing in railway control systems using a rough-set-based approach. In: 2017 XX IEEE International Conference on Soft Computing and Measurements (SCM), St. Petersburg, pp. 248–251 (2017). https://doi.org/10.1109/scm.2017.7970551
8. Butakova, M.A., Chernov, A.V., Shevchuk P.S., Vereskun, V.D.: Complex event processing for network anomaly detection in digital railway communication services. In: 25th Telecommunication Forum (TELFOR), Belgrade, pp. 1–4 (2017) https://doi.org/10.1109/telfor.2017.8249273
9. MacQueen, J.: Some methods for classification and analysis of multivariate observations. In: Proceedings of the Fifth Berkeley Symposium on Mathematical Statistics and Probability, vol. 1: Statistics, University of California Press, Berkeley, pp. 281–297 (1967). https://projecteuclid.org/euclid.bsmsp/1200512992
10. Ester, M., Kriegel, H.-P., Sander, J., Xu, X.: A density based algorithm for discovering clusters in large spatial databases with noise. In: Proceedings of the KDD 1996, pp. 226–231 (1996). https://www.aaai.org/Papers/KDD/1996/KDD96-037.pdf
11. Savvas, I.K., Tselios, D.: Parallelizing DBSCAN algorithm using MPI. In: 2016 IEEE 25th International Conference on Enabling Technologies: Infrastructure for Collaborative Enterprises (WETICE), Paris, pp. 77–82 (2016). https://doi.org/10.1109/wetice.2016.26
12. Tcpdump & libpcap (command-line packet analyzer and library for network traffic capture), April 2018. https://www.tcpdump.org/
13. MPICH (High-Performance Portable Message Passing Interface), April 2018. https://www.mpich.org/
14. Apache ZooKeper (centralized service for maintaining configuration information, naming, providing distributed synchronization, and providing group services), April 2018. https://zookeeper.apache.org/

Diagnosing of Devices of Railway Automatic Equipment on the Basis of Methods of Diverse Data Fusion

Anna E. Kolodenkova[1(✉)] and Alexander I. Dolgiy[2]

[1] Samara State Technical University, Samara, Russia
anna82_42@mail.ru
[2] Rostov State Transport University, Rostov-on-Don, Russia
adolgy@list.ru

Abstract. In the this work it is emphasized that fusion of the diverse data obtained from sources of primary information (sensors, the measuring equipment, systems, subsystems) for adoption of diagnostic decisions at a research of malfunctions of devices of railway transport, is one of the main problems. The generalized scheme of fusion of diverse data reflecting features of this process is considered. Also classification of levels, modern methods of fusion of diverse data in the conditions of incomplete, indistinct basic data is considered. Approach to fusion of diverse data on malfunction of the devices of railway transport received from a set of various sensors with use of the theory of Dempster-Shafer for the purpose of their integration and development of uniform diagnostic decisions for the benefit of end users is offered. Rationing of the weight coefficients reflecting ability of sensors, and fusion of values of mass of probability is the cornerstone of the offered approach. A numerical example for a decision-making illustration at diagnostics of malfunctions of devices of railway transport in the conditions of uncertainty is reviewed.

Keywords: Fusion of diverse data · Normalization of data
Approaches and methods of fusion of data · Railway transport

1 Introduction

The effective centralized fusion of the diverse diagnostic data obtained from various sensors and touch systems is for railway transport of one of the main and serious problems which solution can render considerable effect, as on economic indicators of functioning of railway infrastructure, and on the level of safety of rail transportation in general [1–4]. The analysis of current trends of development of domestic and foreign information and diagnostic systems confirms small efficiency of the traditional approaches and methods to the solution of the specified problem based on the basic data having numerical character. In this regard in the present article for adoption of

---

The work was supported by RFBR grants No. 17-08-00402-a.

A. Abraham et al. (Eds.): IITI 2018, AISC 875, pp. 277–283, 2019.
https://doi.org/10.1007/978-3-030-01821-4_29

diagnostic decisions approach to fusion of diverse data on malfunction of devices of railway transport in the conditions of indistinct basic data is offered.

2 The Generalized Scheme of Fusion of Diverse Data

At the heart of the offered approach to fusion of the diverse data obtained from a set of various sensors use of the following generalized scheme reflecting features of this process submitted in Fig. 1.

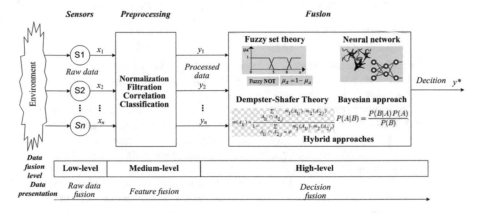

Fig. 1. The generalized scheme of fusion of diverse data

From Fig. 1 it is visible that fusion of data can be classified on three levels [5]:

1. Low level of fusion. This level is often called the level of the raw data (raw data level) or level of signals (signals level). The raw data are considered as entrance data, then unite. As a result of association it is expected to obtain new more exact and informative data, than the raw entrance data. For example, in work [6] the example of low-level fusion to use of the filter of moving average is given.

2. Medium level of fusion. This level is called – the level of signs (attributes, characteristics) (feature level). There is a fusion of signs (a form, texture, edges, corners, lines, situation) as a result of which new objects, or cards of objects which can be used for other problems, for example, of segmentation and recognitions turn out. Also at this level there is a data processing, namely a filtration (fight against noisy data), normalization (transformations to one type of data), correlation, classification of data, to use of methods of "soft calculations" and methods of data mining. Examples of medium-level fusion are given in works [7, 8].

3. High level of fusion. This level is called – the level of the decision (decision level) or level of symbols (symbol level). There is a fusion at the level of the decision as a result of which the global decision turns out. The most traditional and known methods of fusion of data are probabilistic methods (Bayesian networks, the theory of proofs); computing intellectual methods (Dempster-Shafer, theory of indistinct

sets and neural networks). These methods allow to present the coordinated and uniform opinion on diagnostic process to the person making the decision. High level fusion is considered, for example, in work [9].

3 Classification of Levels and Methods of Diverse Data Fusion

Now rather large number of methods is developed for fusion of data. However at the choice of this or that method it is necessary to consider some aspects (what are of fusion methods approach the available data better and will allow to achieve the best results at the solution of objectives?; what is preliminary processing necessary?; how to choose from a data set those which fusion will give the best effect?, etc.).

On the basis of systematization of the review of references in Fig. 2, classification of levels and modern methods of diverse data fusion in the conditions of incomplete, indistinct basic data is presented.

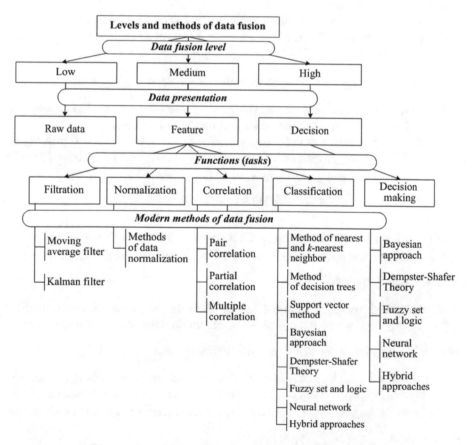

Fig. 2. Classification of levels and methods of data fusion

Let's note that the given division of methods of fusion of the diagnostic decisions given for acceptance as conditional character as in practice they are crossed and interact among themselves.

4 Statement and Algorithm of the Solution of a Task Adoptions of Diagnostic Decisions

Let there are N raw data in the form of various types of malfunctions of devices (x_1, x_2, ..., x_i, where $x_i - i$-th the malfunction type describing a condition of the device $i = 1$, ..., N), received from various S_1, S_2 sensors, ..., S_j ($j = 1,..., M$). Let's assume that all malfunctions are independent from each other, and only one malfunction can occur at any moment. It is required to carry out troubleshooting of a condition of devices, in the conditions of a set of various sensors.

For the solution of this task it is possible to use the following three-stage procedure.

At the first stage the degree of trust $\beta_j(x_i)$ for sensor j-th allowing to estimate conditions of xi devices pays off ($i = 1,..., N$, $j = 1,..., M$), for the purpose of obtaining new information [10].

At the second stage the mass of probability of $m_j(x_i)$ of diagnostics of malfunctions of each device pays off with use of the normalized weight coefficient reflecting ability (sensitivity) of the sensor $w_j^{nor} = [w_{j1}, w_{j2}..., w_{ji}]$ (w_{ji} – a measure of similarity (similarity) between i-th type of malfunction and j-th the sensor $\sum_{j=1}^{M} w_{ji} = 1, 0 \leq w_{ji} \leq 1$), created on the basis of some prior knowledge about each sensor it is also presented in the form of intervals, indistinct triangular and trapezoid numbers.

The mass of probability of $m_j(x_i)$ can pay off with use of the calculated trust degree $m_j(x_i) = w_j^{nor} \beta_j(x_i)$ [10] or with use of assessment of a condition of the device (distance between the S_j sensor and a condition of x_i: $m_j(x_i) = w_j^{nor} \frac{1}{d(x_i, S_j)}$ [11, 12]).

Rationing of weight coefficients can be carried out as follows $w_j^{nor} = \frac{w_{ji}}{\sum_{j=1}^{M} w_{ji}}$ [13]:

If normalized weight coefficient in the form of an interval $w_j^{nor} = [\alpha_{j1}, \alpha_{j2}]$, where α_{j1}, α_{j2} – the minimum and maximum values of an interval, at the same time it is necessary that $\sum_{j=1}^{n} \alpha_{j1} < 1, \sum_{j=1}^{n} \alpha_{j2} > 1$.

If normalized weight coefficient in the form of indistinct triangular number $w_j^{nor} = [\alpha_{j1}, \alpha_{j2}, \alpha_{j3}]$, where $\alpha_{j1}, \alpha_{j2}, \alpha_{j3}$ – the minimum, most expected and maximum values of an interval, at the same time it is necessary that $\sum_{j=1}^{n} \alpha_{j1} < 1, \sum_{j=1}^{n} \alpha_{j3} > 1$.

If normalized weight coefficient in the form of indistinct triangular number $w_j^{nor} = [\alpha_{j1}, \alpha_{j2}, \alpha_{j3}, \alpha_{j4}]$, where α_{j1}, α_{j4} – pessimistic and optimistic estimates of borders of intervals, $[\alpha_{j2}, \alpha_{j3}]$ – the interval of the most expected values, at the same time is necessary that $\sum_{j=1}^{n} \alpha_{j1} < 1, \sum_{j=1}^{n} \alpha_{j4} > 1$.

At the third stage fusion (combination) of values of mass of probability of $m_1(x_i) \oplus m_2(x_i) \oplus ... \oplus m_j(x_i)$ [14].

On the fourth step troubleshooting of a condition of devices is carried out.

In case as weight coefficients numbers were considered, as malfunction of a condition of devices that condition of the device for which the combined value of mass of probability is maximum among similar values is chosen.

If weight coefficients were presented in the form of intervals or indistinct triangular and trapezoid numbers, then as a result of fusion of value of mass of probability are also presented in the form of intervals, or indistinct triangular and trapezoid numbers. In this case troubleshooting of a condition of devices is carried out on the basis of ordering of intervals, or indistinct triangular and trapezoid numbers, according to their ranks. For comparison of indistinct trapezoid numbers the Chyyu-Park, Chiang, Kauffman-Gupt methods are offered [14]. In case of indistinct triangular numbers $m_j(x_i) = [m_j(x_{i1})$, $m_j(x_{i2})$, $m_j(x_{i3})]$, $m_j(x_{i1})$, $m_j(x_{i2})$, $m_j(x_{i3})$ – where the minimum, most expected and maximum values of an interval) the choice is carried out on the expected value of an interval.

As malfunction of a condition of devices that condition of the device which has the greatest expected value is chosen.

5 Example of Troubleshooting of a Condition of Railway Transport Devices

Further we will review an example of troubleshooting of a condition of devices.

There are five conditions of devices $(x_1, x_2, x_3, x_4, x_5)$ and four sensors (subsystem) S_1, S_2, S_3 from which basic data (Table 1) are obtained. It is required to carry out troubleshooting of a condition of devices.

Table 1. Calculated values

Conditions of devices	Sensor S_1	Sensor S_2	Sensor S_3
x_1	3	36	6
x_2	17	57	27
x_3	8	14	18
x_4	12	109	32
x_5	14	84	47

Let's note that sources of primary information on the basis of which diagnostic decisions are passed are the following sensors (subsystems):

1. The subsystem "Tekhnovizor" (S_1) intended for visual diagnosing of malfunctions of elements of the rolling stock. The measured sizes are determined on a basic scale of deviations of the sizes of frictional wedges from the norm expressed in mm [0–20]. Examples of normal values: 1 mm, 3 mm, 4 mm; examples of violations of the mode: 8 mm, 12 mm.

2. The subsystem of laser control of negative dynamics (S_2) intended for identification of fluctuations of bodies and violations of dimensions of the rolling stock the Measured sizes is defined on a basic scale of the distances characterizing amplitude of fluctuations in cm: [1–100]. Examples of normal values: 1 cm, 2 cm, examples of violations of the mode: 50 cm, 120 cm.
3. Subsystem of thermal control of axle knots (S_3). The measured sizes are determined on the basic scale of temperatures in °C characterizing excess of temperature of axle knot over ambient temperature °C: [10–50]. Examples of normal valcs: 3 °C, 12 ° C, examples of an overheat: 19 °C, 27 °C.

The following weight coefficients in the form of indistinct trapezoid numbers are known: $w_1 = [3, 3.5, 4]$, $w_2 = [3, 3.8, 4.5]$, $w_3 = [1, 2, 2.3]$.

Further normalized weight coefficients pay off: $w_1^{nor} = [0.28, 0.37, 0.57]$, $w_2^{nor} = [0.28, 0.4, 0.64]$, $w_3^{nor} = [0.09, 0.22, 0.33]$.

For S_1: $m(\theta) = 0.233$; S_2: $m(\theta) = 0.068$; S_3: $m(\theta) = 0.489$.

Results of calculation of mass of probability and also their fusion for each state on the basis of weight coefficients are presented in Table 2.

Table 2. Calculated values of mass of probability

Conditions of devices	$\beta_1(x_i)$	$m_1(x_i)$	$\beta_2(x_i)$	$m_2(x_i)$	$\beta_3(x_i)$	$m_3(x_i)$	$m_1(x_i) \oplus m_2(x_i)$ $\oplus m_3(x_i)$
x_1	0.115	0.043	0.305	0.122	0.107	0.023	0.07
x_2	0.654	0.242	0.483	0.193	0.482	0.106	0.16
x_3	0.308	0.113	0.119	0.048	0.321	0.071	0.03
x_4	0.462	0.17	0.923	0.369	0.571	0.126	**0.3**
x_5	0.538	0.199	0.712	0.2	0.839	0.185	0.2

Where $m_1(\theta) \oplus m_2(\theta) \oplus m_3(\theta) = 0.24$, $m_1(\varnothing) \oplus m_2(\varnothing) \oplus m_3(\varnothing) = 0$.

From Table 2 it is visible that failure condition of devices of railway transport cannot be found by separate consideration of three sensors. However after fusion of data, comes to light that failure condition is x_4 weighing probability 0.3.

6 Conclusions

In this work approach to fusion of diverse data on malfunction of the devices of railway transport received from various sensors with use of the theory of Dempster-Shafer which cornerstone rationing of the weight coefficients reflecting ability of sensors and fusion of values of mass of probability is offered. Results of the offered approach showed that if the studied conditions of devices are chosen truly, and exact data from sensors are obtained, then approach to diagnostics of malfunctions of devices in the presence of several sensors allows to find malfunction mode, to thereby make the scientifically based diagnostic decision. At the same time the person making the decision does not need to know the principle of work and structure of devices.

References

1. Bevilacqua, M., Tsourdos, A., Starr, A., Durazo-Cardenas, I.: Data fusion strategy for precise vehicle location for intelligent self-aware maintenance systems. In: International Conference on Intelligent Systems, Modelling and Simulation, pp. 76–81 (2015)
2. Dolgiy, A.I., Dolgiy, I.D., Kovalev, V.S., Kovalev, S.M.: Intellectual models of the nonlinear filtration of data in fiber-optical systems of gathering and processing of the primary information. News Volgograd State Tech. Univ. **9**, 63–68 (2011). (in Russian)
3. Reimer, C., Hinüber, E.L.: INS/GNSS/Odometer data fusion in railway applications. In: Symposium Inertial Sensors and Systems, Karlsruhe, Germany, p. 14 (2016)
4. Veloso, M., Bentos, C., Camara Pereira, F.: Multi-sensor data fusion on intelligent transport systems. MIT Portugal Transportation Systems Working Paper Series, p. 18 (2009)
5. Ben Brahim, A.: Solving data fusion problems using ensemble approaches, p. 104 (2010)
6. Polastre, J., Hill, J., Culler, D.: Versatile low power media access for wireless sensor networks, pp. 95–107 (2004)
7. Nowak, R., Mitra, U., Willett, R.: Estimating inhomogeneous fields using wireless sensor networks. IEEE J. Sel. Areas Commun. **22**, 999–1006 (2004)
8. Zhao, J., Govindan, R., Estrin, D.: Residual energy scans for monitoring wireless sensor networks. In: IEEE Wireless Communications and Networking Conference, vol. 1, pp. 356–362. IEEE, Orlando (2002)
9. Krishnamachari, B., Iyengar, S.: Distributed Bayesian algorithms for fault-tolerant event region detection in wireless sensor networks. IEEE Trans. Comput. **53**, 241–250 (2004)
10. Pasha, E., Mostafaei, H.R., Khalaj, M., Khalaj, F.: Fault diagnosis of engine using information fusion based on Dempster-Shafer theory. J. Basic Appl. Sci. Res. **2**(2), 1078–1085 (2012)
11. Mostafaei, H.R., Khalaj, M., Khalaj, F., Khalaj, A.H., Makui, A.: Engine fault diagnosis decision-making with incomplete information using Dempster-Shafer theory. J. Basic Appl. Sci. Res. **2**(1), 105–113 (2012)
12. OtmanBasir, X.Y.: Engine fault diagnosis based on multi-sensor information fusion using Dempster-Shafer evidence theory. Inf. Fusion **8**, 379–386 (2007)
13. Kolodenkova, A.E.: The process modeling of project feasibility for information management systems using the fuzzy cognitive models. J. Comput. Inf. Technol. **6**(114), 10–17 (2016). (in Russian)
14. Dempster, D., Shafer, G.: Upper and lower probabilities induced by a multi-valued mapping. Ann. Math. Stat. **38**, 325–339 (1967)

Technical Aspects of the "Digital Station" Project

Alexander N. Shabelnikov and Ivan A. Olgeyzer[✉]

JSC "NIIAS", Rostov Branch, Rostov-on-Don, Russia
olgeyzer@rfniias.ru

Abstract. JSC NIIAS performed a great work for integration of modern minimally manned technologies into automated hump yards. It is proposed to create a system – "digital station" – which provides generalization of initial information obtained from all devices functioning on a station. Proposed system also provides the consistency verification, elimination of information redundancy and formation of the real-time information models about trains and cars inside marshalling yard. Technical equipment of digital station can provide several aspects. Firstly, it provides complex automation of technical operation control over marshalling yard in real-time mode based on merging of low-level automatics and information planning systems. Secondly, it eliminates the hand input of the information about operations connected with car motion inside station. As well, it optimizes the hand input of the information about technological operations, beginning and end of which can be fixed only by using hand input of information by operational-dispatch or maintain staff via the automated workplaces. Technical aspects of digital station are proposed to be implemented based on system of control and preparation of information about moving of cars and locomotives inside station in real-time mode (SKPI PVL RV).

Keywords: Marshalling yards · Minimally manned technology
Security of breaking-up and makeup of trains · Digital station
Automated systems · Mobile applications for railways
Paperless technologies for railways

1 Introduction

Nowadays, government of JSC "Russian Railways" requires increasing of work productivity, reducing of operational fees, transition to minimally manned technologies in marshalling yard operations together with security increasing during breaking-up and making up of trains [1]. Minimally manned technologies mean the technological operations execution almost without human operator participation. It can be done only using the complete automation of marshalling processes from car entering into yard before it leave it inside a new train.

Assessment of marshalling yards operations is made based on qualitative and quantitative indicators [2], such as number of cars or trains under operation per time unit, time of operation for one car, standstill time of a car, etc.

Herein, increased attention is made for necessity of concentration of sorting operations on large marshalling yards providing operations with high efficiency together with fulfillment of strict requirements for car standstills.

2 Problem Statement

During last years, JSC "NIIAS" have been performing a great work for implementation of modern minimally manned technologies into automated marshalling yards [3, 4].

In this way, it was actually noted about success in digital hump creation at Chelyabinsk-glavniy station in 2017 [5].

The most specific features of technical equipment of such hump are the following:

- Real-time dynamic model of state of all outdoor devices together with rolling stock position [6];
- Wide usage of paperless technologies and hand input reduction;
- Humping control by one operator (including humping from parallel tracks);
- Moving to hump and breaking-up on hump in unmanned mode using Automatic Cab Signalling system (MALS);

To minimize hand input of the information about train accumulations and exchanges on sorting lines, extended information exchange is implemented between integrated automation system for marshalling process control (KSAU SP) and automated system for station operations control (ASU ST).

Example of paperless technology implementation is developed mobile application "electronical dynamic marshalling sheet" for mobile phones and tablets [7] (Fig. 1).

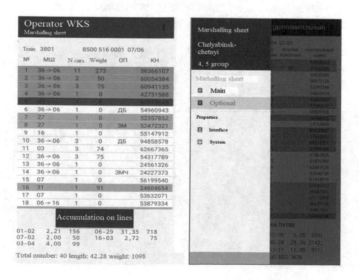

Fig. 1. Electronical dynamic marshalling sheet

Moreover, the trial model of intelligent hump console was presented on Inskaya station in 2016 (Fig. 2). Such console allows to control humping operation at humps of any capacity with any configurations by one operator and provides checking of manual intervention for correctness and permissibility. In Fig. 2, the possibility of interactive control over switches by using the corresponding buttons on display is shown in the upper picture. The bottom picture shows the complete shunting route (when start and terminal points are chosen), which is automatically built by the console.

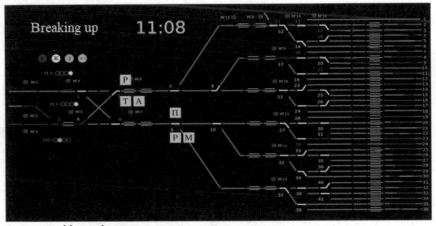

Automated humping
Control over switches, retarders and light signals
Control over shunting operations
Control over devices switching into "automate" mode
Reduction of faultes connected with human factor
Quick reconfiguration

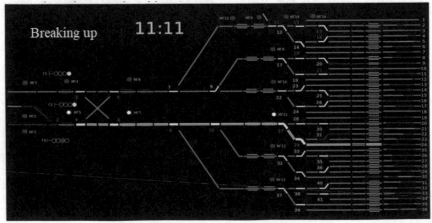

Fig. 2. Screenshots of intelligent hump console

Increasing of marshalling yard capacities together with increasing of traffic volumes require complex analysis of operations of separate technological processes on station. Bottle neck detection in marshalling yard operations allows to correct business-processes, which significantly affect to operation indicators.

Nowadays, systems providing construction of information models about trains and cars on station and formation of main indicators of marshalling operations, in partic-ular, separated standstill data formation, are implemented on Russian Railways. The general drawback of such systems is hand input of the information about hump operations.

Various devices and systems for automation of control are implemented for many marshalling yards (In Russia, these systems are SAI PS [8], subsystems of hump control of KSAU SP [9], MALS [10], ASKOPV [11], set of control and diagnosis systems – ADK SCB, APK DK [12], etc.). Herein, no one of these systems can provide forming of complete and adequate information model about trains and cars in real-time mode. It is caused by lack of completeness and self-sufficiency of information in each of these systems. For example, SAI PS does not control over cars, which do not equipped by RFID sensors and provides only control at input or output of station, MALS is able to control only shunting locomotives mounted with specific on-board equipment, hump subsystems can control only hump zone, etc.

3 Proposed Approach

In aim to increase the efficiency of marshalling operations, it is proposed to start the project for integration of separated systems of marshalling automation into unified complex to create digital station.

It is proposed to create a system, which provides generalization of initial infor-mation obtained from all automation systems, its consistency verification, elimination of its redundancy and formation of the real-time information models about trains and cars.

To construct such system, key principles must be established as the following conditions:

- Station covering by control devices to provide the possibility of digital modelling of station;
- Permanent networking and state monitoring for all devices, mechanisms, rolling stock and human staff (vests, mobile devices) to provide digital model of station;
- Permanent digital modelling at several levels, i.e. actual state of outdoor devices, rolling stock position, locomotive position, staff position.

Technical equipment of digital station can provide complex automation of control over marshalling operations in real-time mode based on merging of low-level auto-matics and information planning systems, hand input elimination for operations con-nected with rolling stock motion inside station and hand input optimization of the hand input of the information about technological operations, beginning and end of which can be fixed only by using hand input of information by operational-dispatch or maintain staff via the automated workplaces.

Therefore, technically digital station should include:

- Object controllers – to diagnose and fix changes and control state of signals and interlocking devices, which are used in processes of receiving, breaking-up and makeup of trains, which are not covered by automated humping systems and MALS.
- Central block of information modelling, gathering and processing – to model the trains, cars and locomotives inside station based on flows of analogue, discrete and digital information from neighboring low-level automatics and information planning systems. As well, modelling is based on referenced data about rolling stock, station parameters, guidelines and operation types on stations.
- Subsystem of analysis and logging – to store data of information models about trains, cars and locomotives on station during last period, to process data and to compute basic indicators of station operations with preparation of analytical reports.
- Automated workplaces (WKS) – to perform online information about current technological situation on marshalling yard for dispatching staff and executives of station and traffic service, to perform online reports about current situation on station by user query, to perform analytical report for last period. Competence limitations and access rights for the certain staff (user of WKS) should define volume of performed information.

4 Experimental Results

Technical aspects of digital station are proposed to be implemented based on system of control and preparation of information about moving of cars and locomotives inside station in real-time mode (SKPI PVL RV) [13], which provides implementation of the following functions:

- Tracking and registration of rolling stock motions inside station;
- Transfer of information about accumulation changes on lines of marshalling yard to the higher-level systems;
- Transfer of information about beginning and end of operations on station lines to the higher-level systems;
- Logging of registered motions on station;
- Logging of messages between the system and other automated control systems;
- Displaying of current location of a rolling stock on WKS of the system and on indicator board of station.

Structure of SKPI PVL RV is illustrated in Fig. 3. In Fig. 3, OC is the object controller, CMB is the central modelling block, DTN is the data transfer network of railway services (particularly, Russian railways).

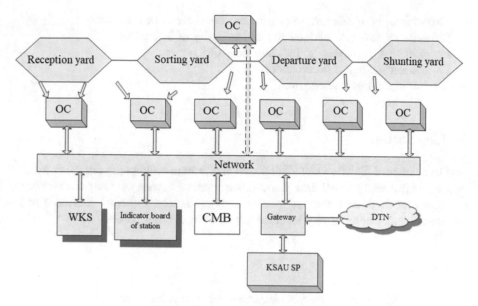

Fig. 3. Structure of SKPI PVL RV

Functions of central modelling block is the following:

- Modelling of state of all outdoor devices on station;
- Modelling of position of rolling stocks on stations for all levels (car, locomotive, shunting group, train). The model should include information about type, dimensions, rolling stock state using data obtained from automated point of receive (PPSS) at the input of marshalling yard;
- Modelling of positions of all staff with electronic tracking devices, which are in dangerous areas;
- Storing of obtained information in database;
- Analysis of statistics and formation of statistic reports providing detection of bottle necks and calculation of backup possibilities [14];
- Transfer of data about moving cars and locomotives, as well as about outdoor devices state, to higher level systems (e.g. automated control system of station) to form the graph reflecting finished operations;
- Diagnosis and displaying of recommendations about required technical service and its times for one or another station process.

Nowadays, SKPI PVL RV is implemented on Bekasovo-Sortirovochnoye yard of Moscow district of Russian railways. There, SKPI PVL RV are built using the following principles:

- Modular structure;
- Possibility of interaction with other information control systems by connection of the corresponding modules;
- Quick access for data with required level of detailing;

- Provision of completeness, reliability and consistency of information;
- Provision of safety, confidentiality and integrity of information;
- Possibility of displaying on indicator board of station;
- Centralized storing of information;
- Electronic workflow integration.

5 Conclusion

Due to dynamical digital model, digital station allows to build graph reflecting finished operations just using actual data. Information input of human operator can be only in correcting mode when automated control tools are absent. Principal difference of the system is impossibility of physical changes in technological process.

References

1. Shipulin, N.P., Shabelnikov, A.N.: Increase in safety of processes of sorting of railway cars. In: Automatic Equipment, Communication, Informatics, №. 8 (2015)
2. Shipulin, N.P., Shabelnikov, A.N.: Complex automation and mechanization of station processes. In: Automatic Equipment, Communication and Informatics, №. 10, Moscow (2017)
3. Shabelnikov, A.N.: Intelligent hump yards. Gudok newspaper, №. 59 (2016)
4. Shabelnikov, A.N., Olgeyzer, I.A.: Management of brake means of hump yards: upgrading and effectiveness. Vestnik RGUPS, Rostov-on-Don, №. 2 (2015)
5. Shabelnikov, A.N., Olgeyzer, I.A., Rogov, S.A.: The modern switchyard. From mechanization to digitalization. In: Automatic Equipment, Communication and Informatics, №. 1, Moscow (2018)
6. Shabelnikov, A.N., Olgeyzer, I.A., Rogov, S.A.: Innovative technologies of the smoothly varying management of brake tools. In: Automatic Equipment, Communication and Informatics. Moscow, No. 3. (2015)
7. Olgeyzer, I.A., Rogov, S.A., Galsky, M.A.: Expansion of opportunities of KSAU SP. In: Automatic Equipment, Communication and Informatics, №. 1, Moscow (2017)
8. Timshenko, A.U.: The Russian development "The system of the automated identification of railway cars" – out of competition. In: Point of Bearing Information, №. 140 (2011)
9. Shabelnikov, A.N., Sokolov, V.N., Sachko, V.I., Odikadze, V.R., Olgeyzer, I.A., Rogov, S. A., Yundin, A.L., Rodionov, D.V.: Service and operation of KSAU SP and KSAU KS. Manual, Rostov-on-Don (2012)
10. Automatic Cab Signalling system. http://mals.su/en.html. Accessed 1 May 2018
11. Soloshenko, V.N.: The automated system of commercial survey of trains and cars. A grant for inspectors of trains. Manual, Moscow (2008)
12. Ivanov, A.A., Legonkov, A.K.: Technical diagnostics and prediction. In: Automation on Transport, vol. 1, no. 3, September
13. Shabelnikov, A.N., Smorodin, A.N.: Complex automation of nodal switchyard. In: Automatic Equipment, Communication and Informatics, №. 4, Moscow (2018)
14. Odikadze, V.R.: Development of technology and development of monitors of functioning system of automation of sorting processes. The thesis for a degree of Candidate of Technical Sciences, Rostov-on-Don (2008)

Evolutionary Development Modelling of the Intelligent Automation Systems for Wagon Marshalling Process from the Standpoint of Smooth Mapping Singularity Theory

Alexander N. Shabelnikov[1,2], Nikolai N. Lyabakh[1,2],
and Yakov M. Gibner[1,2(✉)]

[1] Rostov State Transport University, Rostov-on-Don, Russia
gibner88@gmail.com
[2] JSC «NIIAS», Rostov Branch, Rostov-on-Don, Russia

Abstract. The task of research the intelligent development of marshalling process automation systems from the standpoint of smooth mapping singularity theory is formulated. The types of system potentials and parameters, which determine their change, are systematized. The process of intellectualization is represented by several, qualitatively diverse stages. Mathematical apparatus of identifying dependencies of potentials from their specific attributes is described. The effect of catastrophe initiation is explained, which allows to increase the degree of a research objectivity and efficiency of intelligent technologies introduction into automation systems. It is demonstrated, that catastrophe theory models can be used for description of a series of a difficulty-formalized aspects of intellect initiation: insight, nonlinearity (including liminality) of thinking, symmetry breaking (in the choice of alternatives during decision making), hysteresis in scientific cognition, etc.

Keywords: Marshalling systems · Intelligent
Smooth mapping singularity theory · Catastrophe · Bifurcation points

1 Introduction

Traditionally in the artificial intelligence theory the models and methods are investigated, which reflect the nature of intelligent functioning: process simulation of expected uncertainty rundown about the subject of research, nontrivial solution making. And the problems of intelligent technologies development and effective introduction stay in the background. The paper studies the above problem and illustrates it, using the example of wagon marshalling systems (MS).

The work was supported by Russian Fundamental Research Fund, project No. 17-20-01040.

A. Abraham et al. (Eds.): IITI 2018, AISC 875, pp. 291–299, 2019.
https://doi.org/10.1007/978-3-030-01821-4_31

Marshalling systems enable initiation, advance, and freight traffic distribution among railway network. They determine the traffic frequency and intensity and quality of transportation process.

In this regard, creation of an efficient Integrated Automatic Control System of the Sorting Process (KSAU SP) is an outstanding academic and research problem [1, 6]. The complexity of MS, associated with high-dimensional nature of forecasting and management tasks, expected source data uncertainty and noise, explains the need to develop intelligent functions of KSAU SP [7]. It will provide for the following: system competitive performance in the market, high quality of provided transportation services, increase of production and economic indicators (productivity, profitability, etc.).

The intelligent development of automated systems is a complex process, evolving through several objectively determined stages [6]. The purpose of this study is to reveal the essence and structure of the above process.

Paper [2] describes the mechanism for innovation rating assessment of a product (device, technical-and-process technology system, software), based on the analysis of the innovative product level of fitting the corresponding market segment, on accounting for its distinctive internal attributes. The same mathematic toolkit can be successfully used to assess the intelligence level of the product. A simple assessment of the system intelligence level is important, but it does not reveal the principles of its influence on system performance efficiency.

Serious contradictions develop (financial, organizational, methodological) during implementation of a system development strategy, based on its intellectualization. Resolving the above contradictions requires improvement of the MS automation development theory, with due account of the above aspect. The present study uses the smooth mapping singularity theory [3] as a theoretical foundation of such an improvement.

The smooth mapping singularity theory was actively developed in such subsections as follow: theory of catastrophes, describing the cases of discontinuous, hard-to-predict changes in system status and quality; theory of bifurcation, studying the points of probability branching of development trajectory. The above theories at first stages were applied to technical systems (being simpler and easier for comprehension). But scientific development has allowed to expand the application of formalisms to social and economic processes [4].

In our study the two directions are combined: technical-and-process technology system of KSAU SP is integrated into the economies of industry branch and respective region. The questions under investigation are at the junction of technical and economic problems.

2   Strategies and Stages of Intelligent Systems Development

Each technical-and-process technology system (including KSAU SP) is described by the following:

- vector of parameters changing with time. Thereat, a smooth change of the parameter can lead to discontinuous change of system properties (catastrophe);
- certain potential that varies depending on changes of the above parameters.

Consider a model example: movement of a ball fixed to a weightless spring. The pendulum is characterized by general energy (its potential) E, and the parameters are the weight of ball, m; spring stiffness, k. Depending upon the values of the parameters, the pendulum behaviour will be described by different phase trajectories (Fig. 1):

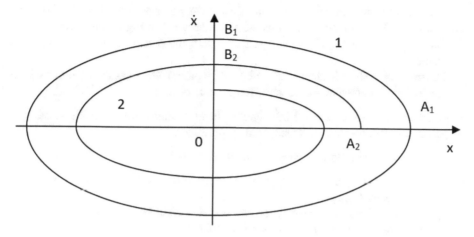

Fig. 1. Phase trajectories of various strategies for the development of intelligent automation systems

- Trajectory 1 is characterized by the initial deviation equal to $0A_1$, lack of friction in suspension (phase trajectory returns to the starting point A_1).
- Trajectory 2 is characterized by a smaller initial deviation of the pendulum $0A_2$ (i.e. less than potential of the system in the first case), availability of friction in suspension (phase trajectory has the shape of a spiral).

Why is this example important in the context of this study? It is typical for many inertially developing processes: the funds invested into the increase of KSAU SP intelligence (in the above example they are simulated by potential energy of the deflected ball, points A1 or A2 in Fig. 1) are exhausted, accelerating the process (points B1 or B2), introduced innovations bring economic effect (funds), which are used for further innovative development. If the reproduction system is not implemented properly (equivalent to friction availability in the pendulum suspension), the process fades (phase trajectory 2 on Fig. 1). If the funds for innovations increase in each cycle, the spiral will unwind. Trajectory 1 is the borderline case between the two described ones above: regressive and progressive strategies of intelligence development.

In general, the phase trajectories of inertial oscillating processes are described by the relationship [5]:

$$\ddot{x} = w^2 x - 2\beta \dot{x}, \tag{1}$$

where

$\ddot{x}$ –   acceleration;

$\dot{x}$ –   speed;

x –   deviation

The relationship parameters (1) characterize the process under investigation. If β equals to zero (no attenuation), then we have Trajectory 1 in Fig. 1, otherwise we have Trajectory 2. Solving Eq. (1) allows you to obtain the change in system parameters over time.

Marshalling system is characterized by many parameters, and also a set of certain potentials. The parameters are in exchange classified into:

- system-general (e.g., level of intelligence, level of innovation, versatility, etc.);
- production (train break-up speed, throughput, etc.).

We will highlight the following potentials for a MS:

- cost of operation (parabola with upward branches) depending on increasing train break-up speed (Fig. 2a);
- operational efficiency depending on the growing train break-up speed (Fig. 2b) – described by parabola with downward branches.
- train break-up safety;
- viability of developed and improved system (KSAU SP);
- quality of provided service (timeliness, absence of cargo breakage and damage);
- environmental performance (noise level, environment pollution);
- system controllability, etc.

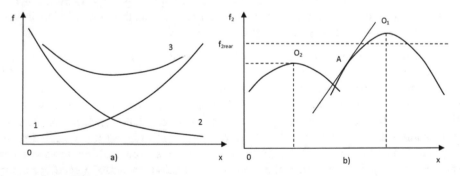

Fig. 2. Graphic illustration of MS potentials' dependency upon the speed of train break-up: (a) efficiency; (b) cost

Minimal and maximal levels, points of inflection of the functions describing the change of potential, are the critical points of the function, which largely define its behaviour. According to Montel [4]: "Functions, as well as living creatures, are characterized by their specific aspects." Figure 3 shows the main stages of intelligence development in technical-and-process technology systems.

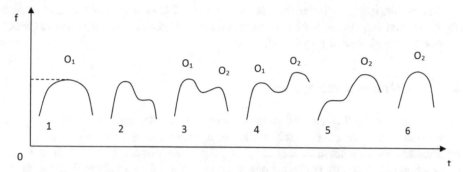

Fig. 3. Stages of intelligence development in technical-and-process technology systems

Curve 1 (Fig. 3) specifies the initial state of the system: point O_1 determines the optimal intelligence level under current conditions. Development of advance technologies and materials, software products (fibre-optic cable, section control switch box, railroad intelligent control system (ICS), development of automatic advice-giving systems for scheduled maintenance, etc.) established conditions for more innovations (Curve 2, Fig. 3 in marshalling processes intelligence. Curve 3 (Fig. 3) corresponds to the development of a new extreme point. Curve 4 reflects the situation of predominance of the new optimal state. Following are the phases of elimination of the first optimum (Curve 5) and consolidation of the new optimal state (Curve 6).

The described scenario is not unique for possible sequences of local transformations, qualitative modifications, or, in the parlance of mathematics, bifurcations. All possible interaction scenarios of two locally optimal modes are described by a universal law [4]:

$$y = x^4 + ax^2 + bx. \tag{2}$$

3   Basic Laws of Smooth Mapping Singularity Theory

The smooth mapping singularity theory is based on a several laws [4]:

1. In the vicinity of a non-critical point, function increment is almost proportional to increment of the argument (see Fig. 2b, Point A).
2. In the vicinity of the maximum (minimum) the typical function increment is almost proportional to square increment of the argument.
3. A typical flat curve is tangent to a line in not more than two points. A three-point tangent can be get rid of by a small change of the curve shape, and a two-point tangent is stable, i.e. does not disappear after small perturbation of the curve.
4. A typical surface is not tangent to any straight line in more than four points.

All the situations reviewed concern the universal laws applicable to any smooth objects (functions, curves, surfaces), except for some special, "atypical" cases (which include both a plane and a cylinder).

4 Catastrophe Description

Let us consider another simplified model example. In the three-dimensional space, which is common for humans, let us allocate a certain space (a plane), in which two subjects are moving towards one another. They can dispart when meeting in the intersection point in only one way: one steps over the other. Imagine that one of the subjects possesses the ability to exit into a three-dimensional space, that is, it can shift aside. Thus it will become "invisible" for the second subject, as it leaves the space of their co-existence. Once their trajectories have disparted, it can return to the initial two-dimensional space.

For the second subject, the catastrophe occurred twice: at disappearing and reappearing of the first subject. It is a hard-to-explain fact in the terms of two-dimensional space. Science fiction and popular science literature gives multiple examples of similar situations: transition to parallel worlds, time travel, etc.

Two-dimensional space is a projection of three-dimensional space that "omits" some details, and events clear and obvious in three dimensions are surprising and unexplainable in two-dimensional representation.

Let us consider Fig. 4, which represents a three-dimensional figure (surface), called the Whitney assembly [4]. The surface under examination is "assembled" of many cubic dependencies. Figure 5 shows several types of them – representatives of a whole

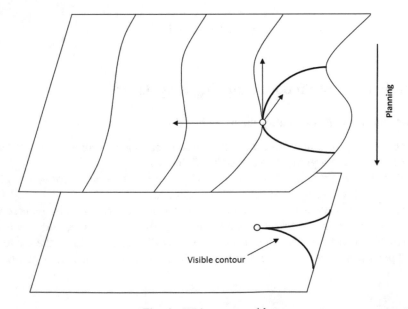

Fig. 4. Whitney assembly

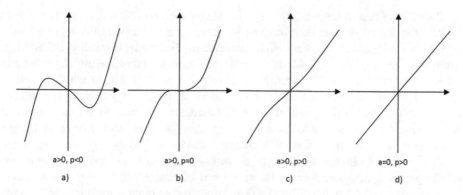

a>0, p<0   a>0, p=0   a>0, p>0   a=0, p>0

a)   b)   c)   d)

Fig. 5. Cubic dependencies synthesizing the Whitney assembly

class of functions, depending on certain parameters (particularly, on time). The axis represents time, the system intelligence level is plotted along the cubic curve.

The values of the potential investigated are shown on the surface with the thick curve – it is a smooth parabolic line, i.e. the potential varies predictably in three-dimensional space.

The universal formula of such a class of functions is a multitude of cubic functions

$$y = ax_3 + px \qquad (3)$$

of variable x, depending at the same time from parameter p. At zero (Fig. 4) there is an "atypical" cubic singularity observed.

As long as the p parameter is negative, the function has the local maximum and minimum in the vicinity of zero. When parameter p tends to zero, the maximum and the minimum draw closer, and at zero value of the parameter a cubic peculiarity arises. At positive values of the parameter both maximum and minimum disappear, "killing" each other. All these phenomena are universal and stable, they are also observed in classes close to the one under consideration. Therefore, such phenomena have a broad scope of applicability.

Let us bring all the previous reasoning together. The customer (user) determines the potential function to be investigated (for example, safety of KSAU SP). He is interested in obtaining a system with values of parameters producing not only maximal security, but providing security with a value not lower than a specified value. That is, if the investigated process is described by Curve 2 (Fig. 2), and the pre-set level of safety is equal to f_{2z}, then reaching the necessary level of safety will be impossible, as a matter of principle. It will be necessary to change the level of system intelligence for transit to Curve 1.

Innovations have successive development in time (see Figs. 4 and 5 along the formed cubic dependencies). The investigated potential of a selected parameter also "smoothly" changes in some (known neither to the customer, nor to the researcher) space (Fig. 4). The researcher chooses a feature space of research (design plane). In case of its unfortunate choice he receives a catastrophe effect (see projection shown in Fig. 4).

Now let us turn to the mechanism of intelligence construction. Let us demonstrate catastrophe theory models may be used to describe a series of difficult-to-formalize aspects of intelligence initiation; insight, nonlinearity (including liminality) of thinking, symmetry breaking (in the choice of alternatives during decision making), hysteresis in scientific cognition, etc. It is known, that insight is a condition in which a person suddenly arrives at a solution of a vital problem or a new projection of a situation. Often such a condition is called an enlightenment and due to bright emotional accompaniment it is referred to the field of a miracle [8]. The fact is that in the human conscience only one projection of a complex cogitative process is reflected, which "conceals details" in the foldings of a multi-dimensional act, revealing only the resulting conception, described by the categories of catastrophe theory.

Any behavior of a subject is finally determined by the brain neural activity. Zeeman [9] supposed, that a brain can be described as a set of tightly bound non-linear oscillators, each of which is modeling behavior of millions of neurons. On the example of a Duffing non-linear oscillator he demonstrated, that the features of non-linear oscillators are very similar to singularities of an "assembly" catastrophe. For this reason it is possible (and necessary) to simulate brain activity (i.e. the process of intelligence construction) using the methods of catastrophe theory.

Wilson and Cowan [10] developed a brain model basing on dynamic equations of interaction of excitory and inhibitory neuron populations. The authors obtained the system of non-linear differential equations for the populations distributed in space, which are subject to investigation using numerical techniques and phase plane method. It is discovered, that foldings in stationary solutions cause the multiple hysteresis phenomenon. Besides, the obtained solutions substantiated the origin of boundary cycles (modeling brain rhythms), in which the oscillations frequency is a monotonic function of stimulus intensity.

5   Conclusions

1. Complex automation processes are developed in large-dimension spaces, which are not always available for analysis. Perspective to these processes from smaller dimension spaces (available for analysis) leads to representations of catastrophe theory.
2. An adequate model for development of automation systems is the Whitney assembly, structuring the dependency of system potentials from its parameters.
3. During a planned transition from one local-optimal mode to the other, a temporary deterioration of the system (economic efficiency, safety, survivability, etc.) is necessary.
4. It is demonstrated, that catastrophe theory models can be used to describe the difficult-to-formalize phenomena of brain activity.

References

1. Shabelnikov, A.N., Lyabakh N.N., Ivanchenko V.N., Sokolov V.N., Odikadze V.R., Kovalev S.M., Sachko V.I.: Sorting hump automation systems on the basis of modern computer technologies. NIIAS, RGUPS, Rostov-on-Don (2010)
2. Kolesnikov, M.V., Gibner, Y.M.: Evaluation of the intelligence degree of systems. In: Abraham, A., Kovalev, S., Tarassov, V., Snasel, V., Vasileva, M., Sukhanov, A. (eds.) Proceedings of the Second International Scientific Conference "Intelligent Information Technologies for Industry", IITI 2017, Advances in Intelligent Systems and Computing, vol. 680. Springer, Cham (2017)
3. Thompson, J.M.T.: Instability and catastrophe in science and technology. Mir (1985)
4. Plotinsky, Y.M.: Models of Social Processes, 2nd edn. Logos, Moscow (2001)
5. Knyazev, A.A.: Lecture on synergetics № 2. Dynamic systems. Phase Space. http://www.sblogg.ru/lekciya-po-sinergetike-2-dinamicheskie-sistemy-fazovoe-prostranstvo.html. Accessed 01 Mar 2018
6. Adadurov, S.E., Gapanovich, V.A., Ljabah, N.N., Shabelnikov, A.N.: Railway transport: on the way to intelligent management. SSC RAS, Rostov-on-Don (2010)
7. Lyabakh, N.N., Shabelnikov, A.N.: Technical cybernetics on railway transport. RGUPS, NCSC HS, Rostov-on-Don (2002)
8. Weisberg, R.W., Alba, J.W.: An examination of the alleged role of «fixation» in the solution of several «insight» problems. J. Exp. Psychol. Gener. **110**, 169–192 (1981)
9. Zeeman, E.C.: Conflicting judgements caused by stress. Br. J. Math. Stat. Psychol. **29**, 19–31 (1976)
10. Wilson R.: Psychology of Evolution. Publishing House, Janus (1998)

Multidimensional Linguistic Variables and Their Application for Resolving the Tasks of Marshaling Processes Automation

Alexandr N. Shabelnikov[1,2], Nikolai N. Lyabakh[1,2(✉)], and Natalia A. Malishevskaya[1]

[1] Rostov State Transport University, Rostov-on-Don, Russia
liabakh@rambler.ru
[2] JSC «NIIAS», Rostov Branch, Rostov-on-Don, Russia

Abstract. Role of multi-dimensional fuzzy sets is updated in the task of railway wagons marshaling processes automation, expanding the resolved tasks range and increasing their solution accuracy. Two identification methods of multi-dimensional functions in tabular form, using univariate membership functions, are developed. Particularly, it is a synthesis based on the operation of fuzzy sets intersection, and a sequential synthesis algorithm for membership functions. An approximation procedure is developed for multi-dimensional membership functions in the points not belonging to table nodes.

Keywords: Automation of marshaling processes · Fuzzy sets
Multi-dimensional linguistic variables · Operations with fuzzy sets
Approximation

1 Introduction

Marshalling yard sorting systems [1, 2] – are the complex objects of railway processes automation. The Rostov Branch of the JSC "Research and Development Institute of Informational Technologies, Automation, and Communication (NIIAS)" developed an Integrated Automatic Control System of the Sorting Process (KSAU SP) [3], introduced at more, than twenty marshalling yards, and utilizing advanced methods of process simulation and decision-making procedures [4]. Complexity of the above systems is described by the following: high dimensions of resolved identification tasks, development and control forecasting; operational statistical data noise; considerable uncertainty of sub-processes under investigation. The last problem is resolved by the means of fuzzy sets theory (FST) [4, 5] through development of intelligent train operation systems [6–9].

Traditionally, the single-dimensional fuzzy variables are used in FST. It limits the class of object under investigation, reduces accuracy, and complicates research and

The work was supported by Russian Fundamental Research Fund, project No. 17-20-01040.

control procedures. In this regard the task of introduction, identification, and use of multidimensional linguistic variables [10] is updated.

2 Fuzzy Sets and Linguistic Variables in the Task of Marshaling Process Control

Formalized procedures of fuzzy sets theory use the notions of linguistic variables [11] and the functions of fuzzy sets membership to certain types of objects. Particularly, when marshaling trains, the important fuzzy variables include:

- "Type of a cut according to rolling properties" [1, 3]. A "cut of wagons" is a group of wagons proceeding in same direction. The differentiated types are as follows: "good runner", "bad runner", etc. It is necessary to consider the type of runner, when solving the tasks of aiming and interval control for cuts, rolling down a gravity hump [3]. In exchange, these complex (compound) values of the linguistic variable "type of a cut" depend upon the following simpler variables (by their structure and identification of membership functions).
- "Weight of a cut", rolling down a gravity hump (GH). Uncertainty is created by inaccuracies of measuring equipment, uneven weight distribution along a cut. In [4] the following values of the corresponding linguistic variable "weight of a cut" are assigned by expertise: "light", "light-medium", "medium", "heavy", "very heavy".
- "Rolling resistance of a cut": depends upon the type of a bearing and the wheel journal box construction design. For example, in paper [4] it was accepted that: roller bearing corresponds to 1, slider bearing corresponds to 0. If a cut of wagons has both types of bearings in various proportions, the value of the variable will change within the range [0; 1]. New types of journal boxes are developed today, which makes the need to consider fuzziness of this variable, characterizing resistance to rolling of a cut, even more critical.
- "Type of wagons in a cut " (platform, open wagon, tank, fully-weighted wagon, etc.), which influences the rate of air resistance while rolling down a gravity hump, and ultimately the running properties of a cut. The cut composition may include wagons of different types, which normalization distributes them along the scale from 0 to 1 according to this property.

Other parameters, influencing the process of a cut rolling down a hump (number of axles, wheel base, length of a cut, etc.) can be given, which are also fuzzy.

Besides the fact of objective data uncertainty, one shall note, that a human person (operator, system developer) uses:

- his/her personal opinion (which takes us further away from the formalisms of probability theory and mathematical statistics);
- means of a natural language (linguistic variables). It makes FST application even more vital.

The corresponding membership functions $\mu(x_i)$ are introduced to describe the above linguistic variables. Membership functions (MF) can be established by description, tables, charts, analytical ratios. The most convenient option for calculation and use in control procedures for the cuts, rolling down a gravity hump by means of KSAU SP is the use of tables:

$$\left\{ \frac{x_i}{\mu(x_i)} \right\}. \tag{1}$$

It allows to simplify computational procedures and enhance visibility of information presentation. Therefore, further reasoning presumes this type of fuzzy object model setting.

Example 1. Let us describe a fuzzy set "weight of a cut" with only two values of variable "light cut" (membership function $\mu_l(x_i)$) and "heavy cut" (membership function $\mu_t(x_i)$) using Table 1.

Table 1. Membership functions for fuzzy sets of "light" (l) and "heavy" (t) cuts.

x_i (mass in tonnes)	20	40	60	80	100	120
$\mu_l\,(x_i)$	1	0.8	0.6	0.4	0.2	0
$\mu_t\,(x_i)$	0	0.2	0.4	0.6	0.8	1

Example 2. Similarly, for fuzzy sets "rolling resistance of a cut", set by two values "high" and "low", it will be represented by the membership functions of Table 2.

Table 2. Membership functions of fuzzy sets "high" (h) and "low" (lo) for rolling resistance of a cut.

y_i (share of rolling bearings)	0	0.25	0.5	0.75	1
$\mu_h(y_i)$	1	0.75	0.5	0.25	0
$\mu_{lo}(y_i)$	0	0.25	0.5	0.75	1

Use of fuzzy set theory tools for mathematic modeling requires to resolve the following tasks:

1. Identification linguistic variables (LV) membership functions. There are several approaches to identification of membership functions [3]:

   - based on the originally known information about the structure and functioning of the object under investigation;
   - statistically (in this case, a membership function, under certain conditions, may be a random variable distribution function);
   - based on subjective opinion of an operator (or a group of expert operators).

2. Developing a model regulating the principle of LV use and comprising a synthesis of the needed operations with fuzzy sets. Thus, a fuzzy algorithm of process technology and decision-making procedure is developed.

In this study, the emphasis is made on the first task.

3 Multi-dimensional Membership Functions

The fuzzy sets reviewed in Examples 1 and 2 were one-dimensional. However, the analysis of rolling control process shows a simultaneous dependency of running properties of a cut from several characteristics: weight, resistance to rolling, type of wagon, number of axles, etc. In this regard, a more generalized characteristic is used – runner type, which assumes the following linguistic values: "good", "average", "bad", and "very bad". The membership functions of the above linguistic values will be the functions of several variables.

They can be identified in different ways. Let us consider the possible variants of multi-dimensional MF synthesis using the example of two conditions. For our convenience let us consider the two following parameters: weight of a cut and resistance to rolling. The two-dimensional fuzzy set in this case will have the following view:

$$\left\{ \frac{(x_i, y_j)}{\mu(x_i, y_j)} \right\}. \tag{2}$$

Further generalization is obvious.

3.1 Multidimensional MF Synthesis Based on Operations with a Fuzzy Set

The theory of fuzzy sets includes various operations with FS. Particularly, the operation of MFs intersection corresponds to logical operator "and". For example, a cut is a "very bad" runner, if it belongs both to the "light" cuts, and to the cuts with "high" resistance to rolling. In this case MF equals to the product of respective partial MFs (3). Besides, if fuzzy sets theory is applied as a generalization of the probability theory, it uses similar laws referring to MFs, considered as the probability of occurrence of any complex event. Thus, the probability of the product of two independent events equals to the product of probabilities of these events – analog (3). Hence, a two-dimensional membership function can be defined as follows:

$$\mu(x_i, y_j) = \mu(x_i)\mu(y_j). \tag{3}$$

Let us return to our examples. There are four combinations possible of linguistic variable values, introduced in Tables 1 and 2 (2 × 2). These are "very bad", "bad", "average" and "good" runners. A combination of "heavy" cuts of wagons and the cuts with "low" resistance to rolling, shall obviously give a "good" runner with a membership function

$$\mu_x(x_i, y_j) = \mu_h(x_i)\mu_{lo}(y_j). \tag{4}$$

A combination of "light" cuts of wagons and the cuts with "high" resistance to rolling, also obviously, gives a "very bad" runner with membership function

$$\mu_{vb}(x_i, y_j) = \mu_l(x_i)\mu_h(y_j). \tag{5}$$

Concerning combinations $\mu_t(x_i)\mu_h(y_j)$ and $\mu_l(x_i)\mu_{lo}(y_j)$, they cannot be by originally known information termed as belonging to "good" and/or "average" runners. They could only be compared basing on the calculation results. In the first case, belonging to a "heavy" cut improves running properties, belonging to the cuts with high resistance deteriorates them. It all depends upon which of the above effects prevails.

Using the relation (5) and data of Tables 1 and 2, we calculate the two-dimensional MF of a "good" runner (Table 3).

Table 3. Two-dimensional membership function of a "good" runner fuzzy set.

$\mu_{lo}(y_j)$	$\mu_t(x_i)$					
	20/0	40/0.2	60/0.4	80/0.6	100/0.8	120/1
0/0	0	0	0	0	0	0
0.25/0.25	0	0.05	0.1	0.15	0.2	0.25
0.5/0.5	0	0.1	0.2	0.3	0.4	0.5
0.75/0.75	0	0.15	0.3	0.45	0.6	0.75
1/1	0	0.2	0.4	0.6	0.8	1

3.2 Sequential MF Synthesis Algorithm

The value of one parameter, characterizing the situation, is fixed, and for the other a membership function is constructed, using one of the presented methods. The procedure is then repeated until all necessary values of the first parameter are considered.

The implementation of this procedure requires solving of two problems.

- Selecting the consequence of parameters' consideration, i.e. it is necessary to determine how to calculate the two-dimensional membership function: by lines or by columns? Theoretically, a question remains open: is this operation commutative?
- Overcoming the high dimension of a problem with three or more parameters considered.

4 MF Approximation

A drawback of the table method of membership function description is the fact, that this table does not provide the MF value at the intermediate values of a cut parameters (the same is true for the membership function described by Table 3). For example, the cuts O_1 and O_2 with the corresponding parameters $x = 60$ and $y = 0.33$ or $x = 70$ и $y = 0.66$

are not included into the Table 3 nodes. For the first cut O_1 (60; 0.33) the task can be simply resolved by linear interpolation of the membership function between points A (60; 0.25) and B (60; 0.5), see Fig. 1, as the points differ only in one parameter y.

From Fig. 1 it readily follows:

$$M(O_1) = M(A) + \frac{y_{O_1} - y_A}{y_B - y_A}(M(B) - M(A)) \tag{6}$$

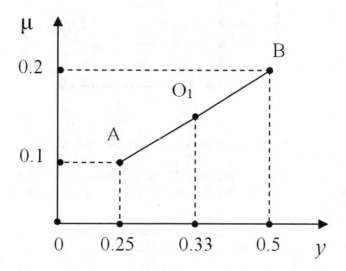

Fig. 1. Renewal of an unknown membership function value by interpolation method

In our example: $y_A = 0.25$, $y_B = 0.5$, $y_{O1} = 0.33$, $\mu(A) = 0.1$, $\mu(B) = 0.2$, whence from (6) it follows that $\mu(O_1) = 0.132$.

For the case of two or more variables (see the example of cut O_2 with parameters (70; 0.33)) the development of interpolation method requires to determine at first the nearest point of procedure. For this purpose, in the space of x and y attributes a proximity measure must be set, with weight ratios accounting for the importance of these characteristics. The minimal distance from the point under investigation to the grid points determines the reference point, to which value the corresponding increments for each parameter are to be added.

Let the point A (see Fig. 2) be the nearest to the investigated point C, and the points B and D are nearest to A and are lying in the nodes of the grid, whence follows:

$$M(C) = M(A) + \frac{y_C - y_A}{y_B - y_A}(M(B) - M(A)) + \frac{x_C - x_A}{x_D - x_A}(M(D) - M(A)). \tag{7}$$

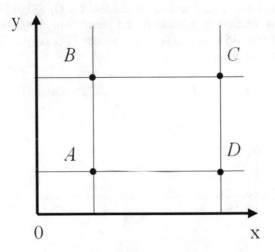

Fig. 2. Approximation of two-dimensional MF

Example 3. It is necessary to calculate μ (C) for cut of wagons C (35; 0.3), using the measure of proximity of the two-dimensional attribute space x_Oy described by the following formula [4]:

$$d(A, B) = \sqrt{2 * 10^{-4}(x_A - x_B)^2 + (y - y_B)^2},\qquad(8)$$

and the interpolation according to (7).

Using (8), we define the distance to adjacent points of Table 3. These are the points: A (20; 0.25), B (40; 0.25), D (20; 0.5) and K (40; 0.5). We obtain: $d\,(A,\,C) = 0.218$; d (B, C) = 0.087; d (D, C) = 0.292; d (K, C) = 0.212. The closest point to C (30; 0.33) is point B. The closest to point B nodes are A and K. The actual location of the points in

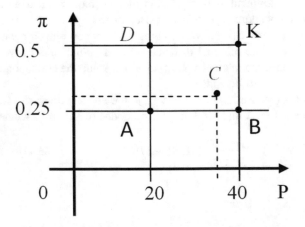

Fig. 3. Geometric illustration of the example conditions

our example is shown in Fig. 3. According to (7), considering the real location of points, we obtain: $\mu(C) = 0.0535$.

5   Conclusions

1. The necessity of multi-dimensional fuzzy sets development for automation of railway marshalling processes is substantiated.
2. Two options of multi-dimensional membership functions synthesis, based on a single-dimensional MFs, are proposed:

   - Synthesis of multi-dimensional MF based on operations over the fuzzy set.
   - Sequential MF synthesis algorithm.

3. A method of multi-dimensional MF approximation is developed.
4. The advanced issues of Synthesis of multidimensional MFs development (confirmation or refutation of commutative property of the sequential MF synthesis, overcoming the high dimension problem under the condition of big number of considered parameters) are formulated.

References

1. Shabelnikov, A.N., Lyabakh, N.N., Ivanchenko, V.N., Sokolov, V.N., Odikadze, V.R., Kovalev, S.M., Sachko, V.I.: Sorting hump automation systems on the basis of modern computer technologies. NIIAS, RGUPS. Rostov-on-Don (2010)
2. Simić, D., et al.: 50 years of fuzzy set theory and models for supplier assessment and selection: a literature review. J. Appl. Logic (2016)
3. Shabelnikov, A.N., Sokolov, V.N.: The latest technologies of marshalling stations automation. Autom. Commun. Inform. № 11 (2007)
4. Lyabakh N.N., Shabelnikov A.N.: Technical cybernetics on railway transport. RGUPS, NCSC HS, Rostov-on-Don (2002)
5. Bray, S., et al.: Measuring transport systems efficiency under uncertainty by fuzzy sets theory based Data Envelopment Analysis: theoretical and practical comparison with traditional DEA model. Transp. Res. Procedia 5, 186–200 (2015)
6. Adadurov, S.E., Gapanovich, V.A., Ljabah, N.N., Shabelnikov, A.N.: Railway transport: on the way to intelligent management. SSC RAS, Rostov-on-Don (2010)
7. Stepanov, M.F., Stepanov, A.M.: Mathematical modeling of intellectual self-organizing automatic control system: action planning research. In: 3rd International Conference « Information Technology and Nanotechnology » , ITNT-2017, Samara, Russia (2017)
8. Pupkov, K.A.: Intelligent systems: development and issues. In: XIIth International Symposium « Intelligent Systems » , INTELS 2016, 5–7 October 2016, Moscow, Russia. Procedia Comput. Sci. 103, 581–583 (2017)
9. Sumalee, A., Ho, H.W.: Smarter and more connected: future intelligent transportation system. IATSS Res. (2017)
10. Finaev, V.I.: Models of Decision-Making Systems. TRTU Publishing, Taganrog (2005)
11. Rosa, M.R., Labella, A., Martínez, L.: An overview on fuzzy modelling of complex linguistic preferences in decision making. Int. J. Comput. Intell. Syst. 9(sup1), 81–94 (2016)

Transport Workers Activities Analysis Using an Artificial Neural Network

Maskim Kulagin[1]([envelope]) and Valentina Sidorenko[1,2]

[1] Russian University of Transport (MIIT), Moscow, Russia
maksimkulagin06@yandex.ru, valenfalk@mail.ru
[2] National Research University Higher School of Economics, Moscow, Russia

Abstract. This article describes modern methods of data processing regarding the task of assessing activities of transportation employees. The main purpose was to find dependencies in data and construct an algorithm for predicting the probability of transport safety violation by employee. The research was conducted for locomotive drivers. The following algorithms were used: neural networks, gradient boosting over decision trees and random forest. Based on the obtained results and drawn conclusions one can think of the perspective for the elaboration and introduction this work for practical use in railway industry, e.g. in "Russian Railways".

Keywords: Machine learning · Neural network
Artificial intelligence · Probability · Random forest · Gradient boosting

1 Introduction

Nowadays, machine learning methods are very popular in the world of computer technology. Machine learning methods are used in various fields of science, since they allow solving problems of analyzing large amounts of data, predicting an adverse event, detecting and recognizing images, recognizing speech effectively and quickly.

Self-driving car systems have been actively developing and popularizing in the world. However, the overwhelming volume of traffic is carried out by people in most large transport companies. The goal of ensuring high quality and safety of transportation requires high level of competences. Most companies train their employees on their own, and then they monitor the work (e.g. "Russian Railways" company).

This article describes a method for a comprehensive activities assessment of vehicles drivers, such as the locomotive driver. The locomotive driver is an employee who controls the locomotive, insures train schedule fulfillment, security requirements, maintains the safety of goods and rolling stock, and maintains a rational mode of driving the train with minimal fuel and electricity consumption.

© Springer Nature Switzerland AG 2019
A. Abraham et al. (Eds.): IITI 2018, AISC 875, pp. 308–316, 2019.
https://doi.org/10.1007/978-3-030-01821-4_33

2   Background

The main objective of Russian Railways as a company is to ensure traffic safety. Today, there is already a system of adaptive management of railway transport infrastructure technical maintenance of Russian Railways (URRAN PROJECT) [1–3], but there is no objective system for evaluating the work of drivers. Therefore, the task of creation of a system for automatic comprehensive evaluation of locomotive drivers is an important scientific and engineering task for Russian Railways.

Human analysis based on machine learning is also used in medicine and banking. For example, a diagnostic system using neural networks, which relies on medical indicators of a person and a mathematical apparatus, is quite popular in medicine [4,5]. The task of credit scoring and credit analysis is solved in banking [6].

The results obtained from this research will help to make the first step in creation of the objective approach for risk management in locomotive drivers work. The study and creation of a unified and objective way of assessing the drivers aims to improve the safety of traffic in rail transport.

3   Task

Nowadays, the activity of locomotive drivers doesn't have an objective evaluation system. Assessment of the locomotive driver's work highly depends on his manager. Accordingly, the weight of each assessment feature depends on the human factor. Therefore, it is expedient to develop an automatic system that will allow to construct an objective evaluation model of the locomotive driver's activity (Fig. 1).

The development of an automatic system goes through several steps:

Step 1. At this step it is necessary to form a list of features that characterize the work of the locomotive driver. The list of feature is determined by the method of expert assessments. At the output of this step, a database must be created.

Step 2. This is one of the most important steps: data processing in order to find patterns. Data processing consists of elimination of emissions, processing of gaps, conversion of categorical features, scaling of numerical features, search for linearly dependent features.

Step 3. At this step, we develop a method for estimating the likelihood of a violation and building a rating system for locomotive drivers [7].

Step 4. This step includes a description of the method of dividing the locomotive drivers into risk groups and creation of a list of measures to reduce the level of risk.

This article outlines a method for estimating the probability of a violation using neural networks.

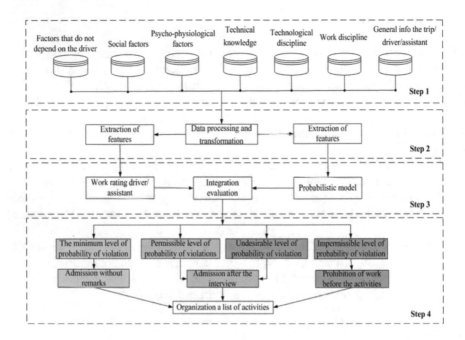

Fig. 1. Block diagram of the system for automatic violation probability determination for locomotive driver (or assistant locomotive driver) and developing a list of measures to prevent violations.

4 Approach

In the article, we used an artificial neural network (ANN) [8] to automate the probability estimation of adverse event occurrence. The task of the ANN is to predict the outcome before the trip (0 - no security violation, 1 - security violation). The task of this research is to determine the probability of a driver breaking a trip. The research was held in several stages:

1. Collecting and study of available information. We used local resources of JSC Russian Railways to obtain the available data. Since there was no access to all data at the moment, we took into account only a limited set of indicators. The obtained indicators are described in details in the next section.
2. Data processing and extraction of features. The object of the research is the trip. A set of features was chosen from a dataset of drivers' trip performance. The amount of data is 180,000 trips for 6 months of work. Each trip includes the following features:

- "Trip start time", "Trip end time", "Driver's experience", "Number of trips committed" - numerical features that passed the elimination phase and were normalized using expression (1):

$$\overline{x}_i^j = \frac{x_i^j - \mu^j}{\sigma^j} \tag{1}$$

where $\overline{x}_i^j$ - normalized i value of j feature; x_i^j - i value of j feature; μ^j - mean j feature that is averaging over all considered trips ($\mu^j = \frac{1}{N}\sum_{i=1}^{N} x_i^j$); σ^j - distributions j feature ($\sigma^j = \frac{1}{N-1}\sum_{i=1}^{N}(x_i^j - \mu^j)$).

- "Safety device", "Railway", "Class" - categorical features that characterize the type of safety device on the locomotive/railway/driver class. These features were processed using OneHotEncoding.
- "Locomotive driver's assistant" - categorical features. This features calculated as follows:

$$x_{wa}^n = \frac{NV_{wa}^n}{NT_{wa}^n}; \tag{2}$$

$$x_{woa}^n = \frac{NV_{woa}^n}{NT_{woa}^n}; \tag{3}$$

where NV_{wa}^n, NV_{woa}^n - the number of violations on the n-th road in the presence or absence of an assistant locomotive driver, respectively;
NT_{wa}, NT_{woa} - the number of trips on the n-th road in the presence or absence of an assistant locomotive driver, respectively.
- "Array of accumulated violations" - this feature describes the content and number of violations committed by the locomotive driver before this trip. The size of the array is 311, where the position of each element in the array corresponds to a particular violation, and the value of this element is the amount of respective violations.

The result was a matrix of 180,000 × 356, where 356 is the total number of features obtained after the conversion.

3. Before building a neural network, we constructed a correlation matrix which shows dependences among different violations. As a result, we obtained a matrix of 311 × 311. The correlation matrix was calculated using the Pearson correlation coefficient. Figure 2 shows the obtained Pearson correlation coefficients distribution.
   According to the graph about 40 (12%) violations have a correlation coefficient between 0.5 and 1. It means that there is a relationship among the violations. For example, correlation between such violations as "Lack of check of brake line density" and "No bleeding of the brake line when receiving the locomotive" is 0.66.
4. Selection of the architecture and parameters of the neural network that was used to find the relationship between the features and the appearance of a violation. During the research for this paper we solved the problem of binary classification [15]. We used an algorithm that based on a fully connected neural network with a single hidden layer. There are few reasons for using neural network:

- a good way to approximate nonlinear functions by input datasets;
- the analysis of human activity requires the use of nonlinear algorithms to build more likely predictions;
- the algorithms of deep learning will identity the relationship between the performance of the locomotive driver and violations in hidden layers.

5. Training of the neural network and assessment of the quality of work on the test database.

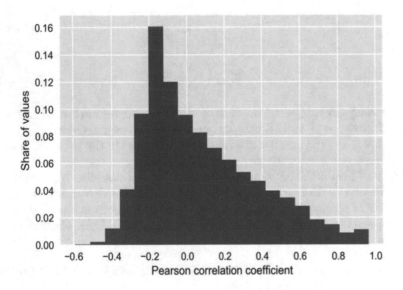

Fig. 2. The distribution of the Pearson correlation coefficient.

The architecture of the neural network with one hidden layer was chosen. A rectified linear function was used in the input and the hidden layers; a sigmoid function was used in the output layer. The training of the neural network was held over 50 epochs with the use of mini-batches gradient descent. The size of the mini-batches for training was chosen experimentally and was equal to 512. The loss function was evaluated using cross-entropy. At the end of the research, we also compared the performance of the neural network, gradient boosting [12, 13] over decision trees and random forest [14, 15].

5   Results

The results of training of the neural network on the training sample are shown in Fig. 3.

According to the graphs the results of the training show good outcome. However, this does not say anything about the actual quality of the algorithm.

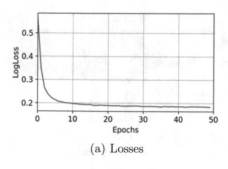

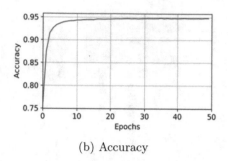

(a) Losses   (b) Accuracy

Fig. 3. ANN training results.

To check the algorithm it is necessary to evaluate its quality on the test sample. Testing was conducted on 40,000 trips. Since we solved the problem of binary classification, we used the following estimates: Precision, Recall, AUC-ROC [9], AUC-PR [10], F-measure [11]. The results of the algorithms are presented in Table 1, Figs. 3 and 4. Table 1 shows the comparative characteristics of algorithms. About 40% of the features in our data are categorical, and it is shown that the gradient boosting and random forests work well enough with these features.

Figure 4 shows confusion matrix of a test sample, by using ANN, Gradient Boosting, Random Forest. In the Fig. 4 of confusion matrix, you can see that high results are shown by two algorithms - Gradient Boosting and ANN. But, the problem of ANN algorithm is the large number of beta errors which may lead to violation's misses. In this case, it is better to determine an excess security breach than to miss it. Thus, the algorithm using gradient boosting over deciding trees showed the best results. This can be seen in both Fig. 5 and Table 1.

Table 1. Comparison of the efficiency of algorithms for the binary classification problem

Metric	ANN	GradientBoost	RandomForest
Accuracy	0.9684	0.9697	0.9106
Precision	1.0	0.9655	0.6213
Recall	0.8916	0.9304	0.7312
AUC-PR	0.9720	0.9796	0.9415
AUC-ROC	0.9690	0.9839	0.9610
F-measure	0.9427	0.9477	0.8268

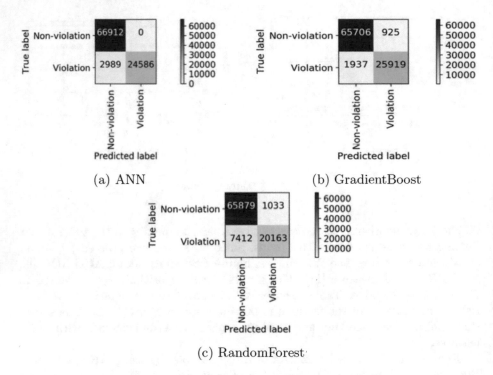

Fig. 4. Confusion matrix for different algorithms.

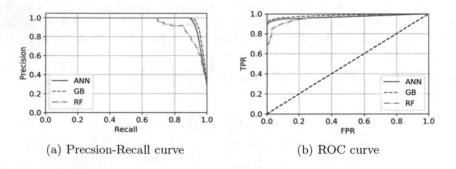

Fig. 5. Functions of the algorithms' work quality.

6   Conclusion

The research showed a sufficiently high quality of results, which does not exclude the possibility that the algorithms have been overfitted. This paper describes the first attempt of applying machine learning methods in Russian Railways. At the present time it is planned to use the results of this research in Russian Railways operational activities and continue further development of the research in the following areas:

1. Collect more data and expanding the number of features.
2. Develop some methods for processing categorical features.
3. Take the appearance of a disturbance into account by analyzing and predicting time series.
4. Search for a universal algorithm to determine the probability of a violation.
5. Establish a risk assessment and risk management system for locomotive teams.

References

1. Gapanovich, V.A., et al.: System of adaptive management of railway transport infrastructure technical maintenance (URRAN project). Reliab. Theory Appl. 10(2)(37) (2015)
2. Gapanovich, V.A., Zamyshlyaev, A.M., Shubinsky, I.B.: Some issues of resource and risk management on railway transport based on the condition of operational dependability and safety of facilities and processes (URRAN project). Dependability 1, 2–8 (2011)
3. Shubinsky, I.B., Zamyshlyaev, A.M.: Main scientific and practical results of URRAN system development. Zheleznodorozhnyi transport 10, 23–28 (2012)
4. Suzuki, K., et al.: Computer-aided diagnostic scheme for distinction between benign and malignant nodules in thoracic low-dose CT by use of massive training artificial neural network. IEEE Trans. Med. Imaging 24(9), 1138–1150 (2005)
5. Nakamura, K., et al.: Computerized analysis of the likelihood of malignancy in solitary pulmonary nodules with use of artificial neural networks. Radiology 214(3), 823–830 (2000)
6. Turkson, R.E., Baagyere, E.Y., Wenya, G.E.: A machine learning approach for predicting bank credit worthiness. In: International Conference on Artificial Intelligence and Pattern Recognition (AIPR), pp. 1–7. IEEE (2016)
7. Kulagin, M.A., Sidorenko, V.G.: The approach to the formation of a drivers rating using different comparison metrics. Electroni. Electr. Equip. Transp. (EET) 1, 14–17 (2018)
8. Goodfellow, I., et al.: Deep Learning, vol. 1. MIT press, Cambridge (2016)
9. Bradley, A.P.: The use of the area under the ROC curve in the evaluation of machine learning algorithms. Pattern Recognit. 30(7), 1145–1159 (1997)
10. Boyd, K., Eng, K.H., Page, C.D.: Area under the precision-recall curve: point estimates and confidence intervals. In: Joint European Conference on Machine Learning and Knowledge Discovery in Databases. Springer, Heidelberg, pp. 451–466 (2013)
11. Powers, D.M.: Evaluation: from precision, recall and F-measure to ROC, informedness, markedness and correlation (2011)
12. Ye, J., et al.: Stochastic gradient boosted distributed decision trees. In: Proceedings of the 18th ACM Conference on Information and Knowledge Management, pp. 2061–2064. ACM (2009)
13. Natekin, A., Knoll, A.: Gradient boosting machines, a tutorial. Front. Neurorobotics 7, 21 (2013)

14. Liaw, A., et al.: Classification and regression by randomForest. R news **2**(3), 18–22 (2002)
15. Müller, A.C., Guido, S.: Introduction to Machine Learning with Python: A Guide for Data Scientists. O'Reilly Media, Inc. (2016)

Analysis of Options for Track Development of a Railway Station Using Graph Theory and Logic Modeling

Vera V. Ilicheva[✉]

Rostov State Transport University (RSTU), Rostov-on-Don, Russia
vilicheva@yandex.ru

Abstract. In this paper we offer use of logic prototyping and methods of the theory of graphs for the analysis of versions of existing and designed transport structures. Questions of maintenance of safety of transformations with preservation of basic functional of stations are considered. The conditions of correctness of the project modifications within the logic prototyping are defined. Logic modelling is used for the diagnosis of logic errors: uncertainty, contradictions, and an impracticability of the set restrictions. The result of prototyping supposes adequate graphic representation. The theory of graphs gives convenient means for the analysis of vulnerability of the obtained structure to occurrence of emergencies and breaking down of connections between nodes of a transport network. We offer methods of allocation of the station framework, revealing divisions of the transport structure that provide its integrity and indicators. Allocating various types of graphs it is possible to organise accident-free traffic on critical intersections of routes with a minimum of delays and preservation of traffic capacity. The method of regulation of movement at intersections of routes is offered. The approach is approved in the task of travelling development of cargo station "Taganrog".

Keywords: Logic prototype · Transport structure · Model transformation
Theory of graphs

1 Introduction

Qualitative designing of complex systems needs a careful analysis of the structure and interrelations of elements of the project as well as estimation of consequences of accepted decisions. Numerous methods and means of prototyping serve this purpose. In [1] you can read about the use of logic prototypes for the analysis of transport systems. The model (prototype) is a result of work of an interpreter of a logic specification that is presented by a class of formulas in the language of calculation of the first order with equality. The interpreter realises a direct logic derivation, diagnosing the following project errors: uncertainty, contradictions, and an impracticability of the set restrictions. The approach allows revealing a project inconsistency at all stages of design.

Nevertheless, the logic approach is of little use for optimisation problems. Development of a transport system can demand the change of travelling infrastructure. The choice of an optimum variant of designing is influenced by the following criteria:

© Springer Nature Switzerland AG 2019
A. Abraham et al. (Eds.): IITI 2018, AISC 875, pp. 317–326, 2019.
https://doi.org/10.1007/978-3-030-01821-4_34

safety maintenance, maximum of simultaneous movement, and minimum of rolling stock delay. Besides logic prototyping, we offer to use the theory of graphs and means of linear optimisation in order to research travelling transformations.

In this paper we investigate the conditions of correctness of the project modifications within the logic prototyping. We also define the correctness criteria of various transformations of the obtained models and their specifications. We carry out project research by means of the theory of graphs to analyze stability and safety of a transport network. Here you can find methods of allocation of the station framework, revealing divisions of the transport structure that provide its integrity and indicators. The method of the analysis of critical intersection of routes is examined. It allows comparing different variants of designing of travelling development and specialisation of tracks according to criterion of minimisation of delay time of a rolling stock. The offered approach has been analyzed by example of a structure of a port cargo station "Taganrog".

2 Modifications of Logic Prototypes

Paper [1] offers use of logic prototypes while designing and analyzing complex transport systems. Let us remind the main idea of the approach. The logic prototype (logic model) of an investigated object or process is the result of executable interpretation of the logic specification that describes a static and/or dynamic system structure and its restrictions. There is a class of formulas used for axiomatics, diagnosed errors of a prototype, and a method of interpretation of descriptions. More precisely, the specification $S = (\Sigma, T)$ of the investigated system is made by the signature Σ and the theory $T = T_{fact} \cup T_{def} \cup T_{rest}$. The signature (i.e. alphabet) $\Sigma = (R, F, C)$ specifies sets of names of relations (predicates R), names of attributes (functions F) and constants C. For C the following is carried out: $C \neq \emptyset, \forall c_i, c_j \in C(c_i \neq c_j)$. Constants cipher objects of a object domain (for example, destinations, parks, directions, tracks, devices, values) that are essential for prototype construction. The theory is represented by axioms which are logic formulas without free variables (all variables are connected by quantifiers or are presented by constants). Axioms T are divided into three groups: T_{fact} is a set of the "real" facts which are atomic formulas with constants instead of variables; the formulas are as follows: $r(\bar{c}), \neg r(\bar{c}), f(\bar{c}) = c$. T_{def} is a set of definitions of relations (dependences between objects) and functions (attributes of objects); the formulas are as follows: $\forall \bar{x}(\varphi(\bar{x}) \rightarrow \psi(\bar{x}))$ or $\forall x \in \theta(\varphi(\bar{x}) \rightarrow \psi(\bar{x}))$, where $\bar{x}$ is a vector of variables, $\varphi(\bar{x})$ is a conjunction of the following formulas $r_i(\bar{t})$, $t_i(\bar{x}) = t_j(\bar{x})$ or their negations ($r_i \in R, t_i, t_j, \bar{t}$ are terms of signature Σ); $\psi(\bar{x})$ is a conjunction of atomic formulas of the following kind: $r_k(\bar{x})$ (definition of the relation $r_k \in R$) or $f_k(\bar{x}) = t_k(\bar{x})$ (definition of the function $f_k \in F$), θ is the finite list. T_{rest} is the set of restrictions on these dependences and values, they are set by any formulas of the first order with the limited quantifiers. The multisortable signature is used to provide totality of functions on a certain set of constants (domain). The specification includes the description of sorts (subsets of constants C), and sorts of their arguments and values are specified for each relation and function. We should notice that the accessory of values of variables to a finite sort (domain) can be considered as quantifier restriction. Here are examples of descriptions of objects of the signature, axioms T_{def} and T_{rest}:

> *Sorts:* OBJECT = FLEET ∪ PORT ∪ PLANT ∪ MAIN.
> *Relations:* is_way (from, to): OBJECT × OBJECT.
> *Functions:* type (of what): OBJECT ∪ TRACK → OBJECT ∪ TRACK_NUMBER.
> *Constants:* 'B', 'G', 'E', 'A', 'Seaport', 'Main', 'Plant': → OBJECT.
> *Definitions:* ∀ x, y, z (is_way (x, y) & is_way (y, z) → is_transit (x, z)).
> *Restrictions:* ¬ ∃ x, y (type (y) = 'Main' & ¬ is_transit (x, y));

The concept of correctness has been defined for specifications; it includes properties of completeness of functions definition, consistency (concerning equality and use of negations). Logic semantics of specifications is represented by initial (minimum as far as homomorphic enclosure is concerned) inductive-computable models from constants C, possessing property of uniqueness to the category of Σ - generated models [2, 3]. This model $M(C, I)$ can be designed according to the description $T_{fact} \cup T_{def}$, its functions and relations are determined by the interpreter in the course of a direct logic derivation. Actually, the interpreter realises function I, which sets predicates from R and functions from F at the set of constants C. It has been proved that there is one-to-one correspondence between correctness of the specification and existence of model of this type for such description. The estimation of correctness of a prototype includes the check of feasibility (validity) of axioms T_{rest} on model M.

An important advantage of a logic approach is possibility, knowing local "laws", dependences and interactions, to receive a global picture of an event as a result: all set of consequences from these local postulates. Design errors are found out at early stages. Prediction errors are diagnosed by the interpreter while checking the axioms T_{rest}, unexpected errors (uncertainty of attributes, contradictions) can be found during interpretation of axioms $T_{fact} \cup T_{def}$. Specifications are easily modified, replenished, and integrated, forming axiom libraries.

During designing there might be situations when it is required to "increase" the model (for example, specification, detailed elaboration of already approved prototype) or to change, transform a prototype, thus saving properties (for example, significant functional, invariant). So, a question arises: how the specification will thus be transformed, and how these impacts can influence its correctness. From the formal point of view, transformation of a logic model can be presented as a combination of some base transformations. We will examine the basic ones.

Expansion by extension of functions and relations at the initial set C. Let $\tau_1, \tau_2, \ldots, \tau_n$ be the sequence of transformations performing transformation of an object (a project, a system) from a state Δ_1 into a state Δ_n. Logic representation of transformation τ_i is specification $S_i = (\Sigma_i, T_i)$, and process of transformation consists in interpretation of S_i that is obtaining the minimal model $M_i = (C_i, I_i)$, representing semantics S_i. We shall notice that extension of a predicate r and functions f means expansion of an area of the truth for r and the domain for f. Then the condition of admissibility of such a transformation are the following statements.

P1: Transformation τ_i is correct if and only if T_i has model M_i. It is equivalent to a correctness of specification S_i.

Condition P1 does not limit the behaviour of sequence of transformations. The correctness of all cycle of transformations can demand the performance of additional restrictions. If for every i $\Sigma_i = \Sigma_{i+1}$, then every I_{i+1} is obtained by extension I_i. Then we have the following:

P2: The sequence $\tau_1, \tau_2, \ldots, \tau_n$ of transformations is correct, if every τ_i is correct and $M_1 \ll M_2 \ll \ldots \ll M_n$ where $\ll$ is the relation of homomorphic enclosure.

Condition P2 is weak for sequence of transformations. It does not demand a correctness of the cumulative specification $S = (\Sigma, T_1 \cup T_2 \cup \ldots \cup T_n)$. As long as formulas with negation are quite possible in axioms, and derivation with the facts obtained at step i is not repeated for specifications S_j, $j < i$, the minimal model for S might exist, but it might not be union of models M_i. For example, in model M_j some predicate $p(\bar{c})$ is fixed to be false, i.e. there is a fact $\neg p(\bar{c})$, but in model M_{j+1} the following consequence is drawn: $p(\bar{c})$. There is homomorphism $M_j \ll M_{j+1}$; however union $M_j \cup M_{j+1}$ is not a model of specification $S_j \cup S_{j+1}$, because it is inconsistent.

Expansion by addition of new elements, relations and attributes (signature expansion). If during conversion $\tau_i \rightarrow \tau_{i+1}$ the signature is $\Sigma_i \subset \Sigma_{i+1}$, the correctness of sequence of transformations is maintained by the following condition:

P3: The sequence of transformations $\tau_i \rightarrow \tau_{i+1}$ is correct if every of them is correct, and model M_i is a subsystem of M_{i+1} in the signature $\Sigma_i \cup \Sigma_{i+1}$, and restriction of interpreting function I_{i+1} on the former carrier coincides with function I_i, i.e. $I_{i+1} \upharpoonright ((R_i \cup F_i \cup C_i) \times C_i^*) = I_i$.

Narrowing of initial logic structure (removal of elements). This generally non-monotonic transformation, nevertheless, can be presented within a considered class of specifications. The variable of a states running a finite interval of «steps» , for example, $\Delta = [\varsigma_1, \ldots, \varsigma_n], \varsigma_j \in C$ is added into axioms T_i: $\forall y \in \Delta, \forall \bar{x}(\phi(\bar{x}, y) \rightarrow \psi(\bar{x}, next(y)))$. Then correctness of transformations τ_i is determined by condition P1, and restriction of the interpreting function should be carried out for their sequence: $I_{i+1} = I_i \upharpoonright ((R_i \cup F_i \cup C_i) \times C_i^* \times \varsigma_{i+1})$. Addition of the variable ς_k into the vector of arguments marks all the elements of the relations that should «be left» . It is not always convenient if the recursive rule works, as in case of transitive closure of the relation. In advance it is impossible to determine the demanded number of steps and to set an axiom of «the last step» . In these cases it is more convenient to use the relation «is deleted» , marking elements which can be «thrown out» .

In some papers the negation is entered into the right part of the formula for the specification of a deleted element of the relation: $\forall \bar{x}(\varphi(\bar{x}) \rightarrow \neg r(\bar{x}))$. Such approach, however, demands determination of non-standard (not logic) semantics of negation since the use of usual semantics will lead to the contradiction (to generation and negation of an element at the same time).

If transformations are carried out asynchronously or in parallel, or there takes place a multi-step recursive transformation, a necessary condition of correctness of transformations is existence for cumulative specification S of the minimal model, isomorphic to union of models M_i.

As a rule, conditions of correctness of transformations are connected not only with a problem of "rightness" of specifications of transformations, but also with a object domain, and with the purposes of transformations that have been done. Preservation of some necessary properties (invariants of transformations) can be specified obviously, in the form of the axioms-restrictions checked on already received model that is a result of transformation, or it can be a consequence of the correctness conditions that are imposed on the sequence of transformations as a whole. So, for example, condition P2 guarantees preservation of the necessary functions.

The obtained set of the logic consequences that is forming a prototype formally forms a new theory T_{fact} that represents a set of the old and new facts, which can be added to another specification or new transformation.

The result of interpretation is the facts. They can be presented in a kind of the attributed graph where nodes of a graph are signature constants, and edges of a graph are predicates, true on these constants. Attributes of the nodes are values of the functions determined on these constants. Graphic representation can be very convenient for study of transformations by means of the theory of graphs.

3   Extracting the Framework of a Station

The scheme of the port cargo station (PCS) can be presented in the form of digraph TG (V, J), where V is a set of nodes, and J is a set of arcs, i.e. ordered couples of elements from V. The basic objects of station (parks, devices) are considered to be nodes of the digraph. The arc is drawn from point v_i to point v_j if it is admissible to move directly from v_i to v_j. During the analysis of any railway station performance it is very important to allocate its basic structure – "framework" where main technological operations and their sequence take place. The theory of graphs [4] allows presenting such a structure visually, and you can carry out research of both separate parts and all system as a whole. The following algorithm is offered.

There is a minimum set of nodes, the other nodes are achievable from them all, i.e. you can find a vertex base of digraph TG. After that you can transit to a more simple digraph TG^* that is condensation of an initial digraph TG. In graph TG you can see strongly coherent generated subgraphs with maximum on inclusion set of nodes – strong components. For this purpose you should construct reachability matrix H (TG), that is determined as follows: $h_{ij} = 1$, if v_j is achievable from v_i; $h_{ij} = 0$ if not. Products H^2 and $H \times H^\tau$, where H^τ is matrix transported to H, allow allocating in digraph TG strong components K_i. In the received graph condensation TG^* is determined by node base B^*. In this case every node base in TG can be obtained from the base TG^*, choosing one node from every strong components TG that belongs to B^*.

The analysis of the obtained digraph of condensation TG^* shows that its structure (nodes and arcs between them) coincides with station "framework" – objects where there are basic operations, and interrelations between them.

Received "framework", as well as any graph, can be presented as a set of facts (logic model) with a unique predicate is_way (x,), set on the set of constants K_i. This set of the facts should be present at each correct transformation, i.e. authentic logic model of the project.

4   Research of Reliability of Transport Network

This problem comprises two stages. Firstly, divisions of the transport structure are determined; their removal leads to destruction of integrity of a network, and technologically important parts of the system will turn out to be unreachable. In graph representation this problem is solved by finding bridges in a digraph TG - the edges $\{v_i, v_j\}$, their delete preserving nodes v_i and v_j leads to a disconnected graph. This procedure also allows finding divisions without bridges that are connected graphs with strong orientation.

For example, in a digraph TG, that corresponds to the structure PCS «Taganrog» , a division without bridges turned out to be subgraph K_1', describing that part of station structure where there is a critical point of traverse of conflicting routes. Then on each arc you can set a direction of edges, so that in the obtained digraph from each node any other node and thus a critical point would be achievable. The following procedure can be carried out for this purpose.

There can be applied a well-known algorithm of finding strong coherent orientation of the graph for obtaining of possible variants of subgraphs G which is a spanning tree of algorithm of search of depth 1. Variants are compared according to criterion of determination of diameter of the obtained digraph, i.e. $\max d(v_i, v_j)$ is calculated on all v_i and v_j, $d(v_i, v_j)$ is length of the shortest track from v_i to v_j. It is determined between what nodes the maximum track takes place; the estimation is made how effective is such variant of orientation of edges for station performance. If this track takes place between technologically important objects, the given variant of design should be rejected.

The second investigation phase includes the analysis of structure of station on stability to occurrence of emergencies and communication disturbance between nodes. It is found out how many connections and which of them can be changed (removed or added) in order not to break traffic safety and to save the basic station functional.

For this purpose it is necessary to find a degree of coherence of strong components in digraph TG and to determine their degree of arc vulnerability as the minimum number of arcs; if you remove them, you can obtain the digraph with a smaller category of coherence. The less this degree is, the more is vulnerability of a transport network. If the degree of arc vulnerability does not exceed 1 as the minimum in-degree of nodes (this result can be obtained, if you have analysed adjacency matrix of the given digraphs), the removal of at least one arc from some set will lead to coherence reduction.

Thus, for example, for station "Taganrog" strong components K_1 and K_2 are strongly coherent digraphs, and connectivity degree equals three. Degree of arc vulnerability in both cases is no more than one, i.e. strong components and all digraph TG as a whole are vulnerable enough. Therefore the available infrastructure of a railway station [5] demands development: increase in number of tracks, their reorganisations, junctions by off-ramps. It will increase not only safety, but also terminal capacity of a station under the conditions of its forecasted increase of cargo processing.

From the point of view of logic prototyping, it is necessary to find out whether there is a model during removal of some facts, whether axioms of "framework" and values of the important functions are saved. In logic model besides conditions P1 (regularization of a predicate is_way) and P3 (removal of the facts), it is necessary to provide presence of necessary tracks and preservation of value of function of terminal capacity of a station.

5   Analysis of Critical Intersections of Routes

Logic prototyping is a previous stage of detailed elaboration of device performance for movement regulation on intersections of tracks. There is a set of the models considering casual character of the investigated phenomena at continuous time [6, 7] or using fuzzy logic, neural networks and fuzzy control in solving the problem of fuzziness in intensive traffic streams [8–10]. However, movement of the scheduled railway transportation has discrete character and is often rigidly determined. Therefore application of the theory of graphs for designing of tracks at intersections can be more adequate means for analysis.

You can use interval graphs for controlling phases of switching on clear signals during movement regulation. The approach allows solving an optimising task of minimisation of general time of a delay of a rolling stock on such intersections. By this criterion you can compare different variants of designing of their travelling development and specialisation of tracks.

Let us illustrate this approach from the point of view of a possible reorganisation of critical intersection of conflicting routes of PCS "Taganrog". The station has the following parks: «P» is a passenger park, "D" is a receiving-departure yard, park "F" is a freight yard, park "E" and «G» are marshalling yards. The following routes are crossed at the yard neck: train A, B, and shunting C (Fig. 1). These intersections increase loading of yard necks and reduce terminal capacity.

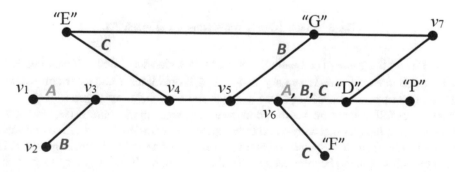

Fig. 1. Intersection of conflicting routes A, B and C (graph TG_1)

For comparison of possible variants of PCS travelling development design we will examine a case of multiple-track intersection: we will add two tracks from park "D". Then route B supposes two compatible variants b_1 and b_2, and route C may have variants c_1 and c_2 (Fig. 2). It means addition of two arcs from node "D" and a new graph TG_2 (Fig. 3) as a result.

Directions which are not crossed, and, hence, do not lead to a crash, and those where simultaneous movement in opposite directions without delays are possible, are compatible.

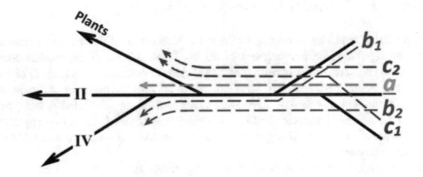

Fig. 2. Multiple-track routes intersection at the critical point at the station

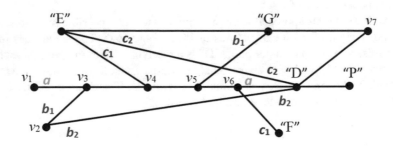

Fig. 3. Graph TG_2 is a transformation of graph TG_1

Let us set the graph of compatibility $S\,(U,\,Q)$ in which set of nodes U are directions of transport movement, and an edge $\gamma\left(u_i, u_j\right) \in Q$ if directions u_i and u_j are compatible. We will construct distribution of arcs for each route on a circumference of a clock, it would be as follows: the arc δ corresponds to a continuous motion interval in a transport direction, and the arcs that correspond to compatible streams are blocked. To each such representation there corresponds a spanning subgraph of arcs of a circumference, that is an intersection graph. For the set graph of compatibility it is necessary to find all spanning subgraphs of arcs of a circumference; then to find the most effective one with minimum time of delay. During one of representations by arcs of a circumference of the graph of compatibility S and a corresponding intersection graph, the transport movement on routes a, b_2, c_2 can be carried out simultaneously. Calculations have shown that delay time has decreased; there is an 8-fold decrease of a shunting route c_1, a 3.5-fold decrease of a train route b_1. The number of the detained movement was reduced: a 4-fold decrease of a shunting route c_1; a 3.5-fold decrease of a train route b_1.

Use of the interval graph for control of a traffic light has allowed solving a problem of minimisation of the general waiting time of transport on an inhibiting signal. It has been found out that this time in case of compatibility of directions a, b_2 c_2 was reduced in 1.7 times. Time of a clear signal for b_1 in the optimum decision has appeared to be equal to zero. It means that movement time in the directions those are least compatible to the others, which also do not perform maximum movement a day, should be minimum in a cycle. Then the basic movements on a route B should be carried out in direction b_2.

The construction of new tracks demands considerable expenses. That is why consideration of variants with smaller quantity of compatible directions is of interest. Transition to interval representation has allowed us to estimate how much delay time will change. The logic description promoted revealing the contradictions connected with impossibility of simultaneous movement on some routes.

6   Conclusion

This paper offers combined use of logic prototyping and elements of the theory of graphs for the analysis of justifiability of existing and projected transport structures. Logic modelling is convenient and useful for the analysis of logic of complex system or process. During prototype construction one can diagnose logic errors and an impracticability of the set restrictions. Modifications of logic prototypes can lead to collapse of the checked up properties. Conditions of a correctness of transformations have been determined.

The result of prototyping supposes adequate graphic representation, therefore further research can be carried out with the use of the theory of graphs. It gives convenient means for the analysis of vulnerability of the obtained structure to occurrence of emergencies and breaking down of connections between nodes of a transport network. Allocating various types of graphs it is possible to organise accident-free traffic on critical intersections of routes with a minimum of delays and preservation of traffic capacity. A computer program has been developed to analyze various variants of traffic control. The program input includes the number of movements in each direction of travel. Output is the transport delay time and the duration of the clear signal. The program confirmed the consistency of model calculations.The approach is approved in the task of travelling development of cargo station "Taganrog".

References

1. Guda, A.N., Ilicheva, V.V., Chislov, O.N.: Executable logic prototypes of systems engineering complexes and processes on railway transport. In: Proceedings of the Second International Scientific Conference Intelligent Information Technologies for Industry (IITI 2017), vol. 2, pp. 161–170 (2017)
2. Ilyicheva, O.A.: Means of effective preliminary treatment of errors for systems of logic prototyping. Autom. Remote Control **9**, 185–196 (1997)
3. Ilyicheva, O.A.: Technology of logic modelling and the analysis of complex systems. Eng. Bull. Don **2**(4) (2012). http://www.ivdon.ru/ru/magazine/archive/n4p2y2012/1234
4. Roberts, F.S.: Discrete mathematical models with application to social, biological and environmental problems. Prentice-Hall, p. 559 (1976)
5. Chislov, O.N., Mamaev, E.A., Guda, A.N., Zubkov, V.N., Finochenko, V.A.: Algorythmic and software support of efficient design of railway transport technological systems. Int. J. Appl. Eng. Res. **11**(23), 11428–11438 (2016)
6. Afanasieva, L.G., Bulinskaya, E.V.: Mathematical models of transport systems based on queuing theory. In: Proceedings of MIPT, vol. 2, № 4, pp. 6–21 (2010)
7. Ahmadinurov, M.M., Timofeeva, G.A.: Models of mass service in the problem of traffic light optimization. In: Vestnik Saratov State Technical University, vol. 3, № 1(57), pp. 217–227 (2011)
8. Yusupbekov, N.R., Marakhimov, A.R., Igamberdiev, H.Z., Umarov, ShX: An adaptive fuzzy-logic traffic control system in conditions of satu-rated transport stream. Sci. World J. **2016**, 1–9 (2016)
9. Hou, R., Wang, Q., Wang, J., Lu, Y., Kim, J.: A fuzzy control method of traffic light with countdown ability. Int. J. Control Autom. **5**(4), 93–102 (2012)
10. Kaedi, M., Movahhedinia, N., Jamshidi, K.: Traffic signal timing using two-dimensional correlation, neuro-fuzzy and queuing based neural networks. Neural Comput. Appl. **17**(2), 193–200 (2008)

Applied Systems

Analyzing Video Information by Monitoring Bioelectric Signals

Natalya Filatova[(⊠)], Konstantin Sidorov[(⊠)], Pavel Shemaev,
Igor Rebrun, and Natalya Bodrina

Tver State Technical University, Lenina Ave. 25, Tver, Russia
nfilatova99@mail.ru, bmisidorov@mail.ru,
pshemaev@rambler.ru, igarrock@mail.ru,
vavilovani@mail.ru

Abstract. The paper considers the problems of creating tools for video information flow automatic analysis through monitoring certain characteristics of bioelectrical signals of an agent (expert, operator).

The authors used spectral analysis methods to obtain time series that illustrate power changes of experimental bioelectric signals (Sa) of an agent during studying a continuous video stream. The fuzzification of spectral features allows moving to fuzzy time series illustrating information evaluation by an agent.

The spectra are calculated using a sliding working window from fragments of two types of bioelectric signals, which are recorded synchronously in two autonomous channels.

Based on fuzzy evaluations of spectral features and Mamdani fuzzy inference algorithm, the algorithm for analyzing video information allows classifying video fragments according to the sign of agent's emotional reaction. Tsukamoto algorithm localizes time markers that determine the beginning and the end of each fragment.

Keywords: Signal analysis · Fuzzy set · Fuzzy signs · Human emotions
Spectral analysis

1 Introduction

The problem of objective evaluation of video information is important for many areas of human activity. Modern technical facilities allow creating large video data archives including the results of monitoring operators, monitoring production and household situations. The problems of automatic analysis of similar objects are considered using various formalization methods. The most common approaches are segmentation [1], automatic referencing [2], semantic indexing [3], annotation [4], ranking and classification of video [5, 6]. In recent years, the interest in content-based video retrieval (CBVR) has been growing. There are publications that raise a question about an

This work was supported by grant from the Russian Foundation for Basic Research (№ 17-01-00742).

emotional semantics of video [7]. This problem seems to be the most difficult for us, since it is related to individual information evaluation, which depends not only on the content, but also on evaluator's state.

In our opinion, one of the ways to solve this problem is using solutions that were obtained in the field of automatic analysis of human emotional reactions [8–11].

Usually, the processes of creating monitoring algorithms and models for interpreting human emotional reactions are autonomous from the research on cognitive processes related to information perception and awareness. Meanwhile, it is obvious that there is some "step" of evaluating the received information before any emotional reaction. In fact, it is a solution of a classification problem ranging the situation in one of the classes known from personal experience. As we do not have knowledge in the structure of such classifiers for now, we can assume that the main result of this process is an emotional reaction sign (a negative or positive reaction). Even if we do not involve the so-called socially induced emotions (conscience, envy, gloating, etc.), we conclude that an emotional reaction requires certain mental processes. It means that emotional and cognitive reactions can be considered as parallel processes [12].

Many authors note the interrelation and mutual influence of these processes [12–14]. Moreover, a number of studies mentions that certain characteristics of intellectual activity (attention, focus on the task, the speed of some mental operations, memory) might change due to the changes in human emotional state. There may be cross-links between an emotional response and cognitive processes [15, 16]. A cognitive process can help reduce the level of emotional reaction, or completely extinguish it. An emotional reaction can accelerate cognitive processes, as well as support associative memories that can generate a solution for an "insight" problem [17].

The enlarged scheme of processing information simulation, which combines emotional and cognitive mechanisms, illustrates the hypothesis of possible using human emotional responses to evaluate information coming from a limited number of channels (Fig. 1).

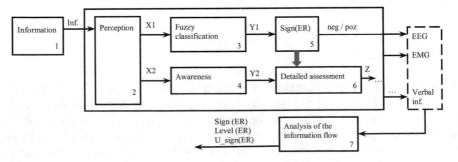

Fig. 1. Simulation scheme of processes related to information processing.

Observation of changes in signal characteristics from the sensors attached to certain points of a human head will allow marking individual video fragments according to a positive or negative information evaluation. The analysis of the array of markers that were united by the results of evaluating one video object by several testees will allow localizing information that causes positive and negative reactions. These data can be considered as objective evaluation of video information, which is not connected to external content.

The report describes the results of testing the possibility of implementing this scheme.

2   An Automated System for Monitoring Operator's Reactions to Information Flow Change

To set up experiments in recording and evaluating human reactions to video content changes, we used an automated system "EEG/Speech/EMG" [9, 18]. The system monitors cognitive activity by recording EEG signals (brain electrical activity), as well as testee's emotional state by recording EMG signals (muscle potentials of facial muscles).

Registration of brain electrical activity signals during video information perceiving by a testee has been performed using the EEGA-21/26-Encephalan-131-03 apparatus. The sensors were fixed on testee's head according to the international system of leads 10–20. The EEG signal sampling frequency was 250 Hz.

The face electromyogram (EMG signals) registration was performed on the left side of testee's face, which is characterized by a clearer mimic reaction [19].

The position of electrodes was chosen based on the experiment purpose and is also related to the work of two muscle groups: corrugator supercilii and zygomaticus major. Usually, when the muscle that wrinkles an eyebrow (corrugator supercilii) moves down, it is associated with the display of anger, sadness, disgust and fear (negative emotions). A big zygomaticus muscle (zygomaticus major) pulls away the corner of the mouth with a smile accompanying positive emotions.

An EMG signal (fEMG) has been recorded according to the Fridlund and Cacioppo method [20]. The recording has been made using Neuro-MEP-4 (Neurosoft) and the Neuro-MEP.NET program. We used bipolar cup Ag/AgCl electrodes with a 3 mm diameter. Testee's skin was degreased with an abrasive paste "Everi" before. To strengthen the contact, the electrodes were covered with an adhesive paste "Ten20". The impedance throughout the experiment did not exceed 10 kOhm. The sampling frequency of the received record was 1000 Hz. To filter the EMG signal, a bandpass filter was used with the parameters $f_l = 20$ Hz and $f_v = 500$ Hz. The choice of the lower limit (f_l) is due to low-frequency noise that occurs when blinking, eye movements, breathing, etc. In addition, there is the Notch filter (f = 50 Hz) to suppress network interference.

3 Marking Information Flows Through a Spectral Analysis of EEG Signals

In order to conduct the experiments, we have prepared two types of files (Sp and Sn). Each of them consisted of several video fragments (s). Moreover, in consultation with testees two rules were observed:

$$Sp = \cup s_i \ (\forall i)s_i \in V_{poz}, \ Sn = \cup s_i(\forall i)s_i \in V_{neg}, \ T(Sp) = T(Sn), \quad (1)$$

where V_{poz}/V_{neg} is a scope of the concepts positive/(negative) evaluations of video examples, $T(S)$ is the length of a file of corresponding type. In the conducted series of experiments $T(S) = 9,5-10$ minutes.

In order to create a basic level of bioelectrical signals, and to speed up the process of recording emotional reactions from viewed information, we used a special file (Sf). It included a frame with a green screen (RGB color model with color parameters: red = 102, green = 153, blue = 0).

During the experiment, a testee has been consistently analyzing information flows: $P1 :: Sf = > Sp = > Sf$ and $P2 :: Sf = > Sn = > Sf$.

In the interval between $P1$ and $P2$, the testee gave a general emotionally motivated evaluation of the presented information (if he liked or disliked).

After the experiment was completed, Sp and Sn files were reproduced once again for the testee who segmented video information into fragments in detail. Each fragment is assigned a marker of an emotional reaction level in points (strong 3, medium 2, weak 1, absent 0, reaction of the opposite sign). Marking results formed three following clusters: (1) fragments with strong emotional reaction, (2) with medium emotional reaction and (3) with weak emotional reaction. A cluster with an absent emotional reaction and a cluster with the opposite sign reaction turned out to be empty.

An electroencephalogram and an electromyogram were recorded for each testee during the perception of information flows.

Since a testee could mark out sections of different duration on a video subjective markup diagram, long fragments were split up into shorter ones (so-called working fragments) when analyzing EEG. At the same time, we tried to avoid including records with eye artifacts in the analyzed section. For all working fragments, we calculated power spectra and found integral estimates of the following form:

$$Azm = \sum_{i=1}^{6} \sum_{j=f_i}^{fk_i} S_j \Delta f, \quad (2)$$

where Azm is absolute power (μV^2), which is determined over the entire frequency range for one lead, i is a frequency interval number, S_j is a value of spectral power at the j-th step, Δf is a Linear frequency step, f_i, (fk_i) is a frequency step number when a frequency corresponds to the beginning (end) of the i-th interval.

According to (2) *Azm* might be split into six components that illustrates brain electrical activity in a certain frequency range (Table 1).

Table 1. Components of *Azm*.

$Azm^1i = 1$	$Azm^2i = 2$	$Azm^3i = 3$	$Azm^4i = 4$	$Azm^5i = 5$	$Azm^6i = 6$
0.2–2 Hz	2–4 Hz	4–8 Hz	8–13 Hz	13–24 Hz	24–35 Hz

Therefore, for each working fragment there is a set

$$CM = \{cm_i\}, i = 1, \ldots n \quad (\forall i)Azm_i = \cup r_{ij}, \quad j = 1, d, \qquad (3)$$

where n is a number of analyzed leads, d is a number of frequency intervals for calculating Azm $(d \leq 6)$ elements.

On the set of *Azm* evaluations, there are three selected subsets that combine the evaluations obtained for time series, which are recorded from the electrodes on the right side of a scalp (*Azm1*), from the left side (*Azm2*) and from the central electrodes (*Azm3*).

It is known that monitoring of brain electrical activity in the frequency range 13 Hz–24 Hz is a way to monitor a sign of human emotional reactions–caused by external video incentives [21].

Figure 2 shows the graphs of *Azm5* change during perceiving a negative video block by a testee. The information flow (*P1*) with positive information has similar results:

$$Azm2^5 > Azm1^5 \text{ and } Azm2^5 > Azm3^5 \qquad (4)$$

The results of the experiments with $Azm2^5$ evaluations illustrate the possibility of separating EEG signals when perceiving information that causes emotional reactions of different sign (Fig. 3).

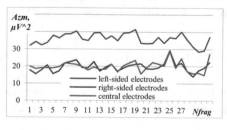

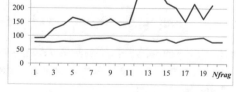

Fig. 2. $Azm1^5$, $Azm2^5$ and $Azm3^5$ change during perceiving a negative video block.

Fig. 3. Azm^5 change during perceiving of *P1* (positive) and *P2* (negative) video blocks.

We used spectral characteristics as a characteristic of brain electrical activity and compared the markings of the information block (for example, *P1*) according to testee's points (*R1*) with *Azm* evaluations for corresponding time intervals (*R2*).

The intersections *R1* and *R2* have been identified for the sections with maximum evaluations. Matching in minimum markups was no more than 30%. Figure 4 shows a markup diagram of a positive video by a testee, as well as averaged estimates of *Azm* for four time intervals.

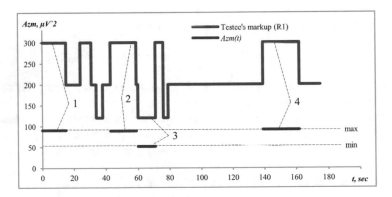

Fig. 4. A testee's markup (*R1*) and averaged estimates of *Azm* for four time intervals of a positive video (*P1*).

Monitoring *Azm(t)* allows selecting video sections that a user marks as positive information. However, it is impossible to divide the video into negative and neutral segments according to *Azm*. To solve this problem, we used EMG signal analysis.

4   Marking Information Flows Using EMG Signal Spectral Analysis

In order to analyze EMG signals, we used a number of *cm* working fragments formed similarly to (3). The absence of artifacts caused by blinking made it possible to introduce an additional restriction:

$$cm_i(\Delta t_j, j), \quad (\forall j)\Delta t_j = 5sek, \quad j = \overline{1, w}, \quad Asm_j = Asm_j^1 \cup Asm_j^2,$$

where *j* is a number of a *i*-th lead working fragments, *w* is a lead number (1 is over an eyebrow and 2 is a zygomaticus muscle), Asm_j^i is an integral evaluation of an *i*-th lead amplitude spectrum.

In the conducted experiments for $i = 1$, the values of the amplitude spectrum in the cm_j fragment illustrated the following dependence:

$$Sn(Tn) \rightarrow cm_i(\Delta t_j, j) \rightarrow Asm_j^1(n), \quad (\forall j)\Delta t_j \in Tn$$
$$Sp(Tp) \rightarrow cm_i(\Delta t_j, j) \rightarrow Asm_j^1(p), \quad (\forall j)\Delta t_j \in Tp$$
$$Asm_j^1(n) > Asm_j^1(p)$$

When information evaluations were negative, the amplitude spectrum level increased significantly (sometimes by 90%) compared with estimates of positive or neutral information fragments (Fig. 5). The situation is reverse for $i = 2$ (Fig. 6).

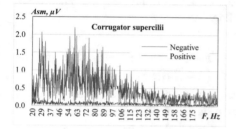

Fig. 5. Amplitude spectra of signals registered at the point ($i = 1$).

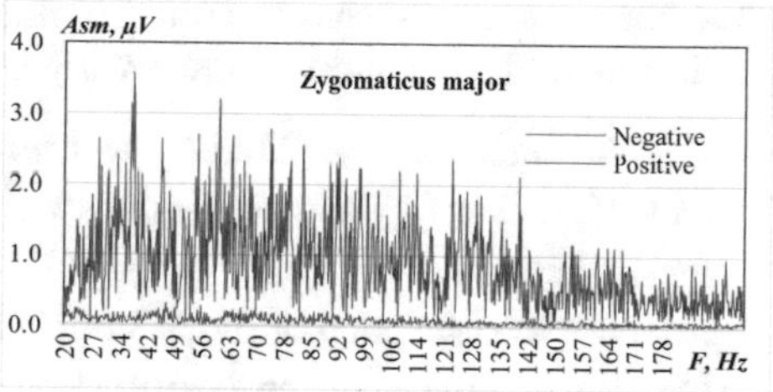

Fig. 6. Amplitude spectra of signals registered at the point ($i = 2$).

Comparison of the marking ($R1$) of the information block ($P1$) according to testee's points with averaged evaluations $\overline{Asm^2}$ illustrates that there is an interrelation between an EMG signal level and a point scale (Fig. 4). The transition from the informational fragment, which a testee rated with a higher score, to a less expressive one (2 points) is accompanied by a 30% decrease in the evaluation $\overline{Asm^2}$, etc. (Fig. 7).

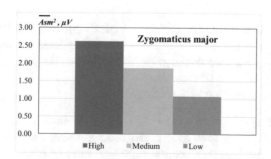

Fig. 7. Comparison of averaged amplitude spectra at a different level of perception of a positive video block.

5 The Algorithm for Analyzing Video Information

The results show that the problem of automatic analysis of video information in order to distinguish fragments with different sign and level of caused emotions can be solved by monitoring spectral characteristics of two types of signals.

Analysis of Asm estimate results in two adjacent calculation windows (j) and $(j+1)$ allows determining the emotional evaluation sign of the information perceived by a user. Moreover, the Azm evaluation is a basis for specifying the time interval boundary for which the found marking is valid.

However, when developing the algorithm, it is necessary to take into account the strong impact of the overall content on our assessments, i.e. the influence of our psychophysiological state on information perception. The experiments showed a scatter in values of spectral characteristics. It leads to the need to move from a quantitative scale for spectral estimates to corresponding linguistic scales.

The algorithm includes the following steps:

1. Determination of amplitude spectrum evaluations on the j-th time interval $(Asm^1(j), Asm^2(j))$ and signal power spectra $(Asm_i^5(j), i = \overline{1, 19})$.
2. Fuzzification $Asm^{1,2}(j)$ and $\{Azm_i^5(j)\}$. Determination of the level of emotional response to perception of the j-th information fragment in the form of two kinds of fuzzy evaluations $\tilde{A}sm^r$ and $\tilde{A}zm_i^5$. Here we use triangular adherence functions and a term set: vary_small, small, mid, big.
3. Determination of the sign of an emotional reaction to perceiving the j-th information fragment $(sign(j))$:

$$
\begin{aligned}
\neg(\tilde{A}sm^1 = vary\_small) \wedge (\tilde{A}sm^2 = vary\_small) &\rightarrow \\
Z = negative(sign = -1) & \\
\neg(\tilde{A}sm^2 = vary\_small) \wedge (\tilde{A}sm^1 = vary\_small) &\rightarrow \\
Z = positive(sign = +1) & \\
(\tilde{A}sm^1 = vary\_small) \wedge (\tilde{A}sm^2 = vary\_small) &\rightarrow \\
Z = neutral(sign = 0) &
\end{aligned}
\tag{5}
$$

4. Using Mamdani algorithm [22] we form a logical conclusion and determine the sign of an emotional reaction to information in the j-th calculation window.
5. Left marker (t_s), position recording; the marker opens the j-th calculation window: $(t_s(j) = t_k(j-1))$.
6. Right marker $t_k(j)$ position determining:

- if $sign(j) = sign(j-1)$, then $t_k(j) = t_k(j-1) + \Delta$, move to 7.
- if $sign(j) \neq sign(j-1)$, then $t_k(j)$ position determining through Tsukamoto fuzzy inference algorithm [22]. We use the following kind of rules:

$$
If (Z_j \neq Z_{j-1}), \quad than \quad y = 1 - C_j(t)
\tag{6}
$$

where $C_j(t) = b_0 + b_1 t$, $\quad b_0 = j$, $\quad b_1 = 1/\Delta$ (Fig. 8).

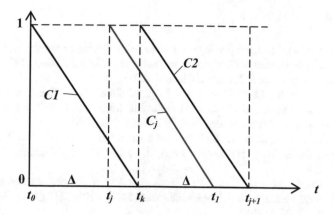

Fig. 8. Correction of the variable match function with consequent rule (6).

7. If video file scanning is finished, then it is the end; if not, we analyze EEG and EMG signals that are synchronous with the $(j+1)$ -th video file interval, then move to the first step.

When determining the position of markers on the time axis that allocate video sections causing reactions with close emotional evaluations, the assumptions are the following:

- there is a lag between the processes of information perception and evaluation by a testee and the changes in integral characteristics of his bioelectrical signals $\tilde{A}sm^r$ and $\tilde{A}zm_i^5$,
- $t_k(j)$ marker position does not coincide with the limits of the interval allocated for calculating $[(t_s), (t_k)]$, $\Delta = t_k - t_s$ signal spectra.

If a change in the fuzzy evaluation $\tilde{A}zm_i^5$ (transition to another fuzzy set) occurs during the transition from $(j-1)$-th to j-th calculated window, then a correction of t_s time marker is necessary for the next calculation window. For this purpose, when moving to the next $(j+1)$-th fragment, the linear dependence $C2$ is replaced by C_j through correction of the parameters $t_s(j+1) = t_j$, $t_k(j+1) = t_j + \Delta$ (Fig. 8).

6   Conclusion

It is possible to consider the obtained results as experimental proof of the possibility to use markers of emotional characteristics of bioelectric signals for marking video information.

The development of this approach will allow introducing bioinspired feedback in automated systems, as well as controlling an information flow achieving a creep of operator's characteristics in a necessary direction.

The proposed approach can be applied in the industry for monitoring operators' state in various technical systems.

References

1. Vijaya, K., Sridhar, I.: Automated tagging to enable fine-grained browsing of lecture videos. In: 2011 IEEE International Conference on Technology for Education (T4E), pp. 96–102. IEEE Computer Society, IIT Madras, Chennai (2011)
2. Wang, M., Hong, R., Li, G., Zha, Z.-J., Yan, S., Chua, T.-S.: Event driven web video summarization by tag localization and key-shot identification. IEEE Trans. Multimedia **14**(4), 975–985 (2012). https://doi.org/10.1109/TMM.2012.2185041
3. Chen, X., Hero, A.O., Savarese, S.: Multimodal video indexing and retrieval using directed information. IEEE Trans. Multimedia **14**(1), 3–16 (2012). https://doi.org/10.1109/TMM. 2011.2167223
4. Zhang, T., Xu, C., Zhu, G., Liu, S., Lu, H.: A generic framework for video annotation via semi-supervised learning. IEEE Trans. Multimedia **14**(4), 1206–1219 (2012). https://doi.org/ 10.1109/TMM.2012.2191944
5. Yu, H.Q., Pedrinaci, C., Dietze, S., Domingue, J.: Using linked data to annotate and search educational video resources for supporting distance learning. IEEE Trans. Learn. Technol. **5**(2), 130–142 (2012). https://doi.org/10.1109/TLT.2012.1
6. Nikitin, I.K.: An overview of complex content-based video retrieval methods. Bull. Novosibirsk State Univ. Inf. Technol. Series **12**(4), 71–82 (2014). (in Russ., Vestnik NGU, seriya: Informatsionnyye tekhnologii)
7. Tamizharasan, C., Chandrakala, S.: A survey on multimodal content based video retrieval. Int. J. Emerg. Technol. Adv. Eng. **3**(1), 69–75 (2013)
8. Abhang, P.A., Gawali, B.W.: Correlation of EEG images and speech signals for emotion analysis. Br. J. Appl. Sci. Technol. **10**(5), 1–13 (2015). https://doi.org/10.9734/BJAST/2015/ 19000
9. Filatova, N.N., Sidorov, K.V., Iliasov, L.V.: Automated system for analyzing and interpreting nonverbal information. Int. J. Appl. Eng. Res. **10**(24), 45741–45749 (2015)
10. Koelstra, S.: Affective and Implicit Tagging using Facial Expressions and Electroencephalography. Ph.d. Dissertation. Queen Mary University of London (2012)
11. Lokannavar, S., Lahane, P., Gangurde, A., Chidre, P.: Emotion recognition using EEG signals. Int. J. Adv. Res. Comput. Commun. Eng. **4**(5), 54–56 (2015). https://doi.org/10. 17148/IJARCCE.2015.4512
12. Baars, B.J., Gage, N.M.: Cognition, Brain, and Consciousness, Introduction to Cognitive Neuroscience, 2nd edn. Elsevier Ltd., Oxford (2010)
13. Marsella, S., Gratch, J.: EMA: a computational model of appraisal dynamics. Cogn. Syst. Res. **10**(1), 70–90 (2009). https://doi.org/10.1016/j.cogsys.2008.03.005
14. Ortony, A., Clore, G.L., Collins, A.: The Cognitive Structure of Emotions. Cambridge University Press, Cambridge (1990)
15. Lazarus, R.S.: Cognition and motivation in emotion. Am. Psychol. **46**, 352–367 (1991). https://doi.org/10.1037/0003-066X.46.4.352
16. Rabinovich, M.I., Muezzinoglu, M.K.: Nonlinear dynamics of the brain: emotion and cognition. Adv. Phy. Sci. **180**(4), 371–387 (2010). https://doi.org/10.3367/ufnr.0180. 201004b.0371. (in Russ., Uspekhi Fizicheskikh Nauk)
17. Isen, A.M., Daubman, K.A., Nowicki, G.P.: Positive affect facilitates creative problem solving. J. Pers. Soc. Psychol. **52**(6), 1122–1131 (1987). https://doi.org/10.1037/0022-3514. 52.6.1122

18. Filatova, N.N., Sidorov, K.V., Terekhin, S.A., Vinogradov, G.P.: The system for the study of the dynamics of human emotional response using fuzzy trends. In: Abraham, A., et al. (Eds.): Proceedings of the First International Scientific Conference Intelligent Information Technologies for Industry (IITI 2016), Advances in Intelligent Systems and Computing 451, vol. 2, part III, pp. 175–184. Springer, Switzerland (2016). https://doi.org/10.1007/978-3-319-33816-3_18
19. Dimberg, U., Petterson, M.: Facial reactions to happy and angry facial expressions: Evidence for right hemisphere dominance. Psychophysiology **37**(5), 693–696 (2000). https://doi.org/10.1111/1469-8986.3750693
20. Fridlund, A.J., Cacioppo, J.T.: Guidelines for human electromyographic research. Psychophysiology **23**(5), 567–589 (1986). https://doi.org/10.1111/j.1469-8986.1986.tb00676.x
21. Muller, M.M., Keil, A., Gruber, Th, Elbert, Th: Processing of affective pictures modulates right-hemispheric gamma band EEG activity. Clin. Neurophysiol. **110**(11), 1913–1920 (1999). https://doi.org/10.1016/S1388-2457(99)00151-0
22. Borisov, V.V., Fedulov, A.S., Zernov, M.M.: Fundamentals of Fuzzy Inference. Hot Line - Telecom, Moscow (2014)

Method of Detecting and Blocking an Attacker in a Group of Mobile Robots

Alexander Basan, Elena Basan$^{(\boxtimes)}$, and Oleg Makarevich

Southern Federal University, Chekov st., 2, 347922 Taganrog,
Russian Federation
{asbasan, ebasan, obmakarevich}@sfedu.ru

Abstract. This article discusses the problem of increasing the resistance to attack by an attacker of a mobile robot groups. The study is limited to the Sibyl attacks and distributed denial of service. An attacker affects the network by sending a large number of packets, or by redirecting the impact on himself. An attacker influences the availability of a group of mobile robots by exerting this impact. This is reflected in the fact that the unit's battery becomes exhausted quickly. Nodes constantly remain in the active mode, receive packets of an attacker and respond to them. This affects the power consumption of the mobile robot. And also, the attacker also affects the network bandwidth. An attacker blocks useful network traffic by sending a large number of packets to the network and flooding it with requests. Due to congestion of wireless communication channels, collisions and queues of packets may occur, and useful packets with important information can be dropped by mobile robots. The developed method for detecting and blocking malicious nodes is based on an analysis of the parameters of the residual energy and the number of sent/received/redirected/dropped network packets. It is possible to identify a node that demonstrates abnormal behavior by analyzing the degree of deviation of these indicators of each mobile robot relative to the indices of the robot group. Probabilistic methods and methods of mathematical statistics are used for the proposed analysis.

Keywords: Mobile robots · Protection · Vulnerability · Anomalies
Attacks · Credibility · Trust · Group management · Security

1 Introduction

The relevance of security problems of mobile robot networks is related to a presence of contradictions between a growing popularity of such networks and a large number of security vulnerabilities. To date, the growth in the popularity of mobile robots, for example, according to the "National Association of Market Participants in Robotics" only in industrial robotics from 2010 to 2014, the average growth in sales per year was 17%. In 2014, 229 thousand robotic complexes were sold for use in industry. The main purpose of a robotic system is to monitor and manage the object, as well as perform the tasks assigned to them. The main vulnerabilities of mobile robot networks are:

© Springer Nature Switzerland AG 2019
A. Abraham et al. (Eds.): IITI 2018, AISC 875, pp. 340–349, 2019.
https://doi.org/10.1007/978-3-030-01821-4_36

- A wireless data medium that allows an attacker to analyze the transmitted data by intercepting them [1];
- Possibility of data modification, as well as the implementation of own data, due to the insecurity of channels and network protocols [2];
- Physical insecurity of autonomous mobile robots, leading to the possibility of reverse engineering;
- The location of mobile devices outside the controlled area, can lead to their interception, as well as the ability of an attacker to influence the environmental parameters, which can lead to disruption of the system.

When developing a system for detecting attacks by an attacker on a mobile robot network, it is also necessary to take into account the fact that non-standard network protocols can be used to transfer data to that network. In addition, robots do not always behave alone; they can implement different behavior scenarios and perform different actions depending on what goal they perform.

2 Related Works

Let's consider some works of authors devoted to the problem of detecting abnormal behavior and malefactors for single mobile robots or groups of robots with centralized control. In [3], authors considered an attack detection system based on the decision tree using the C5.0 algorithm applied to a group of robotic vehicles. The advantage of the presented approach is that for a detection of cyber-attacks, the authors use four characteristics for analyzing the process of communication and information processing, which are called cyber input functions. And also use four parameters to analyze the physical properties of the robot, which the authors call the physical characteristics of the input signal. The authors carried out 5 types of destructive influence on the robot, and get a set of rules for building a decision tree. The disadvantage of the approach is that in this work, attacks on one robot, rather than on a network of robots are conducted. At the same time, the authors considered a limited set of attacks: denial-of-service attacks and attacks aimed at violating physical parameters. Furthermore, for such systems rules must be constantly adding to detect new attacks. This system is aimed at ensuring the availability of the transmitted data.

In [4] an intrusion detection system based on a signature analysis is considered. Authors conducted a series of experiments to create a standard template describing a normal behavior of a robot in the absence of any external influence, as well as random behavioral anomalies. Then a number of situations in which abnormal behavior caused by environmental conditions took place were modeled. The model of normal behavior of the robot, taking into account the weight coefficients, is based on the collected data. The authors conducted a series of attacks that affect a certain set of physical parameters of the node. This approach demonstrates greater efficiency in detecting a malicious node than a simple signature analysis; however, there are some disadvantages: necessity of constant updating and updating of a database of signatures for data control from unaccounted sensors of the mobile robot; conducting analysis of changes only physical parameters of the node and the absence analysis of network data.

The article [5] considers a system anomaly detection network type robots "Internet robots" (Internet of robots). The peculiarity of this system is that it has two subsystems. One is a group of robots that collect data using a sensory system and transmit it to the central node that is connected to an external mobile network. Second it is a mobile network, which includes following modules: a data acquisition module, an anomalies classification module, an orders module. The disadvantage of this system is that it is completely centralized; robots do not communicate with each other and act only through an intermediary. Detection of anomalies occurs using a classifier that is trained with a pre-formed sample. Thus, as a result of examining the works devoted to the topic of detecting attacks on robots, there are three main drawbacks in the existing approaches:

1. Most systems are based on signature analysis, or on a system of rules. In this regard, there are the following limitations: the complexity of detecting new attacks that are not related to the fixed patterns of the attacker's behavior.
2. Systems based on fully distributed detection methods require additional energy costs, computational power costs from nodes and increase bandwidth.
3. With centralized methods, the node performing the basic functions for detecting abnormal behavior is the most vulnerable place of the system.

This article proposes the development of a method based on probabilistic methods for detecting an abnormal behavior of an attacker within a group of mobile robots [6]. The main difference of this method is that it does not require the construction of a standard probability distribution, like other probabilistic methods. The absence of the need to build a reference distribution is due to the fact that the current parameters of a group of nodes are taken to detect abnormal behavior. Then, a function of the normal distribution and the confidence interval are calculated. To assess the behavior of the node Ni, the probability of getting the current indicators of the node in the confidence interval is calculated based on the parameters of all nodes in the group. Thus, it becomes possible to estimate the degree of deviation of the behavior of one node from the behavior of a group of nodes.

3   Development of a Method for Detecting Abnormal Behavior

A feature of the proposed method for a group of robots is that for a normal distribution function it is necessary to obtain data from several nodes performing similar functions. It is necessary that indicators of the group of nodes are in the same range to more accurately determine the degree of deviation of the current node parameters from the group of nodes. This method works most effectively in the case of a group of mobile robots that communicate in the execution of a task. An attacker can influence both a network connection and physical parameters of a network node, realizing the attack. The battery charge parameter is the remaining energy in the battery - e, which allows a device to function in the network. In this study, the residual energy-e parameter analysis was carried out. The parameter L is a total number of packets transmitted on the network. Either a number of packets transmitted through one of the nodes of the network.

A consequence of an attack aimed at depleting the resources of the node will be a decrease in the level of residual energy. In addition, an attacker who carries out this attack may have superiority in the reserves of energy resources. The main purpose of the attacker, in an implementation of the Sybil attack or an attack of redirecting the impact on themselves, is to change the processes and ways of exchanging data on the network. In other words, the attacker achieves such a situation that all or most of the traffic of neighboring nodes is transmitted through his nodes. The normal distribution law is used to calculate a confidence intervals and a probability of a current value falling into the confidence interval for parameters the load L and the residual energy e.

$$L_i(t) \sim N\left(\bar{L}, \sigma_L^2\right),$$
$$Q(E)_i(t) \sim N\left(\overline{Q(E)}, \sigma_{Q(E)}^2\right), \tag{1}$$

where $\sigma_{Q(E)}$, σ_L - the total for a group of nodes, $\bar{L}$, $\overline{Q(E)}$ - the mathematical expectation for congestion and residual energy of the node.

A choice of the normal distribution is due to the fact that it is widely used in queuing networks to represent the distribution of the number of requests in the range of average service duration, similar to the load distribution and residual energy of the nodes. And also an analysis of the quantile diagrams for the residual energy e and for the loading L. Quantile-Quantile graph allows us to compare the distribution of the investigated variable with the theoretical normal distribution. The distribution of e and the load L are located within a straight line, which corresponds to a normal distribution, as seen in Figs. 1(a) and 2(b).

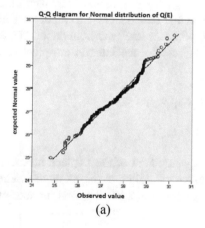
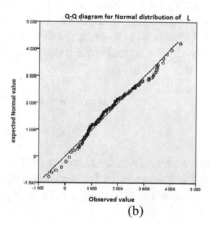

(a)   (b)

Fig. 1. Quantile-Quantile graph diagrams (a) residual energy- e and (b) network load - L.

Trust calculation procedure is performed using the following algorithm.

1. Calculation of confidence intervals for the parameters the network load and the residual energy at the time interval [7].
1.1. Calculation of a variance:

$$D_{Li} - \left(\sum_{i}^{N} (L_i - \bar{L})^2 \right) / n, D_{ei} = \left(\sum_{i}^{N} (e_i - \bar{e})^2 \right) / n, \tag{2}$$

where D_{Li}, D_{ei} is the variance of the parameters L and e of the group of nodes in the current time interval. Calculation of root-mean-square deviation:

$$\sigma_{ei} = \sqrt{D_{ei}}, \sigma_{Li} = \sqrt{D_{Li}}, \tag{3}$$

σ_{ei}, σ_{Li} - is the standard deviation of the parameters L and e of the group of nodes in the current time interval.

1.2. Calculation of upper and lower limits of confidence intervals.

The upper bound of the confidence interval of the parameter e is always equal to the maximum permissible energy value, that is, $a_{max} = initialEnergy$. This is due to the fact that nodes can migrate from one group to another, new nodes may appear with a residual energy value equal to the initial value. This factor must be taken into account in order to avoid mistaken acceptance of the trusted node as malicious. The calculation of a lower bound of the confidence interval is performed only for the residual energy value. Since a lower limit of the parameter L is equal to a minimum required number of packets passed through a node in one time interval L_{min}. These measures are taken for a reason that a mobile robot can exhibit selfish behavior. That is, refuse to participate in the network to save energy, which can artificially "understate" the boundaries of the interval. The formulas for calculating limits of the confidence interval are presented below.

$$a_{min} = \bar{e} - t \cdot \sigma_e / \sqrt{n}, a_{max} = A_{31}, a_{min} < a_{max} \tag{4}$$

$$b_{min} = L_{min}, b_{max} = \bar{L} + t \cdot \sigma_L / \sqrt{n}, \tag{5}$$

where $t * \sigma / \sqrt{n}$ is the accuracy of the estimate; t - argument of the Laplace function; $\Phi(t) = \frac{\alpha}{2}$ where is the Laplace function; α - is a given reliability, in this study the value of the coefficient is equal to $\alpha = 0.98$, so the argument t = 2.34; n - total number of nodes.

2. Implementation of paragraphs. 2.1, 2.2 for each node.
2.1. Determination of the probability of anomalous behavior of a mobile robot.

In order to calculate the mean square deviation and mathematical expectation, it is necessary to shorten the interval for which the value is calculated and only the parameters in the previous time interval L_{i-1}, e_{i-1} and L_i, e_i parameters of the node for the current interval, are taken into account.

$$P_e(a_{\min} < e_i < a_{\max}) = \Phi\left(\frac{a_{\max} - \overline{e_6}}{\sigma_{e_6}}\right) - \Phi\left(\frac{a_{\min} - \overline{e_6}}{\sigma_{e_6}}\right) \tag{6}$$

$$P_L(b_{\min} < L_i < b_{\max}) = \Phi\left(\frac{b_{\max} - \overline{L_6}}{\sigma_{L_6}}\right) - \Phi\left(\frac{b_{\min} - \overline{L_6}}{\sigma_{L_6}}\right) \tag{7}$$

where Φ is a Laplace function; $P_{Q(E)}$, P_L - probability of hit of residual energy of the node and load within the confidence interval; $\overline{e_6} = (e_{i-1} + e_i)/2$; $\overline{L_6} = (L_{i-1} + L_i)/2$ - mathematical expectation of the values of e and L for the sampling interval; $D_{e_6} = \left(\sum_i^N (e_i - \overline{e_6})^2\right)/n$ - variance for the sampling interval for e; $\sigma_{e_6} = \sqrt{D_{e_6}}$ - the standard deviation for the sampling interval for the residual energy; $D_{L_6} = \left(\sum_i^N (L_i - \overline{L_6})^2\right)/n$ - variance for the sampling interval for the node load value; $\sigma_{L_6} = \sqrt{D_{L_6}}$ - standard deviation for the sampling interval L.

2.2. Output the results in a trace log.

3. Implementation of pp. 3.1 for each node.
3.1. Update the following values for each node: the load in the previous time interval, the residual energy at the previous time.

4. Blocking malicious nodes.
4.1. A threshold value of the probability that the node is anomalous is equal to 0.5.
4.2. When the node reaches a value of 0.5, it is necessary to reduce its residual energy level by half, and then the node is considered in an uncertain state.
4.3. If the trust value reaches 0.4, then it is necessary to consider the node malicious and reduce its energy level to zero, so it is excluded from the network [8].

4 Evaluation of the Effectiveness of the Developed Method

To evaluate the effectiveness of the developed method, a model of a group of robots was developed in the simulation environment NS-2.35. Robots communicate with each other via wireless communication and use the TCP/IP protocol stack to transfer information. In particular, the UDP protocol is used for data transmission at the transport level, the ARP protocol is used to transmit control commands at the data link layer, the AODV protocol is used for routing the packets [9]. Procedure for detecting abnormal behavior and output results is called at regular equal time intervals. A group of mobile robots, which includes 10 nodes, was modelled. The N4 node is a base station or a central server. Node N0 is a leader of the group and performs functions of collecting information from robots and redirects it to the central server. The nodes group exchange information with each other and with the group leader [10]. In this case, nodes N6, N7, N8, N9, starting from 50 s will conduct a DDoS attack.

4.1   Implementation and Detection of DDoS Attacks

Detection of a distributed denial of service attack is more difficult task. This is due to the fact that when the number of malicious nodes is greater than the number of trusted nodes, the boundaries of the confidence interval are significantly expanded. The developed method is quite effective in detecting this attack. When the ratio of malicious nodes to genuine nodes is 4 to 5, the method allows immediately detecting all malicious nodes and blocking them already in the second time interval. In Fig. 2 a histogram showing the level of hit of current indicators e and L malicious nodes in the confidence interval. When malicious and trusted hosts are in an equal ratio of 5 to 5, the quality of detection becomes worse. N_9 and N_8 nodes were also detected in the second interval, nodes N_5 and N_6 were detected in the third interval and node N_7 in the fourth time interval, starting from the moment when the attack began. In general, the detection rate is 100%, but the speed has decreased.

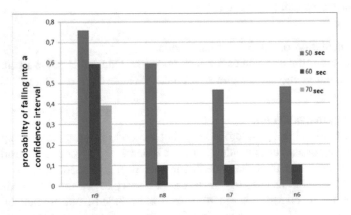

Fig. 2. The detection level of malicious hosts in a distributed denial of service attack for four malicious hosts.

When the number of malicious nodes exceeds the number of trusted in the ratio of 6 malicious to 5 authentic, the detection level is 83%, i.e. one malicious node remains undetected. Implementation and detection of the Sibyl attack. Sibyl attack is that an attacker is represented by several network nodes and tries to redirect most of the traffic to itself [11]. At the same time, he can make a destructive impact on the network by discarding messages, redirecting them to the wrong nodes, violating the routing scheme, or can passively listen for traffic. In this case, it is quite difficult to detect an attack when an attacker does not produce a destructive effect. In [12] there are methods using rigid protection: password protection, cryptographic protection, as well as signature analysis and detection group. These methods are used in MANET, IoF, P2P networks [13], which are not as severely restricted in resources as mobile robot groups. The developed method for detecting abnormal behavior shows the effectiveness of the Sibyl attack detection. Even if the attacker redirects traffic to itself and does not take any action anymore. In this case, detection is possible by changing the degree of load of nodes that conduct an attack on neighboring nodes. In addition, the level of residual energy of the attacking nodes is significantly reduced. In the simulation system NS-2.35, malicious N7–N11 nodes were added to the robot group, which are called (Sybil1-Sybil5), to evaluate the method developed. Figure 3 shows the topology of the network with malicious nodes. The figure shows that the number of malicious nodes and the number of trusted ones, excluding the base station (BS) and the group head (GL), corresponds to half of the network nodes.

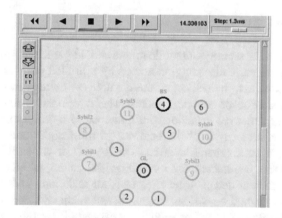

Fig. 3. Network topology for the implementation of the Sibyl attack.

N_7–N_{11} nodes redirect packets, from neighboring mobile robots to itself starting from 50 s, violating the network operation scheme. Initially, mobile robots send packets to the group leader according to a given pattern; the group leader sends packets to the base station. Thus, the method allows identifying 2 malicious nodes N10 and N11 in the first 10 s of the attack. This is due to the fact that these nodes redirect more traffic to themselves. Further, the detected nodes are blocked starting from the 60th s. At the 70th s, nodes N7 and N8 are detected. N9 node was the most difficult to detect. Figure 4 shows a histogram representing the detection level of malicious nodes.

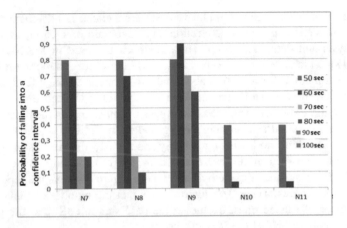

Fig. 4. The detection level of the nodes implementing the Sibyl attack.

5   Conclusion

The issues related to the security of mobile robots and, in particular, group management of mobile robots, are currently being addressed by a limited number of scientists and institutions. Nevertheless, the subject of these studies is quite relevant, in connection with the widespread use of robotic systems. The developed method is a universal means of detecting abnormal behavior with reference to a group of nodes. Since it is possible to conduct an analysis of the behavior of most nodes and identify single or common deviations from general behavior. Due to this, it is possible to increase the number of analyzed parameters for expanding the range of attacks. The method demonstrates a sufficient detection level, detects all malicious nodes within 30–40 s. Limitations of the method consist in the number of malicious nodes that conduct an attack on the network; their number should not exceed 60%, compared to the number of trusted ones. And also this method is applicable to a group of nodes that perform a similar task. Unlike existing methods for detecting abnormal behavior, where it is necessary to compile a signature database or data banks of rules, store them, and then update them, the developed method allows to detect anomalies at the current time and depending on the current situation. This advantage is quite important because mobile robots can be used in different environments and for various tasks. In addition to the protocols of network interaction between mobile robots, the conditions of the medium may change, which in turn also affect the occurrence of anomalies and errors. This method takes into account the threshold values, that is, the permissible level of anomalies, which reduces the number of false positives. In this case, the developed method allows reducing the load on the mobile robot. It does not need to constantly exchange messages, monitor neighboring nodes and update, store the signature database.

Acknowledgment. This research was carried out in the grant of the Ministry of Education and Science of the Russian Federation Initiative scientific projects No. 2.6244.2017/8.9 on the topic "Developing a method for detecting attacks and intrusions, a method for authenticating nodes for a scalable wireless sensor network".

References

1. Basan, A.S., Basan, E.S.: A model of threats for the systems of group management of mobile robots. System synthesis and applied synergetic. In: Proceedings of the VIII All-Russian Scientific Conference, pp. 205–212. Southern Federal University, Rostov-on-DonTownship, Lower Arkhyz (2017)
2. Kim, H.S., Lee, S.W.: Enhanced novel access control protocol over wireless sensor networks. IEEE Trans. Consum. Electron. **55**(2), 492–498 (2009)
3. Vuong, T.P., Loukas, G., Gan, D., Bezemskij, A.: Decision tree-based detection of denial of service and command injection attacks on robotic vehicles. In: Proceedings of 2015 IEEE International Workshop on Information Forensics and Security (WIFS), pp. 1–6. IEEE, Rome (2015)
4. Bezemskij, A., Loukas, G., Anthony, R.J., Gan. D.: Behaviour-based anomaly detection of cyber-physical attacks on a robotic vehicle. In: Proceedings of 15th International Conference on Ubiquitous Computing and Communications and 2016 8th International Symposium on Cyberspace and Security, pp. 61–68. IEEE, Xi'an (2016)
5. Monshizadeh, M., Khatri, V., Kantola, R., Yan, Z.: An Orchestrated security platform for internet of robots. In: Au, M.H.A., Castiglione, A., Choo, K.-K.R., Palmieri, F., Li, K.-C. (eds.) Proceedings of Springer International Conference on Green, Pervasive, and Cloud Computing, vol. 10232, pp. 298–312. Springer, Cetara (2017)
6. Basan, A.S., Basan, E.S., Makarevich, O.B.: The method of counteracting the attacker's active attacks in wireless sensor networks. Izvestia of the Southern Federal University. Tech. Sci. **5**(190), 16–25 (2017)
7. Basan, A.S., Basan, E.S., Makarevich, O.B.: Development of the hierarchal trust management system for mobile cluster–based wireless sensor network. In: Proceeding SIN 2016, Proceedings of the 9th International Conference on Security of Information and Networks, pp. 116–122. Rutgers University, New Jersey (2016)
8. Abramov, E.S., Basan, E.S.: Development of a model of a protected cluster–based wireless sensor network. Izvestiya SFedU. Tech. Sci. **12**(149), 48–56 (2013)
9. Ferronato, J.J., Trentin, M.A.: Analysis of routing protocols OLSR, AODV and ZRP in real urban vehicular scenario with density variation. IEEE Latin Am. Trans. **15**(9), 1727–1734 (2017)
10. Pshikhopov, V.Kh, Soloviev, V.V., Titov, A.E., Finaev, V.I., Shapovalov, I.O.: Group Control of Mobile Objects in Indeterminate Environments. FIZMATLIT, Rostov-on-Don (2015)
11. Abbas, S., Merabti, M., Llewellyn-Jones, D., Kifayat, K.: Lightweight sybil attack detection in MANETs. IEEE Syst. J. **7**(2), 236–248 (2013)
12. Patel, S.T., Mistry, N.H.: A review: sybil attack detection techniques in WSN. In: Proceedings of 4th International Conference on Electronics and Communication Systems (ICECS), pp. 184–188 (2017)
13. Wang, G., Musau, F., Guo, S., Abdullahi, M.B.: Neighbor similarity trust against sybil attack in P2P E-commerce. IEEE Trans. Parallel Distrib. Syst. **26**(3), 824–833 (2015)

Automated Field Monitoring by a Group of Light Aircraft-Type UAVs

Ekaterina Pantelej[1]([✉]), Nikolay Gusev[2], George Voshchuk[3], and Alexander Zhelonkin[3]

[1] Network-Centric Platforms, Ltd., Moskovskoe Shosse 17, Business Center "Vertical", Office 2203, 443013 Samara, Russian Federation
mekachiku-san@mail.ru
[2] Samara State Technical University, Molodogvardeyskaya Str., 244, 443100 Samara, Russian Federation
nikolayagusev@gmail.com
[3] Aeropatrol, Ltd., Moskovskoe Shosse 17, Business Center "Vertical", Office 35, 443013 Samara, Russian Federation
voschuk@gmail.com

Abstract. This paper provides an overview of some existing methods of Earth remote sensing (ERS) using for agricultural needs. Special emphasis is placed on sensing with the help of UAVs. The paper describes the developed software and hardware complex for an aircraft-type UAV group. The proposed solution significantly increases the operating time and automates the process of monitoring agricultural areas. In addition, legislative restrictions on the use of UAVs are considered.

Keywords: Agriculture monitoring · Precision agriculture
Unmanned aerial vehicle (UAV) · Internet of Things (IoT) · Image analysis

The agricultural sector faces many threats, such as pests, plant diseases, climate change and theft. In the Russian Federation, the total area of arable land in 2017 was 116709.5 thousand hectares [1]. Ground-based research and monitoring do not allow for full-value analysis, evaluation and control of such vast agricultural areas. Thus, Earth remote sensing systems are used for timely elimination of pests and diseases, evaluation of fertilizer effectiveness, and protection of agricultural lands.

1 Applications of ERS in Precision Farming

Remote sensing means include satellite systems [2, 3], guided aircrafts and unmanned aerial vehicles (UAVs) [4–6].

The use of satellite imagery is a very common tool for Earth remote sensing. Satellites are constantly making shots of the Earth's surface in conditions of cloudless weather. It is very convenient to use these shots to diagnose the state of objects. One disadvantage of these images is their low spatial resolution. The probe of the Sentinel-2 satellite has from 10 to 60 m in one pixel at a right angle of shooting, depending on the spectrum used. This resolution is enough for identifying the existence of a "problem",

© Springer Nature Switzerland AG 2019
A. Abraham et al. (Eds.): IITI 2018, AISC 875, pp. 350–358, 2019.
https://doi.org/10.1007/978-3-030-01821-4_37

but not enough for establishing its cause and for accurate assessment of the affected area. Another disadvantage of sensing with satellites is the temporal resolution - the time gap before the second (repeated) sensing of the given place. The temporal resolution depends on the trajectory and speed of the satellite. On average, it is between two and seven days for civil satellites that provide access to their images. If it is cloudy or overcast during shooting, the waiting time for the next images increases. In certain cases, more frequent monitoring may be required (up to daily) – for example, during the critical phases of plant growth, testing of new fertilizers, or the harvesting period.

In agriculture, images in the visible range are not of great value. Multispectral images are more important, as they are used for calculating the vegetative index NDVI. Multispectral images have very low spatial resolution. Thus, for full analysis of the NDVI index it is better to use data from several satellites, not just from one. This gives more pixels, and, therefore, more accurate data.

Some services load their images with a delay. There are even free services, for example, Sentinel, but the frequency of image downloading can reach up to 2 weeks, which is not real-time data anymore. All of the above makes the use of satellite imagery useful, but not sufficient for precision agriculture.

The use of manned aircrafts makes it possible to get a high-quality picture at the necessary points of the field in any weather conditions. A significant disadvantage of manned aircrafts is the cost of equipment: not every farm, especially if it is a small one, can afford such expensive machinery as an airplane. Considering the prices for the camera, fuel and the work of pilots and mechanics, the cost of one flight will be significantly higher than the use of UAVs.

UAVs provide more accurate data, in comparison with satellite imagery. Depending on the installed camera, UAVs can provide data of the scale 1:100, 1:2000 and 1:5000, instead of 1:10000 for the "space" satellite imagery. The geodesic accuracy of images reaches 1-3 cm. Using UAVs, data can be obtained in any weather conditions, regardless of the clouds, in contrast to satellite systems that can operate only in clear weather. The cost of one UAV flight is several hundred times cheaper than the flights of "traditional aviation".

Another way to monitor the state of agriculture is real-time systems, consisting of measuring tools and computer centers linked by the Internet of Things [7–9]. Such devices consist of temperature and humidity sensors for soil and environment, soil and light conductivity sensors, sending data via the Internet. Such systems are set up in fields. The obtained local climate data is used to make decisions related to the health of crops. The cost of such monitoring increases with expansion of the investigated territory. The data obtained from the sensors is not sufficient for accurate analysis of uniformity of crop growth and development.

Of course, achieving automation of monitoring and evaluation of agricultural areas is possible only using all of the above-mentioned means, combined via the Internet. Nevertheless, affordable prices of commercial unmanned aerial vehicles, and relative ease of their operation make UAVs the main tool in the tasks of precision agriculture.

2   UAV Group Management System

2.1   Problem Statement

Today many farms use single UAVs in field diagnostics. This practice is common in the Samara and Rostov regions. UAVs of the multi-rotor type have become most widespread. For them, flying over the field costs 50-70 rubles per hectare. In a season it is necessary to make from 5 flights. The cost of monitoring costs 2.5 million rubles a year.

In conditions of small farms, multi-rotor UAVs have proved very successful. However, for large agricultural companies with areas of tens of thousands hectares, and fields often removed from each other, there is a need not just for single UAVs, but for a group of unmanned vehicles controlled from a single center. Implementation and operation of the UAV aircraft type group pays off in 2 years compared to the operation of the single UAV multirotornogotipa. In the case of manual control of the UAVs, special training of the operator is required. Complexity of his work increases with the number of devices. According to rough data, about 60% of accidents in the fields involving UAVs occur due to human factor. Therefore, there is a need for an automated system that would ensure constant monitoring of agricultural fields.

At the moment, the UAV groups are not used in the agricultural tasks. The reason is the absence of a stable working solution for the UAV group. Developments of the UAV group are available from the following companies: Boeing and Lockheed Martin, Naval Postgraduate School, Ehang.

The purpose of the work is to set up an automated control system for a group of UAVs in real time for agricultural tasks.

To fulfill this goal, the complex solves the following tasks: UAV group management, monitoring of the mission satisfaction index, real-time data acquisition and analysis, optimal route construction according to selected parameters, machine learning.

2.2   Used Equipment and Software

The structural diagram of the developed software and hardware complex is presented in Fig. 1, and consists of a group of UAVs with on-board computers, and an operator control center communicating with UAVs via wireless technologies.

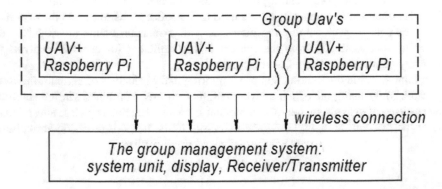

Fig. 1. Structural diagram of the developed software and hardware complex

UAVs are equipped with a camera of the visible and multispectral range. When choosing a UAV, it is necessary to take into account its payload, average flight time, speed and reliability. Tests have shown that the most effective in remote areas would be the use of aircraft UAVs, not the multi-rotor type. Despite active development of the industry of multi-rotor unmanned vehicles, most of them have a flight time limit of 20–35 min for industrial vehicles, less often – 50–60 min. The cost of the latter devices is up to 5 times higher than of drones with shorter flight time. For this cost, it is economically more advantageous to purchase aircraft UAVs providing flight time from two to four hours while maintaining the other functional characteristics. Tests were carried out to compare characteristics of UAVs of aircraft and multi-rotor types. The results obtained are presented in Table 1.

Table 1. Comparison of characteristics of multi-rotor and aircraft types of UAVs

	PHANTOM 3 Advanced (Multi-rotor UAV)	Sovzond Air-Con 4 (Airplane UAV)
Flight time with one battery charge, min.	23	150
Maximum monitoring area on one battery at an altitude of 120 m, ha	~25	~220

These tests were carried out using the DJI PHANTOM 3 Advanced with a camera of 12.4 Mp with 4000×3000 resolution. The territory of 100 hectares was covered at the height of 120 m. This task required 4 flights, while the landings were carried out to change the battery. Spatial resolution is tens of centimeters. With an aircraft UAV, only one flight is required for a similar mission, which saves the operation time and eliminates the need to purchase a large number of spare batteries. In addition, a multi-rotor UAV can develop the speed of about 16 m/s, while an aircraft UAV develops the speed of up to 25 m/s. Figure 2 shows results of UAV comparison for flights at maximum speed at the height of 120 m. The "Number of flights" means the flights on one battery charge. Depending on requirements for image resolution, it is possible to fly at a higher altitude, and, consequently, to cover a larger area. At the height of more than 500 m, spatial resolution decreases so much that the quality difference with satellite images disappears. Therefore, the greatest efficiency can be achieved when operating a system with aircraft UAVs.

Raspberry Pi was chosen as on-board computer, since it has the necessary functionality and relatively small weight.

The operator control stand is shown in Fig. 3. The stand has a number of displays, receiving and transmitting devices, a personal computer and peripheral devices.

Connection between the group and the control stand can be established in different ways, depending on the mission and operating conditions. Possible options include Wi-Fi, mobile network, or radio communication. Depending on the chosen option, the communication range will change.

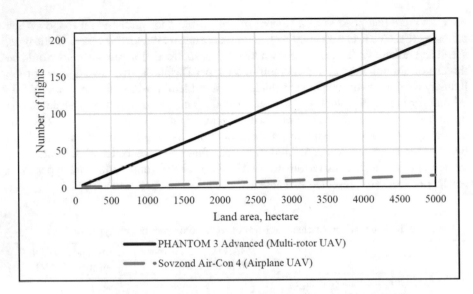

Fig. 2. Comparison of characteristics of multi-rotor and aircraft UAVs

Fig. 3. The swarm control system stand

Multi-agent technology is used to solve the tasks of group monitoring and continuous automated patrolling of territories [10].

2.3   Multi-agent Scheduler

Initially, operator draws up the mission and team plan, which contains defined areas for the flight, optimal flight parameters, departure times and mission objectives. Commands are sent to the UAV group. In standard flight conditions, the general task is decomposed into sets of subtasks (territories for flight) comparable to characteristics of the group members. The system creates optimal plans to cover the selected observation squares, as well as the task schedule for each UAV. The operator receives data from UAVs in real time on the control stand, having the ability to modify, extend and terminate missions at any moment.

The scheduler balances action plans of the group members in response to dynamically occurring events. For example, when flying over a square at an altitude of 500 m, one UAV detects non-uniformity in crop growth (Fig. 4). There is not enough spatial resolution for accurate assessment of the affected area and for identifying the cause of this problem at the selected altitude. The system decides to lower the altitude to the optimal value of 100 m, with recalculation of parameters, taking into account the remaining battery life of this UAV, the required area of detailed flight and the remaining uninspected territory. The scheduler decides that the second UAV is required for the detailed flight, and the rest of the territory is distributed among the remaining members of the group. As a result, the group inspects all the fields and diagnoses slug invasion on field A.

Fig. 4. Analysis of images taken by satellites and UAVs. Pink color indicates the area of the crops destroyed by pests (slug).

Depending on the time of year, the flight schedule and task distribution can vary. Thus, during the ripening season, the following situation is possible (Fig. 5). The data is obtained by stitching NDVI images. Red color shows the territories with mature crops, and green color – with ripening crops. Different cultures mature at different time intervals. Besides, in the fields, it is possible to point out accelerated or slow maturation: along roads, pipelines, on the hills and lowlands. All these areas require detailed monitoring during the harvest period. For example, if the crops over a pipeline are not harvested on time, they can begin rotting by the time the main part of the field is harvested.

Fig. 5. Analysis of images taken with the help of a satellite. Red areas indicate ripe harvest.

Information about the cultivated land (culture, processing technology) and problem areas is loaded into the system by the operator and is formed on the basis of experience of fulfilling such missions in the previous years. All this data forms the knowledge base. Based on this information, special missions are formed for the group of unmanned vehicles, with marked areas of increased monitoring importance.

The knowledge base also contains object classes, relations between them, and UAV satisfaction indices. The use of a swarm of unmanned vehicles combined into a single software complex makes it possible to automatically fly over agricultural lands, assess the information received, distribute missions among all UAVs, identify problem areas in a timely manner and take measures to eliminate them, monitor effectiveness of the fertilizers used, issue full reports and upload the obtained information into the cloud.

The described software and hardware complex was developed by our team. It successfully passed the tests with a group of multi-rotor UAVs. The use of the complex on aircraft UAVs will make it possible to cover larger areas in less time and solve not only the tasks of local patrolling, but also monitoring of rather extensive agricultural areas.

2.4   Existing System Restrictions

The use of UAVs in Russia is inhibited by the new amendments to the Air Code [11], according to which all aircrafts over 250 g are subject to mandatory registration, and for each flight it is necessary to obtain a permit. Formally, each landing is considered as completion of the flight and, consequently, cancellation of the current permit. For each next take-off it is necessary to obtain a new permit, even if the landing was done only for battery replacement. Under the current code, aircraft UAVs that can stay in the air for up to 4 h have advantages over drone (multi-rotor) UAVs with flight duration up to only about 30 min.

In agriculture, the flight usually takes place over large areas with no secret or guarded facilities and without crowds of people. Thus, it would be good to simplify the flight procedure and, for example, issue flight permits in advance for the entire agricultural season only for the territories of the particular farm. This would facilitate the use of the proposed software and hardware complex and UAVs, as flights will depend only on the current needs and weather conditions, not on legislative requirements and bureaucratic runaround.

We suppose that it is highly advisable to consider amendments to the Russian Air Code for the needs of the agricultural industry.

3 Conclusions

In today's world with constantly growing population, it is important to develop precision agriculture using remote sensing technology. The most promising and requested trend is the use of a group of aircraft UAVs, combined into a single system, consisting of the UAVs with installed Raspberry Pi on board, the software developed by our company and a single flight center. The hardware and software complex can plan flight tasks, automatically analyze the footage during the flight, make independent decisions on monitoring tasks, and issue reports and recommendations based on the performed analysis. The system has proved itself in tests on a group of multi-rotor drones. In order to extend the capabilities of the system and save time for end users, we recommend using the system on aircraft UAVs. The key factor currently limiting the civil application of UAVs of any kind is the legislative restrictions in Russia.

Acknowledgements. The work was supported by the Ministry of Education and Science of the Russian Federation within the contract agreement #14.574.21.0183 - unique identification number RFMEFI57417X0183.

References

1. Collection: Agro-industrial complex of Russia in 2016. Rosinformagrotech, p. 704 (2017)
2. Uwizera, D., McSharry, P.: Forecasting and monitoring maize production using satellite imagery in Rwanda. In: Technological Innovations in ICT for Agriculture and Rural Development (TIAR), pp. 51–56. IEEE (2017)
3. Yang, C., Everitt, J.H., Du, Q., Luo, B., Chanussot, J.: Using high-resolution airborne and satellite imagery to assess crop growth and yield variability for precision agriculture. In: Proceedings of the IEEE, vol. 101, No. 3, pp. 582–592 (2013)
4. Li, W., Yuan, H., Li, W., Song, L.: Prediction of wheat gains with imagery from four-rotor UAV. In: 2nd IEEE International Conference on Computer and Communications, pp. 662–665 (2016)
5. Perez-Ortiz, M., Gutierrez, P.A., Pena, J.M., Torres-Sanchez, J., Lopez-Granados, F., Hervas-Martinez, C.: Machine learning paradigms for weed mapping via unmanned aerial vehicles. In: 2016 IEEE Symposium Series on Computational Intelligence (SSCI) (2016)
6. Vehicl, J.N., Prado, J., Lino, M.: Low-cost multi-spectral vegetation classification using an unmanned aerial. In: IEEE International Conference on Autonomous Robot Systems and Competitions (ICARSC), pp. 336–342 (2017)
7. Hwang, J., Shin, C., Yoe, H.: Study on an agricultural environment monitoring server system using wireless sensor networks. Sensor **10**, 11189–11211 (2010)
8. Mekala, M.S., Viswanathan, P.: A survey smart agriculture IoT with cloud computing. In: 2017 International conference on Microelectronic Devices, Circuits and Systems (ICMDCS) (2017)

9. Yonghong, T., Bing, Z., Zeyu, L.: Agricultural greenhouse environment monitoring system based on internet of things. In: 3rd IEEE International Conference on Computer and Communications, pp. 2981–2985 (2017)

10. Skobelev, P., Budaev, D., Brankovsky, A., Voshuk, G.: Multi-agent tasks scheduling for coordinated actions of unmanned aerial vehicles acting in group. Int. J. Des. Nat. Ecodynamics **13**(1), 39–45 (2018)

11. The Air Code of the Russian Federation. http://docs.cntd.ru/document/9040995. Accessed 5 April 2018

Visualization of Hydrogen Fuell Cells Laboratory

Zdenek Slanina[1(✉)], Filip Krupa[1,2], Jakub Nemcik[1,2], and Daniel Minarik[2]

[1] Faculty of Electrical Engineering and Computer Science, VSB-TU Ostrava,
708 33 Ostrava, Czech Republic
zdenek.slanina@vsb.cz
[2] Centre of Energy Utilization of Non-traditional Energy Sources ENET,
708 33 Ostrava, Czech Republic
daniel.minarik@vsb.cz

Abstract. This article deals with fuel cell hydrogen laboratory analysis for re-implementation of control and visualization software blocks because of technology delay problems connected with used control hardware and software tools and units. This laboratory is part of a broader cluster of energy sources to enable bi-directional energy flow and its regulation both in off-grid and on-grid mode. It's connection with visualization is described as well as the issue of safety and security of visualization via operating modes and user accounts.

Keywords: Fuel cell · Off grid · Visualization · Control

1 Introduction

The fuel cell Laboratory is a part of the energy unit (CENET - Center for Energy Utilization of Non-traditional Energy Sources at VSB-TU Ostrava), which is responsible for the development and research on storage and generation of electricity from renewable and other specific energy sources (see [1–4,6,8]). The facility is run as a research center for the design and development of energy technologies. Laboratory is equipped with five NEDSTACK series fuel cells, named FCS 8-XXL. These are trays each containing 64 fuel cells. These fuel cells can generate up to 8 kWe (230A approximately). The working voltage ranges from 25.6 V to 64V DC. The maximum current is 300A.

The individual cells are equipped with Proton-Exchange Membrane. The membrane fuel cell uses as an electrolyte an acidic polymeric membrane with platinum-based electrodes. PEMFC cells operate at relatively low temperatures (below 100°C) and can adjust electrical performance to meet dynamic performance requirements. Due to the relatively low temperatures and the use of precious metal electrodes, these cells have to work on pure hydrogen. PEMFC cells are currently the leading technology for light commercial vehicles and material handling vehicles, and to a lesser extent for stationary and other applications.

© Springer Nature Switzerland AG 2019
A. Abraham et al. (Eds.): IITI 2018, AISC 875, pp. 359–368, 2019.
https://doi.org/10.1007/978-3-030-01821-4_38

A PEMFC fuel cell is also sometimes called a polymeric electrolyte membrane fuel cell (also PEMFC).

Hydrogen fuel is processed on an anode where the electrons are separated from the protons on the surface of a platinum-based catalyst. Protons pass through the membrane on the cathode side of the cell, while the electrons move in the outer circuit and generate electrical power. On the cathode side, an electrode with precious metals, combines protons and electrons with oxygen. This process produces water which is excluded as the only waste product. Oxygen can be provided in purified form or extracted directly from the air.

In our case pure hydrogen gas is used as fuel, and humidified air as the oxidant. However, technology is ready to test other gases such as chlorine, nitrogen, oxygen and more.

2   Technology Overview

The main technological part are fuel cells. Their working condition is provided by compressor turbines that bring the air under pressure into the oxygenation chamber. This air is moistened with steam, which is made up of other laboratory equipment. The supply of fuel, in this case hydrogen, is provided by a set of pneumatic valves that are driven by nitrogen. Therefore, it is necessary first to manually open the gas bombs with nitrogen in order to control the valves designed to conduct other gases. Last but not least, there is cooling of the fuel cells, which ensures the optimum temperature of the membrane. Ideal operating temperature ranges from 55 °C to 70 °C, ideally 65 °C. The diagram of the technologies can be seen at Fig. 1.

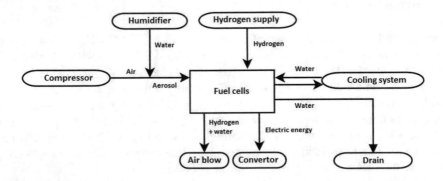

Fig. 1. Diagram of supporting technologies.

Compressor turbines serve to inject air and water vapor into the oxygen cell of the fuel cell. In the laboratory there is one turbine for each fuel cell. The control of these turbines is solved using Siemens Sinamics G120C USS/MB. This drive has output power of up to 15 kW at low load and 11 kW at high. Turbine switching is addressed here by switching contactors to this inverter before it is started. When the inverter is started, it detects the load on the terminals and outputs the required power. It is very critical that the inverter configuration, even in standby mode, does not change the configuration of the connected turbines. It could cause irreversible damage to both the inverter and the turbines themselves.

Part of the power generation chain is the humidification of the working membranes in the fuel cells on the side of the flowing air, in order to avoid the drying and porosity of the membranes and thus their destruction. This activity is provided by the steam generator FOG 100 Ambiente. The steam formed is then mixed with the inlet air. Intensity, and hence air humidity, is now regulated by two different flow nozzles for each turbine. Nozzle opening is provided by PLC-controlled solenoid valves. The humidifier has its own pressure and communication control, but we do not use it now. The switching of the humidifier unit is provisionally solved by connecting it to the drawer housing and then switching the power supply contactor on to this cabinet.

The whole technology is connected to the Profibus bus, which connects all devices to a central PLCs. Next, the data is passed through S7 protocol and OPC communications to the server. From there, the data is further distributed through the Ethernet network to individual clients (Fig. 2).

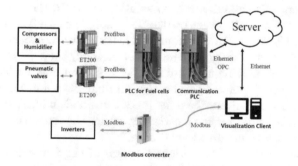

Fig. 2. Communication scheme.

3   Fuell Cell Lab Visualization

The original visualization is based on the ControlWEB system (Moravian instruments). The main reasons for the new design of visualization were the time lag and reaction times of the system. The response of the system to the sent command ranged from 2 to 15 s depending on the run time of the visualization program. Because it is critical, where responses within hundreds of milliseconds are required, up to a maximum of one second, the concept has to be changed.

The main reason for the delay was the concept of technology layout in individual PLCs and the determination of the flow of data sent. Therefore, there are more PLCs for other technologies in the cluster. The problem was congestion of the communication line and overload of the communication PLC, referred to as Accumulation and Production of Electricity see Fig. 3. Technology controlled from the main control room is comprised of six redundant PLCs, all of which were fed into one main PLC, which further sent data to the OPC communications server. At this node, a delay of up to two seconds was recruited.

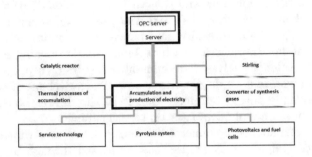

Fig. 3. Connection scheme.

Next problem was the visualization solution itself. In the two image panels, elements of all technologies were overlapped using the SHOW method and the HIDE method. Counters limiting communication have also been inserted here. Although objects were hidden by the HIDE method, they were still physically placed on the panel and requested data. With the number of individual technologies and elements themselves, there was a huge data flow between the OPC server and visualization. Since the redesign of the old visualization would be very demanding, and it would never be certain whether the response would really be reduced, we decided to create a new design.

After further analysis of the original design of the fuel cell system, a gross violation of the rules of automated control systems has been identified. A U50 invertor drive (designed for this special purposes) plays a key role in the power generation control chain. In the original design of the entire system concept, this drive control was solved without any treatment on old visualization, without any link to the PLC controller (the inverter communicates with the Modbus bus itself).

4   New Technology Visualization

A significant change was made by the topology of the network's own PLC linking, technology and visualization itself, see Fig. 4. There was a PLC for fuel cells apart from the existing PLC assembly. The communication blocks have been preserved for the time being, but the whole energy system topology is planned for the future. The above-mentioned U50, which is essentially generating power, has received the right Modbus communication transmitter on Profibus. Thanks to this converter, the PLC can now fully control the operation of the fuel cells.

There was also a part of fuel cells with its own Modbus - Profibus converter. Thanks to this solution, an internal fuel cell internal diagnostic unit is now available. This diagnostics allows you to view the temperatures of each cell, the specific values of the voltage and the efficiency of the article.

The visualization itself is now divided into two parts - server and client. Therefore, the communication between the server and individual clients runs on the client-server architecture. The server part contains communication tables designed to be connected to the PLC using the S7 protocol, the visualization tables that provide data partitioning on individual screens, and last but not least, database communication that provides storage, diagnostics, and representation of graphical system actions.

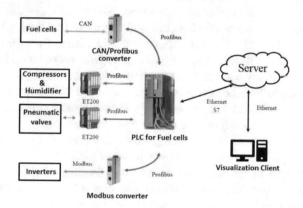

Fig. 4. New system topology.

To increase the speed of the communication line, a communication type other than OPC was chosen between the PLC and the server, namely the S7 protocol. This type of communication is optimized for Simatic, it provides both a one-sided and two-way configuration of the connection between the devices, allows sending of dynamic message lengths and, last but not least, confirmation of reception and length of the message, which is important for ensuring the validity of messages.

The client-to-server architecture of Promotic allows almost countless clients. Since a large number of clients and the volume of data transferred could greatly exploit the communication line, it was decided to divide the transmitted data into individual structures created on the basis of the data used on the actual display screens of the individual clients themselves. Therefore, the server section contains four types of tables that are linked directly to data blocks in the PLC. These tables contain values, which are system feedback, parameters that need to be set before starting the system, commands that the system controls, and eventually a fault table that indicates the error states of each system element. The data from these tables is then divided into Recieve and Send groups. Group names are always tailored to match the direction of communication from the given data sending point.

This means that if the values are received from the PLC that the server sends to individual client stations, then these values are in the Send group. In the case of commands received from the operator towards the server and then into the PLC, this data is in the Recieve group. This logic is also used in the client part of the visualization where the received data are in the Recieve groups and the commands sent and the parameters are in the Send groups. The tables in these groups are marked with the appropriate markings of their counterpart on the client side. These groups are then subdivided into individual data blocks based on the screen type.

The basic types of screens are panels. These screens include a summary of the technologies or individual parts of the technology. It serves mostly informative purposes and orientation between individual elements. The second type of screen is so-called popups. This type of screens serves to display more detailed information on individual technological elements and individual control. The connection itself, however, invokes the PmWeb component, a feature of the PmData feature. This component has a unique tag for each table that serves as a link identifier.

Because the original control system used a completely different view of the controlled system, it was necessary to create a completely new structure of the technology layout. This structure is partly based on the original concept, but only the parts that were important to be preserved.

Creating the concept of screens was preceded by an analysis of the whole technology as a whole, of the technological subsystems and of the individual elements, so that the structure of the screens is effectively distributed ([5, 7, 9, 10]).

Control panels for hydrogen technology control in automatic mode are located in the control panel of this screen. If the control system allows this, it is possible to choose between operating modes and if the fuel cell technology can be switched to automatic mode then its input parameter will be the required power. Control is also available for electrolysis technology. Its input parameter then in the automatic mode is the run time. In the information section, an indication of the energy consumed and the amount of hydrogen generated will be generated.

The other parts of the hydrogen technology visualization are the H2 screen. Its name is derived from the original visualization of the operator request. This screen is only an information section and therefore does not contain any signals for the control sequences. Its background is a gas scheme taken from the original visualization system. This scheme is enriched with the display elements of the pressure and flow values in the pipeline, and the control elements of the taps for forming the gas distribution routes.

This automatic mode display is informative only because the control system itself defines the routes of the gases used based on pressures in the gas cylinders, the use of particular fuel cells, based on their efficiency, the hourly hours and the need for the supplied energy. If the entire hydrogen technology is in manual mode, this screen will serve to manually set the hydrogen distribution path or other gases.

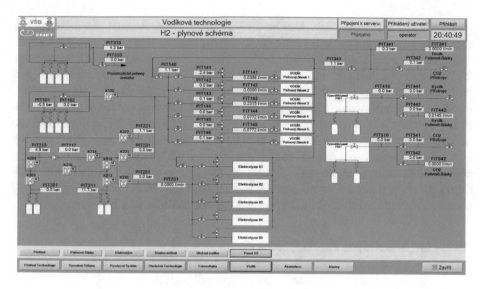

Fig. 5. Hydrogen technology information screen (only czech localization is available now).

One of the most important hydrogen fuel cell visualization panels is the screen with the same name (see Fig. 5). As with all other technology screens, this screen is primarily designed for manual control. This overview is again divided into the information and control section. This part of the visualization contains almost everything about the management of fuel cells and their operating technologies. The only thing that is not shown on this screen is the U50, which is located on the viewing screen and the cooling circuit, which is controlled by another part of the power unit and thus by other PLCs. In the future, reorganization of the signals of the technological elements between the control systems is considered.

In the information section there is a scheme of interconnection of five fuel cells among other service technologies. This scheme includes measurement of pressures and flows of gases entering the fuel cells marked PIT and FIT, passive elements of manual and reduction cocks and fuel cells themselves. In addition, we can see the BTW fuel cell membrane dampening valve icons, the steam inlet valves in the YV oxidation circuit, and the compressor unit icons that supply air to the fuel cells. Last but not least, the information section contains two icons of controlled elements. This is the icon of the inverter operating turbine units, marked G1 and the humidifier icon, for controlling humidification intervals, marked G2.

There are several partitions in the control section of the screen. In the first are the buttons with the indication to select the operating mode. In the second section there is a control unit for switching the humidifier unit. It should be noted here that the laboratory is not yet optimally operationally and electrically optimized. This results in the humidifier being powered by a drawer that is controlled by a contactor switch. Therefore, this button switches the contactors to power the socket cabinet located in the laboratory. Another element of the control part is power control of the G1 inverter, compressor units and YV valves, the control and power of which are fed into the same cabinet. Since it is important for fuel cell fuel cell parameterization to be in the ready state, the screen also includes U50. This control triggers the drive power and sets it to the ready state until it is parameterized. The fifth section contains the connection of the fuel cells to the terminals of the U50. Since this drive contains large capacitors, it needs to be pre-charged. Therefore, the load connection can be connected until the drive is ready. Although this part of the screen is reserved for control purposes, there is an indication of the flow and humidity of the intake air into the fuel cells. This information is common to all fuel cells. Last but not least, the control section contains an indication of an emergency stop and failures confirmation.

5   Visualization Security

The hydrogen laboratory, which includes fuel cell technology, is one of several elements of the energy unit for the generation and accumulation of electrical energy. This object can be moved by people without sufficient qualifications to manage this system, such as students. That is why you need to ensure visualization and configuration or system setup. To this end, Promotic offers a user account system and so-called operating modes have been created in the application.

Operating modes, or operating modes, serve to divide traffic into certain situations, affect user access to the system, runtime security and, in particular, to protect operators themselves. Four operating modes indicate the states in which the managed system can be located. Individual elements of a technology unit may have their own operating modes, depending on the complexity and necessity of changing them. These elements may be in different states, but they always have traceability.

Automatic mode is used to fully separate the part or whole technological unit. In this mode, it is essential that all inferior elements of the technology used by the system are in the ready state for automatic operation. If everything is ready for automatic operation, the system waits for an input parameter or impulse from the parent system or operator or responds directly to the input operating conditions in which the system is operating. If the automatic work sequence ends by completing the task, not meeting all termination conditions, the system stays in the ready auto mode. If a request to switch the operating mode occurs during the automatic sequence operation mode, the system is required to safely terminate its run prematurely and meet the request. If the automatic mode is lost, even if it is the only subset of the system that is part of it, the work process is terminated and the operating mode is switched.

If the fuel cell system is fully automatic when properly set up, the Hydrogen Laboratory is part of the research center. Also in full operation, it is possible to use the manual mode of this part of the power unit to work with the voltage directly in the laboratory. This option results from the electrical connection and the bringing of special contacts back to the laboratory.

The working mode itself is designed for a manually operated work process supervised by the operator. If the system is not ready for any reason for automatic mode, it is in manual mode. In addition to manually entering values for a system request, this mode can change certain parameters, sequencing, or diagnostics. Running the system into working mode in this operating mode is the responsibility of the operator. Of course, the control system checks the critical conditions, howeverit can bypass, for example, the conditions for the correct startup and startup of the system into working mode. It is therefore necessary for the operator to be trained to run a hydrogen laboratory.

Local or service mode is primarily designed to process significant changes to the system and put the entire system into operation. In this mode, the system allows the input of parameters, measuring ranges, testing of individual control signals without the use of traffic control. Steps in this mode can only be done by an administrator or a system developer, not a regular operator. Local mode can be closely tied to hardware elements on racks, such as key mode switches. These switches provide security for a particular part of the technology, for service adjustments, or for control. In our case, these are key switches on the U50 inverter cabinet, which is not located directly in the laboratory. It is therefore necessary to switch these switches while working in these switchboards so as not to endanger the service technician by unexpected workflow initialization. Local mode can also be invoked on specific elements from visualization if it is a feature that does not have its own hardware switches. Once any element of the system is in the local mode, the control system blocks the execution of the work process. When one or more elements switch to this mode, the rest of the system remains in manual or off mode. At the same time, it is not possible to remove the local mode by switching the mode from a different location than was invoked. This means that if any switchboard mode is switched to local mode, it can not be changed from visualization until the key is re-rotatable, or by pressing a button on that rack indicating the change of the working mode.

Off mode is used to indicate inactive elements that are not currently live, so they are not in any working mode. If the system is in manual mode, it is not a requirement for all subsystems to be turned on.

6   Conclusion

The original aim of the work on the hydrogen laboratory was to remove the visualization and system reaction delay. After analysis, significant system errors were detected in both the topology, the control program, and the visualization itself. As part of measuring time lags between technology elements, a problem in technology topology has been found. Delayed issues were also caused by building a PLC control program. The second problem was the original visualization, which was poorly designed, contained a lot of timers and read all the data constantly, which led to several seconds delay. For this reason, the new visualizations described in this article were created. The new visualization is processed into smarter screens and removes direct command control from the elements of the technology, but creates a space for controlling the commands that trigger the device elements in the PLC safety sequences. The new topology and communication principle helped to remove the essential component of the delay. Next step is interconnection with other energy blocks within center to find correct working set point for all technologies combinations.

Acknowledgement. This work was supported by the project SP2018/160, "Development of algorithms and systems for control, measurement and safety applications IV" of Student Grant System, VSB-TU Ostrava.

References

1. Chesalkin, A., Minarik, D., Moldrik, P., Otevrel, V.: Metal hydrides as an energy storage systems in rreal operating conditions. In: Proceedings of the 8th International Scientific Symposium on Electrical Power Engineering, Elektroenergetika 2015, pp. 507–510 (2015)
2. Minarik, D., Horak, B., Moldrik, P., Slanina, Z.: An experimental study of laboratory hybrid power system with the hydrogen technologies. Adv. Electr. Electron. Eng. **12**(5), 18–528 (2014)
3. Vantuch, T., Misak, S., Jezowicz, T., Burianek, T., Snasel, V.: The power quality forecasting model for off-grid system supported by multiobjective optimization. IEEE Trans. Ind. Electron. **64**(12), 9507–9516 (2017)
4. Kosmak, J. Misak, S.: A power quality parameters setup for power quality management model as an integrate part of active demand side management. In: Proceedings of the 9th International Scientific Symposium on Electrical Power Engineering, Elektroenergetika 2017, pp. 523–528 (2017)
5. Idzkowski, A., Walendziuk, W., Borawski, W.: Analysis of the temperature impact on the performance of photovoltaic panel. In: Proceedings of SPIE - The International Society for Optical Engineering, vol. 9662 (2015)
6. Kumar, A., Alice Hepzibah, A., Cermak, T., and Misak, S., Analysis and simulation of PEM fuel cell with interleaved soft switch boost converter for stand-alone load conditions. In: Proceedings of the 8th International Scientific Symposium on Electrical Power Engineering, Elektroenergetika 2015, pp. 199–202 (2015)
7. Idzkowski, A., Leoniuk, K., Walendziuk, W., Budzynski, L.: Monitoring and control system of charging batteries connected to a photovoltaic panel. In: Proceedings of SPIE - The International Society for Optical Engineering, vol. 9662 (2015)
8. Horak, A., Prymek, M., Prokop, L., Misak, S.: Economic aspects of multi-source demand-side consumption optimization in the smart home concept. Acta Polytech. Hung. **12**(7), 89–108 (2015)
9. Kaczmarczyk, V., Bradac, Z., Fiedler, P.: A heuristic algorithm to compute multimodal criterial function weights for demand management in residential areas. Energies **10**(7) (2017)
10. Fendrychova, M., Fiedler, P.: Challenges in ARC fault detection for high power DC applications. IFAC-PapersOnLine **49**(25), 552–556 (2016)

Processing of Conceptual Diagrammatic Models Based on Automation Graphical Grammars

Alexander Afanasyev, Anatoliy Gladkikh, Nikolay Voit,
and Sergey Kirillov$^{(\boxtimes)}$

Ulyanovsk State Technical University, Ulyanovsk, Russia
{a.afanasev,n.voit}@ulstu.ru, kirillovsyu@gmail.com

Abstract. The presented work is devoted to conceptual design in area of creation and processing of diagram models. Authors offer a family of automatic graphical RV-grammars which allows to analyze, control and interpret diagrammatic models of automated systems presented in the widely used visual languages as UML, IDEF, BPMN. The estimation of time costs of the analysis is executed and the graph is constructed.

Keywords: Conceptual design · Automated systems · Visual languages
Diagrammatic models · Workflows · Automaton grammars

1 Introduction

Diagrammatic models developed in visual languages (UML, IDEF, eEPC, BPMN, SharePoint, ER, DFD, etc.) are actively used in the practice of designing complex automated systems (CAS), especially at conceptual stages. CAS includes, among others, e-learning environments. In particular, the open technology platform for distance learning Moodle contains more than 1.5 million lines of source code. For example, in master technology of RUP (Rational Unified Process) [1], UML is the basis for representation of architectural and software solutions, and when using the ARIS (Architecture of Integrated Information Systems) - eEPC. The use of diagram models allows to increase the efficiency of design solutions, to avoid "expensive" mistakes, to improve the understanding of the project by executors, to organize interaction between customers and project executors, to improve project documentation and in some cases to automate the process of obtaining program code. The solution of these problems is connected with the problem of "success" in the creation of CAS. According to research by Standish Group [2], only about 40% of projects meet the requirements of the declared functionality, implemented within the specified time and within the specified budget.

Due to the specifics of modern CAS design, the considered models are dynamic and distributed, which is determined by the changing business processes and the work of remote teams of designers in the environment.

© Springer Nature Switzerland AG 2019
A. Abraham et al. (Eds.): IITI 2018, AISC 875, pp. 369–378, 2019.
https://doi.org/10.1007/978-3-030-01821-4_39

The research of structural (primarily topological features) and semantic (connectivity of diagrams, possibly presented in various visual languages, in terms of the text component) is an actual task, which has great practical value.

2 Related Works

In the modern theory of graphic visual languages, a logical model is used to represent diagrams, which contains graphical objects and connections between them. To process such models, graphic grammars are used. John L. Pfaltz and Azreil Rosenfeld offered web grammar [3]. Zhang and Costagliola [4, 5] developed a positional graphical grammar related to context-free grammar. Wittenberg and Weitzman [6] developed a relational graphical grammar. Zhang and Orgun [7] described a graphical grammar preserving in their works. However, the mentioned graphical grammars have the following disadvantages:

1. Positional grammars. It doesn't involve the use of connection areas and cannot be used for graphical languages, whose objects have a dynamically changing number of inputs/outputs,
2. The authors of the relational grammars tells about the imperfection of the neutralization mechanism, specifically the incompleteness of the generated list of errors.
3. There is no control of semantic integrity (text attributes of complex diagram models presented in different visual languages), as well as semantic consistency in terms of structural issues of diagrams between themselves and the conceptual model as a whole.
4. The common disadvantages of the above grammars are: the increase in the number of products in the construction of grammar for unstructured graphic languages (with a constant number of primitives of the graphical language to describe all variants of unstructured there is a significant increase in the number of products), the complexity of grammar construction, the large time costs of analysis (analyzers built on the basis of the considered grammars, have a polynomial or exponential time of graphic diagrams analysis.

In [8], colored Petri nets are used for the dynamic semantic analysis of workflows, and in [9] the pi-Calculus approach formalizing workflows into algebraic statements of first-order logic. However, there is no analysis of the text component of the diagrammatic models in these methods.

In the most common tools for creating and processing diagram models, such as Microsoft Visio [10], Visual paradigm for UML [11], Aris Toolset [12], IBM Rational Software Architect (RSA) [13] analysis and control of diagram models is performed by direct methods, requires several "passes" depending on the type of error. There is no control over the structural features of the complex diagram models, and also the semantic control of the integrity and consistency of the structural and text attributes of the associated diagrammatic models of dynamic workflows.

3   Problem

Diagrammatic models of the conceptual design of automated systems have a number of features, the analysis of which allows us to identify a number of mistakes related to the most "expensive" ones.

From the structural (syntactic) point of view, these visual languages have a number of graphic objects with a variable number of inputs/outputs, through which complex structures associated with the use of "AND", "OR", "XOR" splitters can be organized. Splitters can "stand" at a considerable distance from each other, forming structures with a remote context. The methods and tools discussed above are not in a position to control such features. The results may include diagrams that contain freezes, deadlocks and deadlocks.

The next structural feature of the considered class of models is their complexity. This is especially true for UML and IDEF. For example, the top of the UML usage diagram can be deployed into an activity diagram, the individual blocks of which can correspond to class diagrams, etc. Implementing such a hierarchical dynamic end-to-end analysis is an important task.

Let us consider the semantic features of the diagrammatic models. Diagrams should not contain conflicting information.

Contradiction of the model can cause serious problems in its implementation and subsequent use in practice. For example, the presence of closed paths when mapping the aggregation or composition relationship leads to errors in the code that will implement the corresponding classes. Having elements with the same name and different property attributes in the same namespace also leads to ambiguous interpretations and can be a source of errors.

In the collective development of automated systems, the work of designers opens up additional sources of difficult to diagnose, distributed over a multitude of error diagrams.

When working in a team, it is important to monitor the consistency of concepts used in complex diagrams. Semantic control of complex diagrams allows you to monitor errors distributed across different diagrams. With semantic control, the complex idea created about an automated system becomes consistent and consistent. Early diagnostics of such errors can reduce development time by reducing the iterations associated with their correction.

Thus, the solution of these problems is related to the development of mechanisms that allow analysis and control of the syntactic and semantic features of conceptual models. The article proposes a family of automatic graphical grammars that allow efficient processing of such models.

4 RV-Grammar

RV-grammar of the language L (G) is the ordered five non-empty sets

$$G = \left(V, \Sigma, \widetilde{\Sigma}, R, r_0 \right) \tag{1}$$

where $V = \{v_e, e = \overline{1,L}\}$ is auxiliary alphabet; $\Sigma = \{a_t, t = \overline{1,T}\}$ is terminal alphabet graphic language; $\widetilde{\Sigma} = \{\widetilde{a_t}, t = \overline{1,T}\}$ is quasi terminal alphabet; $R = \{r_i, i = \overline{0,I}\}$ is grammar G schema (set of names of products complexes, each complex r_i consists of a subset P_{ij} of products $r_i = \{P_{ij}, j = \overline{1,J}\}$; $r_0 \in R$ is RV-axiom grammar.

RV-grammar is automate-based grammar. It operates with inner memory to control correctness of diagrams. Finite state machine is used to choose rule by memory changing. A number of different structures are used as memory: stacks, lists, queries, etc. and combination of them. These conditions make possible to construct grammar with linear dependence from element count. All operation takes limited time, because reading and writing data is constant time operation. In each tack of diagram analysis operate with only one element. On Fig. 1 is presenter RV-grammar in graph form.

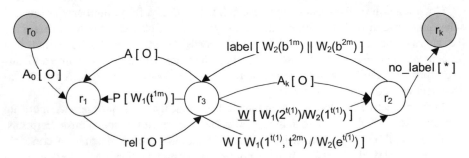

Fig. 1. RV - grammar graph form for the PGSA language after minimization

5 Hierarchical RVH-Grammar

The construction of a hierarchical RV grammar is based on grammatical models in the form of a recursive Woods network. Suppose that not only terminal (quasi terminal) symbols are allowed as symbols marking the arcs of the graph of RV grammars, but also nonterminal symbols denoting complex syntactic constructions whose presence is necessary so that the transition along the arc to the next complex is permissible. As nonterminal symbols, the names of production complexes can act, and as complexes themselves - other RV-grammars (we will call them subgrams). To organize work on the hierarchical RV grammar, a store mechanism is used, similar to that used when calling subroutines from the main program and then returning to the main program. Using RVH grammars allows you to reduce the amount of grammar, it is easy to modify it, effectively construct a grammar from existing subgrams (Fig. 2).

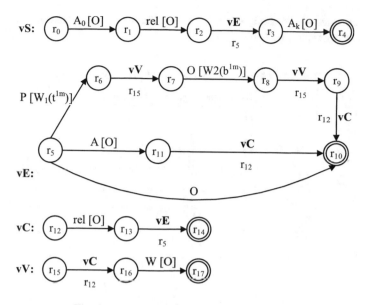

Fig. 2. Hierarchical RVH-grammar example

6 Temporal RVT-Grammars

Flows of design works are a powerful tool for analyzing the business processes of the enterprise and contain design tasks that are addressed to specific departments and executors of the enterprise.

Such design works can be performed simultaneously in different structural divisions and different executors, so the issues of synchronization, resource blocking, deadlocks, bottlenecks, etc., encountered in the theory of informatics, arise in the field of business process management enterprise. Such streams of design work can be represented in the temporal basis of "Before", "During", "After", putting the time schedule for the execution of the enterprise's work in accordance with the parameter.

Temporal RVT - grammar of a language L (G) is an ordered eight non-empty sets

$$G = \left(V, \Sigma, \widetilde{\Sigma}, C, E, R, T, r_0 \right) \tag{2}$$

where $V = \{v_e, e = \overline{1,L}\}$ is auxiliary alphabet; $\Sigma = \{a_t, t = \overline{1,T}\}$ is terminal alphabet graphic language; $\widetilde{\Sigma} = \{\widetilde{a}t, t = \overline{1,T}\}$ is quasi terminal alphabet; C is set of clock identifiers; E is the set of temporal relations "Before", "During", "After" (initialization of the clock $\{c := 0\}$, relations of the form $\{c \sim x\}$, where x the variable (the identifier of the clock), a is a constant, $\sim \epsilon\{=, <, \leq, >, \geq\}$); $R = \{r_i, i = \overline{0,I}\}$ is grammar G schema (set of names of complexes of products, each complex r_i consists of a subset P_{ij} of products $r_i = \{P_{ij}, j = \overline{1,J}\}$); $T \epsilon \{t_1, t_2, ..., t_n\}$ is a set of time stamps; $r_0 \epsilon R$ is RV-axiom grammar.

RVT-grammar was developed for basic elements of BPM notation. Table form of this grammar presented in Table 1.

Table 1. Grammar example for simple BPMN diagram

N	State	Quasi term	Next state	Operation with memory
1	r0	A0	r1	o
2	r1	rel	r3	o
3	r2	labelEG	r3	$W_2(b^{1m}, b^{t(6)})$
4		labelPG	r3	$W_2(b^{2m}, b^{t(6)})$
5	r3	Ai	r1	o
6		Aim	r1	o
7		Ait	r1	$W_1(t_s^{t(6)})$
8		Akl	r2	$W_3(e^{1m}, e^{2m})$
9		Ak	r4	o
10		A	r1	$W_1(t_s^{t(6)})$
11		EGc	r1	$W_1\left(t^{1m^{(n-1)}}\right)/W_3(k = 1)$
12		EG	r2	$W_1(1^{t(1)}, k^{t(2)})/W_3(e^{t(2)}, k\ ! = 1)$
13		_EG	r2	$W_1(\mathrm{inc}(m^{t(1)})/W_3(m^{t(1)} < k^{t(2)})$
14		_EGe	r1	$W_1\left(t^{1m^{(n-1)}}\right)/W_3\left(m^{t(1)} = k^{t(2)}, p!\ = 1\right)$
15		_EGme	r1	$o/W_3(m^{t(1)} = k^{t(2)}, p = 1)$
16		PGf	r1	$W_1\left(t^{2m^{(n-1)}}\right)/W_3(k = 1)$
17		PG	r2	$W_1(1^{t(3)}, k^{t(4)})/W_3(e^{t(3)}, k\ ! = 1)$
18		_PG	r2	$W_1(\mathrm{inc}(m^{t(3)})/W_3(m^{t(3)} < k^{t(4)})$
19		_PGe	r1	$W_1\left(t^{2m^{(n-1)}}\right)/W_3\left(m^{t(3)} = k^{t(4)}, p!\ = 1\right)$
20		_PGje	r1	$W_1\left(t^{2m^{(n-1)}}\right)/W_3\left(m^{t(3)} = k^{t(4)}, p = 1\right)$
21	r4	no_label	r5	*
22	r5			

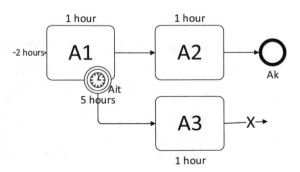

Fig. 3. Temporal BPMN diagram example

The example of diagram, which can be analyzed by RVT-grammar presented in Fig. 3. The main feature is time labels for every object from terminal alphabet.

7 The Method for Controlling the Semantic Integrity of Flow Diagrams

The control of the semantic integrity of the models under consideration is connected with the diagnosis of the consistency and consistency of the text component of the diagrams. An ontological approach is proposed for these purposes. Ontological analysis can be carried out during the process of analyzing the diagram or in the complete construction of an ontology upon completion of the analysis of the diagram. In the first case, it becomes possible to immediately point to a specific element that led to an error. However, you will have to spend more time, since ontology analysis should be performed on each element. In the second case, the process of combining ontologies must be carried out only once. Next, the second option will be considered. The general scheme of the ontological analysis of the diagrams is shown in Fig. 4.

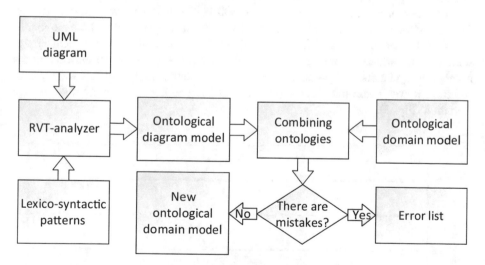

Fig. 4. Main scheme of diagram ontological analysis

To control the semantic integrity of different types of diagrams, we will make changes to the RV-grammar rule:

$$a_t(\chi,\varsigma) \xrightarrow{W_v(\gamma_1,...,\gamma_n)} r_m \tag{3}$$

where χ is the procedure for extracting semantic information, and ς is the procedure for retrieving the temporal information. The procedure of χ is to find the appropriate rule in the list. The rule consists of two parts - replacement and pattern. The pattern part describes the lexico-syntactic pattern.

The replacement part specifies the location of this text unit in the partial semantic tree of the diagram.

8   Time Estimation of RV-Grammars

The RV analyzer has a linear time characteristic of the control, i.e. The time spent analyzing the input list of chart elements is determined by the linear function of its length and is calculated by the formula:

$$t = c * L_s \tag{4}$$

or

$$t = c * k * L_k, k = L_s * L_k \tag{5}$$

Where L_s - the number of transitions over the states of the automaton, L_k - the number of elements of the diagram and C - the algorithm implementation constant, which shows how many commands (operators) are spent on analyzing one object when implementing a control algorithm on a specific PC.

An experiment was conducted in which 500 PGSA diagrams were generated, then for each of them the number of generated elements was determined and the number of states of the analyzing machine was calculated. The result of the experiment is presented in the form of a dot diagram in Fig. 4, it confirms the linear dependence of the number of applications of grammar rules on the number of graphic objects and links. In addition, it was determined the expectation of the coefficient of equal and average deviation (Fig. 5).

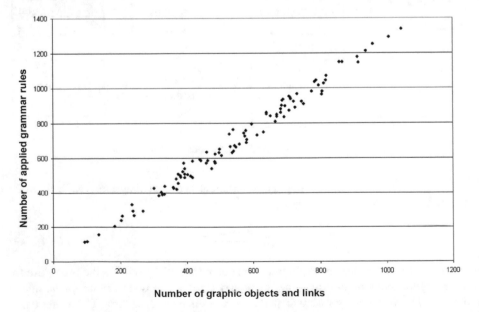

Fig. 5. The values of the experimental linearity coefficients for the PGSA language

9 Conclusion

A family of automatic graphical grammars has been developed that make it possible to extend the class of diagnosed errors in the process of developing conceptual models of automated systems. RVN - a grammar offering mechanisms for neutralizing errors, RVH is a grammar that assumes its construction on the basis of sub grammars, as well as an easy modification of the basic grammar and re-use of sub grammars, and RVTI is a grammar allowing to take into account time characteristics. Developed analyzers for most visual languages in the form of plug-ins for MS Visio.

Further directions of work are the expansion of the possibilities of semantic analysis of diagrammatic models from the point of view of coordinating text attributes of diagrams with project documentation.

Acknowledgment. The reported study was funded by RFBR according to the research project № 17-07-01417. The research is supported by a grant from the Ministry of Education and Science of the Russian Federation, project No. 2.1615.2017/4.6. The reported research was funded by Russian Foundation for Basic Research and the government of the region of the Russian Federation, grant № 18-47-730032.

References

1. Fowler, M.: New Programming Methodologies. http://www.maxkir.com/sd/newmethRUS. html. Accessed 14 May 2018
2. The Standish Group. http://www.standishgroup.com/outline. Accessed 14 May 2018
3. Fu, K.: Structural Methods of Pattern Recognition. Mir, Moscow (1977)
4. Zhang, D.Q., Zhang, K.: Reserved graph grammar: a specification tool for diagrammatic VPLs. In: IEEE Symposium on Visual Languages, Proceedings, pp. 284–291 (1997)
5. Costagliola, G., Lucia, A.D., Orece, S., Tortora, G.: A parsing methodology for the implementation of visual systems. http://www.dmi.unisa.it/people/costagliola/www/home/papers/method.ps.gz. Accessed 14 May 2018
6. Wittenburg, K., Weitzman, L.: Relational grammars: theory and practice in a visual language interface for process modeling. http://citeseer.ist.psu.edu/wittenburg96relational.html. Accessed 14 May 2018
7. Zhang, K.B., Zhang, K., Orgun, M.A.: Using graph grammar to implement global layout for a visual programming language generation system. In: Feng, D., Jin, J., Eades, P., Yan, H. (eds.) Conferences in Research and Practice in Information Technology, vol. 11, pp. 115–122. Australian Computer Society, Sydney (2002)
8. Jalali, A., Wohed, P., Ouyang, C.: Aspect oriented business process modelling with precedence. In: Mendling, J., Weidlich, M. (eds.) Business Process Model and Notation. BPMN 2012. Lecture Notes in Business Information Processing, vol. 125, pp. 23–37. Springer, Heidelberg (2012)
9. Hasso Plattner Institut. http://bpt.hpi.uni-potsdam.de. Accessed 14 May 2018
10. Roth, C.: Using Microsoft Visio 2010. Pearson Education, United States of America (2011)
11. Paradigm V. Visual paradigm for UML. Hong Kong: Visual Paradigm International. http://www.visual-paradigm.com/product/vpuml/. Accessed 14 May 2018

12. Santos Jr., P.S., Almeida, J.P.A., Pianissolla, T.L.: Uncovering the organisational modelling and business process modelling languages in the ARIS method. Int. J. Bus. Process Integr. Manag. 2(5), 130–143 (2011)
13. Hoffmann, H.P.: Deploying model-based systems engineering with IBM® rational® solutions for systems and software engineering. In: Digital Avionics Systems Conference (DASC), 2012 IEEE/AIAA 31st. – IEEE, 2012. – C. 1–8 (2012)
14. Sharov, O., Afanasev, A.: Syntactically - implementation-oriented graphical language based on automatic graphical grammars. Programming 6, 56–66 (2005)

Intelligent System for Assessing Organization's Possibilities to Achieve Sustained Success

Inna Nurutdinova$^{(\boxtimes)}$ and Liubov Dimitrova

Don State Technical University, Rostov-on-Don, Russian Federation
nurut.inna@yandex.ru, kaf-qm@donstu.ru

Abstract. The questions of creating intelligent information system designed for assessing an organization's maturity level in the direction of achievement of stable development are considered. An unbiased evaluation of the possibility of achieving a stable success is an important task of any organization. The urgency of this problem is growing in the conditions of contemporary dynamically changing world and national economies. The choice of a linguistic approach to the solution of this problem, based on the application of the theory of fuzzy sets, has been substantiated. The architecture of the expert system and subsystem interrelations, including standard blocks, and also original subsystems, has been described. The problem solving algorithm on the basis of expert assessments of the main criteria, contributing to elimination of obstacles in the organization activity, has been suggested. We have considered the problems of presentation and quality analysis of the fuzzy expert information, determined linguistic variables, and developed membership functions. The base of knowledge has been created, a technique of fuzzy inference has been explained, an example of assessment of the organization's maturity level on the basis of expert assessments of the main criteria has been provided. Application of the suggested intelligent system of monitoring the organization's state makes it possible to react immediately to quickly changing conditions, in which the organization is functioning, correct both tactical methods and strategic aims.

Keywords: Intelligent system · Organization maturity level
Linguistic approach · Linguistic variable · Membership function
Fuzzification · Knowledge base · Production rule · Fuzzy inference
Defuzzification

1 Introduction

Achieving a sustained success is an urgent task for each business entity (enterprise, organization, firm, etc.). In the following, we will use a general concept, organization, implying any of the mentioned entities. Providing a sustained success of an organization involves optimal interrelation of high indices of satisfaction of all the parties concerned: consumers, owners, shareholders, suppliers, business partners, and communities. Contemporary quickly changing economy requires a prompt response, therefore, this aspect of the activity analysis takes on particular significance. Qualitative and quantitative characteristics of all aspects, influencing the organization functioning, must be balanced, and this serves as the main criterion for assuring success. Self-assessment is

© Springer Nature Switzerland AG 2019
A. Abraham et al. (Eds.): IITI 2018, AISC 875, pp. 379–388, 2019.
https://doi.org/10.1007/978-3-030-01821-4_40

one of the main tools of the organization functioning analysis and provides continuous monitoring of the organization state. Self-assessment technique suggests the use of 5 maturity levels, each of them corresponds to a kind of activity, regulated by the standard [1]. In the process of self-assessment, strong and weak aspects are determined, activity priorities are ranged, a valuable array of information for each of the aspects, involved in the organization functioning, is formed. The result of self-assessment is the possibility of taking urgent tactical measures and correction of strategic actions.

In accordance to [1] 6 main groups of criteria, intended to encourage elimination of obstacles in the organization's activity: management for achieving sustained success of the organization; strategy and policy; resources management; processes management; monitoring, measurement, analysis and study; improvements, innovations, and learning.

2 Problem Solution

2.1 Problem Statement

The technique specified by the standard [1] doesn't permit us to determine a maturity level of the organization as a whole, as the purpose in hand gives rise to uncertainty even at the level of a criterion analysis. It is stipulated, first of all, in that the document doesn't have a distinct algorithm focused on definition of the maturity level even within the framework of the mentioned elements. At the same time, to define the activity priorities and correction of development strategy, it is also important to define a maturity level of the organization as a whole. The purpose of this paper is to develop an intelligent system for assessing a maturity level of the organization in the direction of achieving a sustained success, and providing a continuous monitoring of the organization state.

The difficulty of making decisions while assessing a maturity level of the organization is connected with defining a direction of actions, the final result of these actions will be a successful solution of the problem. A fuzzy character of the expert information, which is subject to processing, gives the problem even a larger level of complexity.

Application of the theory of fuzzy sets [2, 3] enables us to formulate a formal description of all the factors being beyond the quantification, due to use of fuzzy linguistic concepts. The result of introducing linguistic variables is a complete qualitative description of the problem, taking into account conditions of uncertainty.

Description of actions, aimed at making the only truth decision in complex multilevel systems including quality management systems [4], with the help of the apparatus of the theory of fuzzy sets, makes it possible to use an adequate formal language. Methods and concepts of the theory of fuzzy sets (fuzzy constraints, linguistic variables) allow us to identify the essence and directional effect of the process of decision making in a multilevel system in fuzzy conditions. A mathematical apparatus of the theory of fuzzy sets is successively applied for solving problems in the spheres of economy, logistics, marketing, etc., such as, for instance, estimation of production processes quality [5] and service quality [6], estimation of relations with customers and choice of the supplier of logistic services [7, 8], estimation of innovative development of regions [9] and level of interregional integration [10], analysis of customer queries and confidence estimation in electronic commerce [11] etc. The suggested approach to modeling the processes of decision making on the basis of self-assessment with the use of mathematical apparatus of the theory of fuzzy sets makes it possible to remain the consideration unity of the complex hierarchical system on the basis of the main elements and processes being part of it, and relations between them, and also take into consideration the main factors, apart from the way of their assessment. Practical application of fuzzy modeling in the present problem will make it possible to obtain an objective assessment of the maturity level of the organization.

2.2 Methods

To formalize real systems a linguistic approach is used, and dependence between sets of input and output variables is described in the qualitative level as statements in the form of production rules. Fuzzy and linguistic variables, fuzzy relations make it possible to move from verbal presentations of problems' elements and relations between them to numerical ones.

Fuzzification, composition, and defuzzification [2, 3] are the stages, generating the basis of the procedure of the process fuzzy modeling, when they are successfully performed, then the main aim is achieved – getting a reliable assessment.

Description of the problem statements in a linguistic form takes place at the stage of fuzzification. Input linguistic variables (LV) are determined in accordance with 6 groups of criteria presented in [1]. Membership functions (MF) of all terms of input LV, constructed with the use of the preset exact values from universes of input LVs, elucidate degrees of confidence that the input LV takes on a value, namely a certain term.

The process of grouping and defining a unified fuzzy set of all fuzzy sets, determined for all terms of each input LV, takes place at the stage of composition. Being obtained as a result of grouping, a unified fuzzy set is a value for each derived LV. As a result of manipulation on application of collection of rules – fuzzy knowledge base – a truth level is determined to presuppose each rule. The reason for this process is a certain fuzzy operation, corresponding to conjunction or disjunction of terms in the left part of the rules.

The main point of the defuzzification stage in this problem consists in getting exact values of the organization's maturity level, these values being based on fuzzy inference. Fuzzy inference looks like an approximation of dependence "input – output" on the basis of linguistic statements of the kind "IF, THEN", and also logic operations on fuzzy sets. The output in this problem is discrete likewise the problems of pattern recognition, prediction etc. The problem decision diagram for assessing a maturity level of the organization is presented in Fig. 1.

Let us consider the problem where, depending on possible values of input information A_j, an expert arrives at conclusion on the resulting assessment B_j (on the maturity level of the organization). Let $\{X\}$ denote a set of values of input features i.e. a population of evaluations of the main criteria, determining the value of the output estimate B – maturity level. A successful solution of the problem is impossible without resolution of the problem of modeling expert information on relations of the considered features and procedures of decision making.

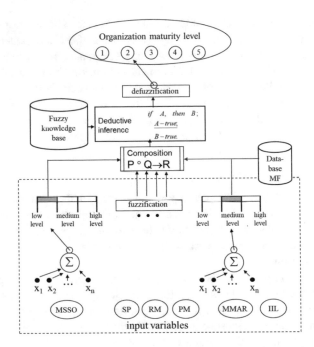

Fig. 1. General diagram of the maturity level assessment method.

The models of input X and output B features are represented in the form of semantic spaces and corresponding them MFs. The generalized model of the object domain "assessment of an organization's maturity level" has the form of composition of fuzzy relations of the considered semantic spaces $R = X \rightarrow B$. Relation R is a fuzzy set on the direct product $X \times B$ of the complete space of predictions X and complete space of conclusions B.

As a result of fuzzification of the features under examination we have constructed MFs of linguistic variables of the groups of criteria and output parameter – an organization's maturity level. Selection of the MF kind is based on different assumptions on the characteristics of these functions (symmetry, monotonicity, continuity of the first derivative etc.) with regard to uncertainty specific nature. MFs of the standard kind are most convenient for expert assessment, and also for data input-output and storage.

An important point is definition of the basic term-set and the extended ones of the LV. Selection of the optimal amount of terms is made both on the basis of a priory analysis of the object domain and a posteriori one on the basis of the analysis of the models consistency [12].

Obtaining a particular solution provides for fulfillment of certain sequence:

– construction of the generalized LV on the set of input features influencing the value of the output feature (i.e. aggregation of left parts of the rules);
– calculation of MF for the generalized LV;
– definition of truth degree of fuzzy rules, on the basis of which selection of values of the output LV is performed.

At the heart of the inference mechanism of the intelligent information system there is a model of the object domain, representing a composition of fuzzy relations of semantic spaces of assessments of the main criteria and the resulting assessment of the maturity level. An expanded form of the fuzzy inference for the system of knowledge of this kind is similar to [13]:

$$\mu_{B'} = \bigvee_{x \in X} \left(\mu_{A'}(x) \wedge \mu_R(x, y) \right).$$

One of the most widespread methods used at the stage of defuzzification is the method of "barycentre" [13].

2.3 Linguistic Description

The linguistic description of the subject domain is made on the basis of the judgments of five experts who have extensive experience in the practical work on the analysis of quality management systems and self-assessment of the state of the organization. Since maturity level is determined on the basis of all the values, let us represent each of them in the form of a LV. For input LVs we introduce the following terms: low (corresponds to levels 1 and 2), middle (corresponds to level 3), high (corresponds to levels 4 and 5). Introduction of three terms given that there are five maturity levels meets the requirement of minimal uncertainty for experts. According to the names of the main maturity criteria, quoted in [1], let us introduce LV.

LV tuples "management for achieving sustained success of an organization" (MASSO), "strategy and policy" (SP), "resources management" (RM), "processes management" (PM), "monitoring, measurement, analysis, and study" (MMAS), "improvements, innovations, and learning" (IIL) have the form:

<MASSO, score {low, middle, high}, [4–20]>;
<SP, score{low, middle, high}, [3–15]>;
<RM, score{low, middle, high}, [8–40]>;
<PM, score{low, middle, high}, [2–10]>;
<MMAS, score{low, middle, high}, [6–30]>;
<IIL, score{low, middle, high}, [3–15]>.

For the output LV "organization's maturity level" (OML) let us select a 5-term model. LV tuple OML has the form:

<OML, % {1 level, 2 level, 3 level, 4 level, 5 level}, [0–100]>.

As a result of implementation of the fuzzification method, MFs of input and output LVs were constructed with the help of typical functions of trapezoidal form [14], the parameters of which are obtained on the basis of average expert estimates.

Figure 2 shows several graphics of the generalized MFs.

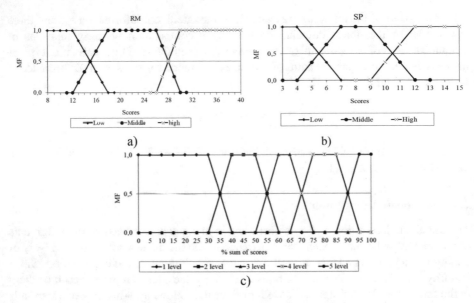

Fig. 2. Membership functions of LV: (a) RM; (b) SP; (c) OML.

In accordance with the developed algorithm [15], general consistency of many models of a feature assessment by an expert is determined by additive k_a and multi-plicative k_m indices [16]. Consistency indices k_{ij} and difference indices d_{ij} (Hamming linear distance) between the models of i-th and j-th experts form accordingly matrices of pair consistency K and fuzziness indices D, both for each of the terms, and for the whole population of terms [16].

All consistency characteristics are obtained. For the purpose of illustration Table 1 gives characteristics of consistency for LV of MASSO.

Table 1. Characteristics of expert information consistency.

K				D				k_a	k_m
1	0,892	0,892	0,761	0	0,07	0,072	0,168		
0,892	1	0,895	0,709	0,072	0	0,072	0,216	0,694	0,687
0,892	0,895	1	0,791	0,072	0,07	0	0,144		
0,761	0,709	0,791	1	0,168	0,216	0,144	0		

After determining satisfactory quality of expert information we go to the stages of composition and defuzzification.

2.4   Fuzzy Inference

On the basis of hierarchy analysis method [17], scores of significance coefficients for six criteria were obtained (Table 2).

Table 2. Values of significance coefficients for input LVs

MASSO	SP	RM	PM	MMAI	IIL
0,166	0,155	0,158	0,2	0,155	0,166

A base of knowledge for assessment of the organization's maturity level, that contains 729 rules, was created. As an illustration, let us present some rules:

1. IF < MASSO is «low level» and SP is «low level» and RM is «low level» and PM is «low level» and MMAS is «middle level» and IIL is «low level»> THEN < OML is level 1 >;
2. IF < MASSO is «low level» and SP is «low level» and RM is «middle level» and PM is «middle level» and MMAS is «low level» and IIL is «low level»> THEN < OML is level 2 >;
3. IF < MASSO is «middle level» and SP is «middle level» and RM is «middle level» and PM is «middle level» and MMAS is «middle level» and IIL is «middle level»> THEN < OML is level 3 >.

Fuzzy inference is based on maximin composition as a composition rule of fuzzy inference and operation of taking minimum as fuzzy implication. The problem was solved in MatLab medium with the help of Fuzzy Logic Toolbox application software package. Figure 3 presents the surfaces "inputs-output" corresponding to a synthesized fuzzy system of production rules. When constructing, input variables were selected by pairs, and the rest ones were fixed at the middle levels.

On the basis of the surfaces kind in Fig. 3 we can make a conclusion that the presented model of relations among features of the object domain adequately reflects real interrelations.

Let us illustrate an inference of the maturity level exact value for the next collection of input factors: MASSO = 10, SP = 10, RM = 20, PM = 10, MMAS = 15, IIL = 10. As a result, an assessed value of the organization's maturity level was obtained, it being equal to 80%, and that corresponds to level 4 of the organization maturity by the standard [1].

The presented technique is intended to create an intelligent system for assessing the state of an organization to achieve a sustained success. A structural diagram of this system is given in Fig. 4. The use of such system under the conditions of constant monitoring will make it possible to quickly reveal and work out the proper managerial decisions promoting achievement of sustained success of the organization.

IIS contains, besides conventional blocks of the expert system, subsystems of problems solution with the use of key elements, criteria groups, definition of features weight.

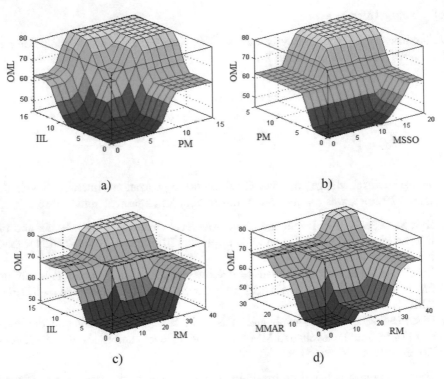

a)   b)

c)   d)

Fig. 3. Response surfaces of organization's maturity level vs (a) processes management and management for achieving sustained success of the organization; (b) processes management and improvements, innovations, and learning; (c) resources management and improvements, innovations, and learning; (d) resources management and monitoring, measurement, analysis and study

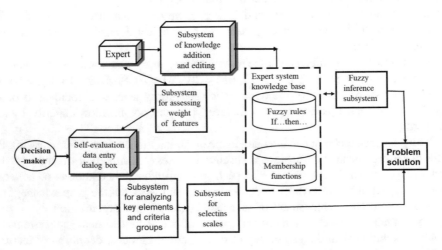

Fig. 4. Structural diagram of the intelligent information system (IIS).

3 Conclusion

The technique for obtaining objective assessment of an organization's maturity to achieve a sustained success has been presented, it is based on the application of the mathematical apparatus of the theory of fuzzy sets. The object domain has been studied and its formalized description is given, linguistic variables have been introduced in accordance with the criteria of the standard [1], the basic term-sets have been determined, their membership functions have been constructed. The knowledge base has been created, and fuzzy inference of assessment of the organization's maturity level is based on it. The stage of defuzzification was illustrated in MatLab with the help of Fuzzy Logic Toolbox application software programs, exact value inference of maturity level for certain assessments of criteria has been presented. The generated knowledge base and the suggested technique make the basis of the intelligent information system to analyze the condition of the organization on the basis of the available self-assessment, performed by experts. A well-timed assessment of the organization's maturity level is necessary for making decisions and rapid response in dynamically changing economic conditions.

References

1. Nacional'nyj standart RF GOST R ISO 9004-2010. Menedzhment dlya dostizheniya ustojchivogo uspekha organizacii (Managing for the sustained success of an organization. A quality management approach). Standartinform, Moscow (2011). (in Russian)
2. Zadeh, L.: Fuzzy sets. Inf. Control **8**, 338–353 (1965)
3. Zadeh, L.: Knowledge representation in fuzzy logic. In: Yager, R.R., Zadeh L.A. (eds.) An Introduction to Fuzzy Logic Applications in Intelligent Systems, The Springer International Series in Engineering and Computer Science, vol. 165, pp. 1–27. Springer, New York (1992)
4. Nacional'nyj standart RF GOST R ISO 9001-2015. Sistemy menedzhmenta kachestva Trebovaniya (Quality management systems. Requirements). Standartinform, Moskow (2015). (in Russian)
5. Hrehova, S., Vagaska, A.: Application of fuzzy principles in evaluating quality of manufacturing process. WSEAS Trans. Power Syst. **7**(2), 50–59 (2012)
6. Borisova, L.V., Dimitrov, V.P., Nurutdinova, I.N., Serbin, D.M.: Osobennosti ehkspertnogo kontrolya kachestva v sfere obsluzhivaniya (Features of expert quality control in the service sector). In: sbornik nauchnyh trudov Mezhdunarodnoj molodezhnoj nauchno-prakticheskoj konferencii Kachestvo produkcii: kontrol', upravlenie, povyshenie, planirovanie, pp. 110–113, ZAO "Universitetskaya kniga", Kursk (2014). (in Russian)
7. Ding, J.-F.: Assessment of customer relationship management for global shipping carrier-based logistics service providers in Taiwan: an empirical study. WSEAS Trans. Syst. **11**(6), 198–208 (2012)
8. Eremina, E.A., Vedernikov, D.N.: Informacionnaya sistema vybora postavshchika na osnove metoda nechyotkogo logicheskogo vyvoda (Information system for vendor selection based on fuzzy inference method). Sovremennye problemy nauki i obrazovaniya. **3**, 294 (2013). (in Russian)
9. Zaharova, A.A.: Integral'naya ocenka innovacionnogo razvitiya regionov na osnove nechyotkih mnozhestv (Integral assessment of innovative development of regions based on fuzzy sets) Sovremennye problemy nauki i obrazovaniya. **3**, 25 (2013). (in Russian)

10. Nurutdinova, I.N., Borisova D.V.: Nechetkoe modelirovanie ocenki mezhregional'noj ehkonomicheskoj integracii (Fuzzy modeling of the interregional economic integration assessment). KANT **4**(25), 226–231 (2017). (in Russian)
11. Wang, G., Chen, S., Zhou, Z., Liu, J.: Modelling and analyzing trust conformity in e-commerce based on fuzzy logic. WSEAS Trans. Syst. **14**, 1–10 (2015)
12. Dimitrov, V., Borisova, L., Nurutdinova, I.: Modelling of fuzzy expert information in the problem of a machine technological adjustment. In: MATEC Web of Conference 13. Ser. 13th International Scientific-Technical Conference "Dynamic of Technical Systems", DTS-2017, P. 04009 (2017)
13. Asai, K., Vatada, D., Sugeno, S.: Prikladnie nechetkie sistemi (Applied fuzzy systems). Mir, Moscow (1993). (in Russian)
14. Kofman, A.: Vvedenie v teoriyu nechyotkih mnozhestv (Introduction in the theory of fuzzy sets). Radio i svyaz', Moscow (1982). (in Russian)
15. Borisova, L., Dimitrov, V., Nurutdinova, I.: Algorithm for assessing quality of fuzzy expert information. In: Proceedings of IEEE East-West Design & Test Symposium (EWDTS 2017), Novi Sad, Serbia, pp. 319–322 (2017)
16. Averkin, A.N., Batyrshin, I.Z., Blishun, A.F., Silov, V.B., Tarasov, V.B.: Nechetkie mnojestva v modelyah upravleniya i iskusstvennogo intellekta (Fuzzy sets in the models of management and artificial intelligence). Nauka, Moscow (1986). (in Russian)
17. Saaty, T.L., Kearns K.P.: Analiticheskoe planirovanie. Organizaciya sistem (Analytical Planning. The organization of Systems). Radio i svyaz', Moscow (1993). (in Russian)

The Application of MATLAB for the Primary Processing of Seismic Event Data

Anatoly Korobeynikov[1,2], Vladimir Polyakov[2], Antonina Komarova[2], and Alexander Menshchikov[2(✉)]

[1] Pushkov Institute of Terrestrial Magnetism,
Ionosphere and Radio Wave Propagation of the Russian Academy of Sciences
St.-Petersburg Filial, Moscow, Russia
korobeynikov_a_g@mail.ru
[2] St. Petersburg National Research University of Information Technologies,
Mechanics and Optics, St. Petersburg, Russia
v_i_polyakov@mail.ru,
piter-ton@mail.ru, menshikov@corp.ifmo.ru

Abstract. This paper discusses the use of wavelets for the preprocessing of data from seismic observation stations. The authors analyze the application of wavelets for seismic event data processing, discuss natural disasters prediction trends and make a decision about importance of performing additional seismic data analysis. We provide a new approach of seismic event preprocessing based on different methods combination and MATLAB usage. This approach can be used during educational process in the fields of digital data processing, wavelet analysis, natural disasters research and MATLAB study.

Keywords: MATLAB · Wavelets · Earthquake · Digital data processing
Seismic event

1 Introduction

The MATLAB system is currently used to solve problems in various subject areas [1–16]. This is not least due to the presence of well-implemented algorithms for solving a very large set of applied mathematical problems. This paper discusses the use of wavelets for the preprocessing of data from seismic observation stations. Any user can perform his own seismic event analysis because of the processing. This approach, in our view, leads to the increased interest in MATLAB usage and application.

Seismic events with a magnitude of 5 or more, no matter where in the world they occur, are currently recorded by seismic stations. Seismic events fixation with a magnitude less than 5 is possible where there is a presence of highly sensitive seismic observation points.

Such observation points are available, for example, in the USA and Europe. But if a seismic event occurs, say, on the seabed or ocean floor, in Africa, in Afghanistan, or in the newest states of Central Asia, it may not be registered. Incompetent mass media often provide information about the increase in the number of earthquakes on o ur planet.

In fact, the increase in the number of recorded seismic events depends on improving the sensitivity of the equipment and increasing the number of highly sensitive seismic

© Springer Nature Switzerland AG 2019
A. Abraham et al. (Eds.): IITI 2018, AISC 875, pp. 389–398, 2019.
https://doi.org/10.1007/978-3-030-01821-4_41

observation points. This statement is confirmed by the statistics of recorded earthquakes with magnitudes 6 and 7 from 1900 to 2017. The small growth can be explained by the usage of more effective tools.

If earthquakes with a large magnitude are taken into consideration, then their number does not increase, in other words, the proximity of average annual numbers is observed.

Furthermore, a fairly bad trend has been found, which shows an annual increase of the number of people affected by natural disasters. Statistical data analysis shows that most of the average annual human losses are due to the weather disasters.

The next disaster by the number of losses is an earthquake. The earthquake that occurred in Southeast Asia on December 26, 2004, is considered as the most tragic of the beginning of the 21st century. The earthquake and the tsunami caused by it claimed the lives of more than 232 thousand people.

But it is necessary to understand that a weak earthquake can also be very dangerous.

In most cases, this is associated with a poorly seismic understanding of the area where it occurred.

For the prediction of a strong seismic event, a certain set of features is currently used, allowing to calculate the formation of an earthquake with a sufficiently high probability [17]. But for weak earthquakes these signs, unfortunately, work very badly [18–21]. This creates a rather tense situation. For example, intensive mining and minerals processing require such an organization of enterprises, which often include dangerous production cycles. If such an enterprise is created on a seismically unexplored territory, then it is quite natural that there will be greater risks. Thus, to create conditions for damage reducing, specialists need to have as much information as possible about seismic events in the area.

In this article, we considered methods of data preprocessing of seismic events, occurred in early 2018, using wavelets implemented in MATLAB.

2 Application of Wavelets for Seismic Event Data Processing

Mathematical modeling and data processing are often performed using the MATLAB system [1–3]. In this system, a large number of digital data processing methods are implemented to solve different problems in different subject areas. For the processing of non-stationary signals, the Wavelet Toolbox extension package, which is one of the new and powerful tools, is attracted quite often.

The MATLAB system and accordingly the Wavelet Toolbox are dynamically developing. The difference between the new versions of the Wavelet Toolbox is, first of all, in the appearance of new versions and the modification of the available functions. The paper presents the results of the modified *cwt* function (Introduced in R2016b). By default, this function works with *the Morse* wavelet which means that the parameter of the wavelet name is '*morse*'.

We can consider results of this function for data processing of different seismic events. Each event has its own data set. In the source code, after reading, it is denoted by X.

On Tuesday, January 23, 2018, an earthquake with a magnitude of 7.9 points occurred on the coast of Alaska. The earthquake was noticed at 09:31:42 UTC. The epicenter was at a point with coordinates of 56.046° N and 149.073° W at a depth of 25 km below sea level. Geographically this is 280 km to the south-east of the city Kodiak.

We have processed the data obtained from different points of seismic observations.

1. Kodiak Island, Alaska, USA. Data about the station are presented in Table 1.

Table 1. Data about the station of seismic observations KDAK: Kodiak Island, Alaska, USA

Network	Station code	Latitude	Longitude	Elevation
II	KDAK	57.78°	−152.58°	152 m

Source code in MATLAB may look as follows:

```
clc
clear
% Calculation of CWT coefficients from data on the
earthquake in the Gulf of Alaska
% KDAK: Kodiak Island, Alaska, USA
% Network    Station Code    Latitude    Longitude
Elevation
%   II    KDAK    57.78°    -152.58°    152 m
my_name_file='II.KDAK.00.BH2.M.2018-01-
23T093122.019538.txt';
[F,mes]=fopen(my_name_file,'rt');
if F == -1
    disp('Error opening file');
    disp(mes);
    quit cancel;
end
my_string=fgetl(F);
X=[]; my_sec=[];
my_string=fgetl(F);
while my_string~= -1
    my_sec=[my_sec
3600*str2num(my_string(12:13))+60*str2num(my_string(15:16
))+str2num(my_string(18:27))];
    X=[X str2num(my_string(29:end))];
    my_string=fgetl(F);
end fclose(F);
count=numel(X);
% Sampling frequency
my_frequency=40;
%Chart of initial data on the earthquake.
figure;
plot(my_sec/60.0,X);
xlabel('mins');
ylabel('nm/s^2');
title('The of Earthquake Data in Gulf of Alaska
23.01.2018');
grid on
figure;
[wt,f,coi] = cwt(X,my_frequency);
% Plot the data, including the cone of influence.
cwt(X,my_frequency);
grid on
```

The results are shown in Figs. 1 and 2.

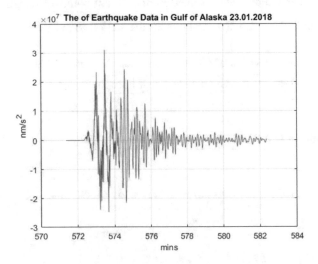

Fig. 1. Initial data chart on the earthquake in the Gulf of Alaska 23.01.2018 (KDAK)

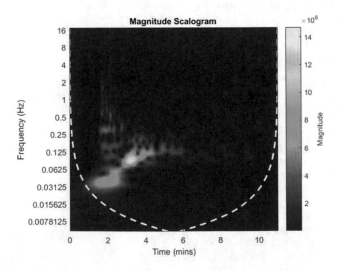

Fig. 2. Calculated wavelet coefficients (KDAK)

Figure 1 shows the initial data in absolute time. Since the earthquake is very powerful, its beginning can be seen even on the raw data.

In Fig. 2 the calculated wavelet coefficients in relative time are presented. It can be seen that the values of some wavelet coefficients start to change after the third minute, which can be interpreted as the beginning of a seismic event. This result correlates with the onset of the seismic event presented in Fig. 1.

We have processed the same seismic event using data from another observation point.

2. PET: Petropavlovsk, Russia. Data about the station is presented in Table 2.

Table 2. Data about the station of seismic observations PET: Petropavlovsk, Russia

Network	Station code	Latitude	Longitude	Elevation
IU	PET	53.02°	158.65°	110 m

The source code in MATLAB is almost the same except the file name and comment about the point of observation. The results are presented in Figs. 3 and 4.

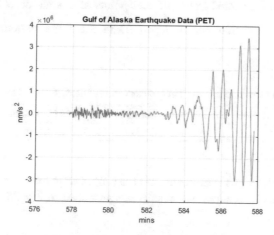

Fig. 3. Initial data chart about the earthquake in the Gulf of Alaska 23.01.2018 (PET)

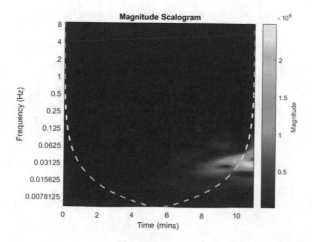

Fig. 4. Calculated wavelet coefficients (PET)

The analysis of Fig. 4 shows that the values of the wavelet coefficients begin to change significantly at the 11th minute. This fact can be interpreted as the beginning of a seismic event.

3   Seismic Events with a Small Magnitude

In this section, we consider an example where it is impossible to say anything definite, looking at the chart of the initial data, i.e. seismic events with a small magnitude.

We analyze data on the earthquake that occurred on the Kuril Islands on 30 January 2018 with a magnitude of 4.4 points. The earthquake was fixed at 03:07:48 UTC. The epicenter was at a point with coordinates 45.1455° N and 147.449° E at a depth of 134.35 km below sea level.

We process data obtained from different points of seismic observations.

1. ERM: Erimo, Hokkaido Island, Japan. Data about the station are presented in Table 3.

Table 3. Data about the station of seismic observations ERM: Erimo, Hokkaido Island, Japan

Network	Station code	Latitude	Longitude	Elevation
II	ERM	42.01°	143.16°	40 m

The source code in MATLAB is almost the same except the file name and comment about the point of observation. The results are presented in Figs. 5 and 6.

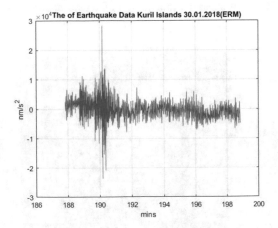

Fig. 5. Initial data chart data about the earthquake in the Kuril Islands 30.01.2018 (ERM)

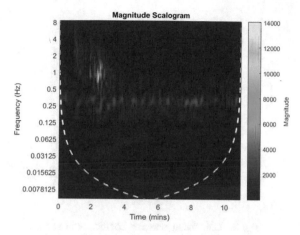

Fig. 6. Calculated wavelet coefficients (ERM)

The analysis of Fig. 6 shows that the values of the wavelet coefficients are maximal on the 3rd minute. This fact can be interpreted as the beginning of a seismic event.

2. YSS: Yuzhno Sakhalinsk, Russia. Data about the station are presented in Table 4.

Table 4. Data about the station of seismic observations YSS: Yuzhno Sakhalinsk, Russia

Network	Station code	Latitude	Longitude	Elevation
IU	YSS	46.96°	142.76°	150 m

The source code in MATLAB is almost the same except the file name and comment about the point of observation. The results obtained are presented in Figs. 7 and 8. The analysis of Fig. 8 shows that the values of the wavelet coefficients are maximal on the 7th minute. This fact can be interpreted as the beginning of a seismic event.

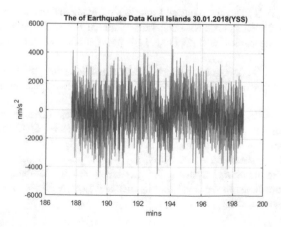

Fig. 7. Initial data chart about the earthquake in the Kuril Islands 30.01.2018 (YSS)

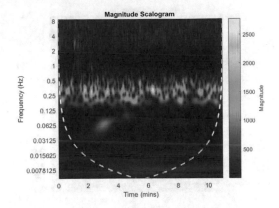

Fig. 8. Calculated wavelet coefficients (YSS)

4 Conclusion

The article demonstrates the processing by wavelets of non-stationary signals realized in MATLAB. These signals are data on earthquakes. This approach can be used, for example, in the educational process to achieve several goals - training work in the MATLAB system, the application of digital data processing methods, the usage of wavelet analysis in solving data processing problems, the study of earthquake science. In addition, the skills of correct interpreting of the data obtained can bring future benefits. For example, seismic observatories sometimes register foreshocks before the main shock in the focal zone of an earthquake [8, 9]. If strong earthquakes have not occurred in this area for a long time, this may be a prerequisite for assessing the threat as real.

References

1. Korobeynikov, A.G.: Development and analysis of mathematical models using MATLAB and MAPLE – St. Petersburg: St. Petersburg National Research University of Information Technologies, Mechanics and Optics, 144 pp. (2010). https://elibrary.ru/item.asp?id=26121333
2. Korobeynikov, A.G.: Designing and researching mathematical models in MATLAB and Maple environments. – SPb: SPbSU ITMO, 160 p. (2012). https://elibrary.ru/item.asp?id=26120684
3. Korobeynikov, A.G., Grishentcev, A.Yu.: Development and research of multidimensional mathematical models using computer algebra systems. – SPb: NIU ITMO, 100 p. (2014). https://elibrary.ru/download/elibrary_26121279_54604165.pdf
4. Velichko, E.N., Grishentsev, A., Korikov, C., Korobeynikov, A.G.: On interoperability in distributed geoinformational systems. In: Lecture Notes in Computer Science (including subseries Lecture Notes in Artificial Intelligence and Lecture Notes in Bioinformatics), vol. 9247, pp. 496–504 (2015)

5. Grishentcev, A.U., Korobeynikov, A.G.: Interoperability tools in distributed geoinformation systems. J. Radio Electron. (3), 19 (2015). http://jre.cplire.ru/jre/mar15/7/text.pdf
6. Tung, K.K.: Topics in Mathematical Modeling. Princeton University Press, Princeton (2007). van de Koppel, J., Huisman, J., van der Wal, R., Olff, H.: Patterns of herbivory along a prouductivity gradient: an empirical and theoretical investigation. Ecology **77**(3), 736–745
7. Pianosi, F., Sarrazin, F., Wagener, T.: A Matlab toolbox for global sensitivity analysis. Environ. Model. Softw. **70**, 80–85 (2015)
8. Korobeinikov, A.G., Ismagilov, V.S., Kopytenko, Yu.A., Petrishchev, M.S.: The study of the geoelectric structure of the crust on the basis of the analysis of the phase velocities of ultra geomagnetic variations. Cybernet. Programm. (2), 36–43 (2013). https://doi.org/10.7256/2306-4196.2013.2.8736. http://e-notabene.ru/kp/article_8736.html
9. Korobeinikov, A.G., Ismagilov, V.S., Kopytenko, Yu.A., Petrishchev, M.S.: Processing of experimental studies of the Earth crust geoelectric structure based on the analysis of the phase velocities of extra-low-frequency geomagnetic variations. Softw. Syst. Comput. Methods (3), 295–300 (2013). https://doi.org/10.7256/2305-6061.2013.3.10381
10. Korobeynikov, A.G., Fedosovsky, M.E., Zharinov, I.O., Polyakov, V.I., Shukalov, A.V., Gurjanov, A.V., Arustamov, S.A.: Method for conceptual presentation of subject tasks in knowledge engineering for computer-aided design systems. In: Proceedings of the Second International Scientific Conference "Intelligent Information Technologies for Industry" (IITI 2017), vol. 2, pp. 50–56 (2017). https://link.springer.com/chapter/10.1007/978-3-319-68324-9_6
11. Korobeynikov, A.G., Fedosovsky, M.E., Gurjanov, A.V., Zharinov, I.O., Shukalov, A.V.: Development of conceptual modeling method to solve the tasks of computer-aided design of difficult technical complexes on the basis of category theory. Int. J. Appl. Eng. Res. **12**(6), 1114–1122 (2017). ISSN 0973-4562. http://www.ripublication.com/ijaer17/ijaerv12n6_46.pdf
12. Velichko, E.N., Korikov, C., Korobeynikov, A.G., Grishentsev, A.Y., Fedosovsky, M.E.: Information risk analysis for logistics systems. In: Lecture Notes in Computer Science (including subseries Lecture Notes in Artificial Intelligence and Lecture Notes in Bioinformatics), vol. 9870, pp. 776–785 (2016)
13. Wang, Z.: Application of Matlab in the university mathematical experiment course. Comput. Digit. Eng. (2013)
14. Biao, W.: Brief analysis on application of Matlab in higher mathematics teaching. Comput. Digit. Eng. (2013)
15. Velichko, E.N., Grishentsev, A.Y., Korobeynikov, A.G.: Inverse problem of radiofrequency sounding of ionosphere. Int. J. Mod. Phys. A **31**(2–3) (2016). ISSN0217-751X. http://www.worldscientific.com/doi/abs/10.1142/S0217751X16410335
16. Korobeynikov, A.G., Aleksanin, S.A., Perezyabov, O.A.: Automated image processing using magnetic defectoscopy. ARPN J. Eng. Appl. Sci. **10**(17), 7488–7493 (2015). ISSN 1819-6608. http://www.arpnjournals.com/jeas/research_papers/rp_2015/jeas_0915_2586.pdf
17. Xanthakis, J.: Possible periodicities of the annual released global seismic energy (M > 7.9) during the period 198-1971. Tectonophysics. **81**(1–2), T7–T14 (1982)
18. Ashit, K.D.: Earthquake prediction using artificial neural networks. Int. J. Res. Rev. Comput. Sci. (IJRRCS) **2**(6), 2079–2557 (2011)
19. Moustra, M., Avraamides, M., Christodoulou, C.: Artificial neural network for earthquake prediction using time series magnitude data or seismic electric signals Expert Syst. Appl. **38** (12), 15032–15039 (2011)

20. Wang, Y., Chen, Y., Zhang, J.: The application of RBF neural network in earthquake prediction. In: Third International Conference on Genetic and Evolutionary Computing, pp. 465–468 (2009)
21. Panakkat, A., Adeli, H.: Neural Network model for earthquake magnitude prediction using multiple seismicity indicator. Int. J. Syst. **17**(1), 13–33 (2007)

Grid-Tie Inverter Intellectual Control for the Autonomous Energy Supply System Based on Micro-gas Turbine

Pavel G. Kolpakhchyan[1(✉)], Vítězslav Stýskala[2], Alexey R. Shaikhiev[1], Alexander E. Kochin[1], and Margarita S. Podbereznaya[1]

[1] Rostov State Transport University, Rostov-on-Don, Russian Federation
`kolpahchyan@mail.ru`
[2] VŠB - Technical University of Ostrava, Ostrava, Czech Republic
`vitezslav.styskala@vsb.cz`

Abstract. The article discusses the development of an intellectual control system for an electric power converter (OnGrid Inverter) of an energy complex based on a micro-gas turbine and a high-speed electric generator. Viewed energy complex is designed for operation in the power supply system with distributed generation. Therefore, the intelligent control system of the OnGrid Inverter must provide an analysis of the generation and consumption balance in the electrical grid, distribute electricity production between sources, and ensure high quality of electricity. The article describes the structure and principle of the OnGrid Inverter synchronization system with the electrical grid. Regulation principle described, shows the structure of the OnGrid Inverter intellectual control system. An example of intellectual power supply control system consisting of two OnGrid inverters and a load connected to an external electrical grid is considered. The results showed the correctness of the above article the approach to the formation of a OnGrid Inverter intellectual control system.

Keywords: OnGrid Invertor · Hi-speed electric generator
Phase locked loop · Micro-grid

1 Introduction

The use of energy complexes with distributed generation is one of the ways to improve the efficiency of energy systems. Its use in autonomous power supply systems makes it possible to unite diverse energy sources into a single system for the production, distribution and consumption of various types of energy [1–3]. The main source of energy in such systems are installations that uses organic fuel (gas, for example). Devices that use alternative sources of energy (wind, solar radiation) operate in the cogeneration mode and reduce the consumption of primary fuel.

© Springer Nature Switzerland AG 2019
A. Abraham et al. (Eds.): IITI 2018, AISC 875, pp. 399–408, 2019.
https://doi.org/10.1007/978-3-030-01821-4_42

The application of autonomous power supply systems with a capacity of 1 to 2 MW is effective [4–6]. The generating capacities of up to 100–200 kW are required for such systems [6,7]. The natural or liquefied gas (as well as biogas) is increasingly used as a fuel [4–6,8,9]. Gas costs lower than other types of fossil fuels, its application is more environmentally friendly. In some range of capacities, the use of gas turbines has several advantages over gas-piston engines, therefore the micro-gas turbine units are most common for the systems under consideration [1–3,10,11].

The effective operation of a gas turbine is possible at a speed of not less than 60 000–100 000 rpm and at capacities up to 200 kW. In this case, it is necessary to use a high-speed electric generator to generate electricity. The most well-known manufacturers of micro gas turbine, such as Capstone Turbine Corporation (USA), Calnetix Power Solutions (USA), Turbec (Italy) and others are doing so [8].

The output frequency of high-speed electric generator is 1000–2000 Hz, which does not allow it to be connected directly to a load or an electrical network with an industrial frequency voltage. Therefore, the creation of a power conversion system produced by a high-speed electric generator is required, which will allow obtaining output parameters corresponding to the industrial standard of electricity supply. In addition, the electricity conversion system must be compatible with the electric grid and similar power plants.

The key problem of the efficiency of the energy system with multiple, heterogeneous sources of electricity is a coordinated control of the processes of generation and consumption of energy. The OnGrid Inverter regulates the transmission of electricity generated by the micro-gas turbine plant to the electrical grid. OnGrid Inverter implements not only the power conversion function to the form corresponding to the requirements of the power supply system, but also the implementation of the joint work with other power plants and electricity consumers. The use of traditional approaches to control by a OnGrid Inverter does not allow to effectively solving the task. Therefore, the development of intellectual OnGrid Inverter control system to provide analysis of the balance of generation and consumption of electricity in the electrical grid carry out the distribution of electricity generation between sources, to provide high quality power, synchronization with the supply network.

2   Problem Formulation

The high-speed electric generator, which located on a common shaft with a micro-turbine, has a frequency of output voltage up to 2 kHz. It is necessary to use a static semiconductor converter having the ability to be synchronized with the grid to work in conjunction with an electrical grid of industrial frequency (50 Hz). The sudden changes in load from the electric generator are unacceptable for a gas turbine, therefore the presence of a rheostat brake in the converter circuit must be provided. The DC link simplifies the realization of the rheostat brake in comparison with matrix converters [13,14]. In addition, the connection of several electric generators to the common DC voltage link for increasing the reliability of power supply and redundancy is possible.

More efficient management of power system with distributed generation requires the creation of intellectual OnGrid Inverter control system. Its functions should include the synchronization of the output voltage to the electrical grid, providing high-quality electricity at the expense of managed suppress harmonic current in the grid, analysis of the production and consumption of electricity, condition monitoring, process control of reactive power compensation. Solving these problems requires a network selection scheme of the OnGrid Inverter, the principles of synchronization with the electrical grid and management of operation of the OnGrid Inverter, the development of algorithms effective management of power generation and consumption.

3   Circuit Design of the Inverter Synchronized with the Grid

An On-Grid inverter is a three-phase stand-alone voltage inverter. In our case, the use of a two-level bridge inverter is rational. The use of more complex schemes, in this case is not justified, since for operation at a voltage of 0.4 kV the two-level stand-alone voltage inverter can be assembled on available power semiconductor devices for a voltage of 1200 V and there is no need to complicate the converter circuit for applying a power semiconductor devices to a lower voltage. In addition, the use of multi-level stand-alone voltage inverters requires special measures to equalize the voltage across capacitors in the case of an asymmetric load.

It is necessary to have an adjustable rheostat brake in the DC link. It protects the turbine from uncontrolled acceleration in cases of significant fluctuations and sudden loss of load. The rheostatic brake is used with voltage fluctuations in the DC link. In a continuous mode, the power of the rheostat brake should be about 10% of the rated power of the On-Grid inverter with the possibility of short-term (up to 1–2 min) operation at rated power. Therefore, it is reasonable to use a power semiconductor device of the same type as the stand-alone voltage inverter as a breaker for the rheostat brake.

The higher harmonics in the output voltage is one of the important indicators of the operation of the On-Grid inverter. Time Harmonic Distortion (THD) estimates this indicator. In accordance with All-Union standard 32144-2013 (EN 50160:2010, NEQ – Electric energy Electromagnetic compatibility of technical equipment. Power quality limits in the public power supply systems), the THD value for 0.4'kV distribution grids should not exceed 8%. The voltage at the output of the stand-alone voltage inverter is obtained by using PWM (pulse-width modulation). To reduce the level of the higher harmonics to the required value, it is necessary to use a three-phase sine-filter of the second order at the output of the inverter. The filter parameters are calculated by the method described in [15].

A significant voltage from the main harmonic of the output current falls on the inductance of the sine filter. This fact must be taken into account when choosing the voltage in the DC link. With the mentioned parameters of the sine filter, the voltage in the DC link should be increased to 700 V for operation of an On-Grid inverter on an electrical grid with standard parameters.

The mains inverter is connected to the distribution grid. The control system generates a current of the set value at the output of the inverter. It is synchronized with the grid voltage at the connection point. Next we will consider the synchronization of the On-Grid inverter with the grid and the control principles.

4   Synchronization of the On-Grid Inverter with the Grid

Operation of the inverter in parallel with the requires synchronization in phase and frequency with the voltage at the connection point. For this purpose, the Phase Locked Loop (PLL) is used [16,17]. The application of this method for a three-phase grid provides the accuracy and synchronization time, which is sufficient for the effective operation of the control system.

This synchronization method is based on the introduction of a coordinate system d–q rotating in an electrical space with variable velocity. The projections of the stress vector on the axis of the fixed coordinate system α–β are calculated using the Clarke transform from the measured values of the mains voltage at the connection point. The projections of the stress vector on the axis of the rotating coordinate system d–q are determined using the Park transformation.

If the angular velocity of rotation of the stress vector and the coordinate system d–q coincide, the projections of the stress vector on the axis of this coordinate system are constant values. The synchronization task is to orient the coordinate system d–q in such a way that the grid voltage vector is located along the d axis, and its component along the q axis is equal to zero. This can be achieved by acting on the rotation speed of the coordinate system d–q.

When the coordinate system d–q rotates synchronously with the voltage vector, the projections of this vector are constant. The magnitude of the projection on the q-axis determines the phase mismatch. This projection must be zero for synchronization. Therefore, it is fed to the regulator input acting on the rotation speed of the coordinate system d–q.

Figure 1 shows the PLL block structure.

The projections will have oscillations with respect to the average value with an insignificant difference in the angular velocities of rotation. Therefore, it is necessary to extract a constant component from the signal about the magnitude of the projection of the vector onto the q-axis by means of a filter to eliminate oscillations at the output of the regulator. The same filter suppresses high-frequency components of the signal, which arise because of the presence of higher harmonics in the voltage of the grid. Therefore, the considered synchronization system can be used in the case of high-frequency interference in a grid.

There are two main modes of this PLL. One of them performs synchronization from the mismatched position, the second – supports synchronization. The

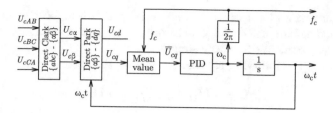

Fig. 1. Phase locked loop structure

OnGrid inverter control is impossible until the end of the synchronization procedure; therefore, the current control at the output of the converter must start only after synchronization. The synthesis of the regulator in the PLL was performed using the approach outlined above [16,17]. Since we consider the grid frequency as known, we take a value of 1 to 2 Hz greater than the estimated value as the initial frequency to speed up the phase synchronization procedure. In this case, the synchronization process is performed in a time corresponding to 3 to 4 half-periods of the mains voltage (0.03 to 0.04 s). After that, we use the signal from the output of the PLL to control the operation of the OnGrid inverter, and the unit goes into the synchronization maintenance mode.

5   Principles of Building a Grid Inverter Control System

The control of the electric power generation process is carried out by an intellectual control system of the OnGrid Inverter by forming a current value synchronized with the grid to which the inverter is connected. The magnitude of the current, the phase shift relative to the voltage and the harmonic composition are determined by the power output from the generating unit, the need for reactive power compensation, and the requirements for ensuring the quality of electricity. Intellectual control system determines these values depending on the selected control strategy (to achieve minimum power losses, optimizing the cost of electricity for consumers, aligning resource generating plants, etc.). To solve these problems, the intellectual control system of the OnGrid Inverter can receive information from the electrical system management system. An autonomous operation of the device is possible, in which the amount of power given by the OnGrid Inverter is determined by the voltage at the point of connection to the grid. In this case, the intelligent control system generates an external characteristic of the OnGrid Inverter with a predetermined slope and current limitation. This principle ensures the distribution of electricity, in which generation is carried out by sources closest to consumers, responding to changes in production or consumption.

The current formation at the output of the On-Grid inverter occurs at the filter choke. We set the second-order sine filter at the inverter output. Therefore, we do not take into account the capacity of the sine filter when synthesizing the current generation system at the output of the inverter and we represent a filter as a choke having an inductance and an active resistance. On the one hand, the mains voltage acts on the choke, on the other hand – the voltage of the inverter which acts as a controlled voltage source.

In this case, the following equation describes the choke current:

$$\mathbf{U}_N = (R_f + j\omega L_f)\,\mathbf{I_c} + \mathbf{U}_c, \tag{1}$$

where $\mathbf{U}_N$ – vector of the inverter output voltage; $\mathbf{U}_c$, $\mathbf{I}_c$ – vectors of the voltage and current in the place of connection to the grid; R_f, L_f – resistance and inductance of the filter choke; $\omega = 2\pi f$ – angular frequency of the electrical grid.

The Eq. (1) in the projections on the axis of the coordinate system d–q rotating synchronously with the voltage vector at the point of connection to the grid has the following form:

$$\begin{cases} U_{Nd} = R_f I_{cd} - \omega L_f I_{cq} + U_{cd}; \\ U_{Nq} = R_f I_{cq} + \omega L_f I_{cd} + U_{cq}. \end{cases}$$

Formation of the choke current is carried out by projections on the axis of the coordinate system d–q.

The expressions for the projections of the choke current on the axis of the selected coordinate system have the following form:

$$\begin{cases} I_{cd} = \dfrac{1}{R_f}\,(U_{Nd} + \Delta U_{Nd})\,; \\[2mm] I_{cq} = \dfrac{1}{R_f}\,(U_{Nq} + \Delta U_{Nq})\,, \end{cases} \tag{2}$$

where, the cross-coupling compensation voltages are determined from the following expressions:

$$\Delta U_{Nd} = \omega L_f I_{cq} + U_{cd}; \quad \Delta U_{Nq} = \omega L_f I_{cd} + U_{cq}. \tag{3}$$

Equations (2) describe the transfer functions of the regulatory object. With considering (2) and (3) we synthesized the structure of the OnGrid inverter automatic control system (Fig. 2).

In accordance with (2) regulators of the current components along the axes d and q must be integral. However, we used proportional-integral regulators in the system under consideration to compensate the effect of the voltage and current projection filters at the connection point.

6   Mathematical Modeling of Processes in the Power Supply System with Inverters Synchronized with the Grid

To illustrate the use of an inverter synchronized with the grid, we considered an electrical power system consisting of two OnGrid inverters, a load and a

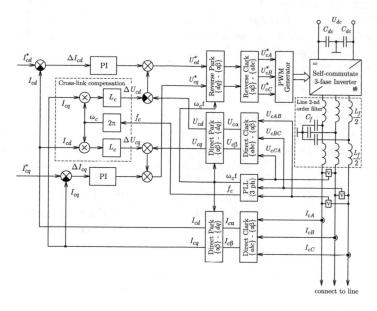

Fig. 2. Structural diagram of the On-Grid inverter automatic control system

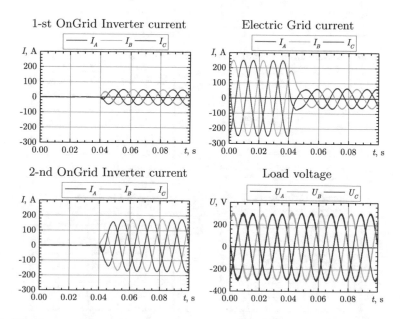

Fig. 3. The first and second grid inverters currents and the load voltage during the connection of inverters to the grid

distribution grid connected to an external electrical grid. A three-phase AC distribution grid with a frequency of 50 Hz and a voltage of 380 V is connected to the electrical grid. Two power inverters with a capacity of 100 kW and a load of 100 kW are connected to the distribution grid.

The calculations were carried out under the following conditions. At the initial time, the OnGrid inverters are disconnected from the distribute grid. The load is fed from the external grid. OnGrid inverters are connected to the grid in 0.05 s. During this time the PLL blocks complete the synchronization procedure with the grid voltage at the connection points and they are ready to work. At 0.04 s, the on grid inverters are simultaneously connected to the grid. The first inverter generates a current with an amplitude of 50 A, the second one - with an amplitude of 170 A.

To demonstrate the possibility of load redistribution between inverters over a time interval of 0.2 s to 0.3 s, the current amplitude of the first inverter increases to 170 A, and of the second one decreases to 50 A. From 0.35 s to 0.4 s, the

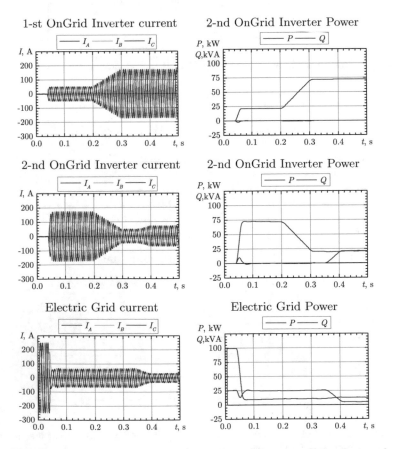

Fig. 4. Electric grid currents, active and reactive grid power of the first and second grid inverter (P – active power; Q – reactive power)

current setting of the second inverter along the q-axis changes from 0 A to −50 A. This mode corresponds to the generation of reactive power.

Figures 3 and 4 shows the results of modeling. Figure 3 shows the dependence of the electrical grid current, the currents at the connection point of the first and second inverter and the load voltage after connecting the inverters to the grid. Figure 4 shows the same values over the entire calculated time interval, as well as the active and reactive power of the electrical grid of the OnGrid inverters.

An analysis of the results shows that the principles of synchronization with the grid and the inverters operation control described above correspond to the specified goals. As an example, we considered the simultaneous connection of inverters to a grid with a current reference step change. In this mode, the current is controlled without oscillation and overshoot. The control system maintains stability and good control quality. The regulation of On-Grid inverters in the mode of generating reactive power has also high quality. The analysis of the load voltage showed that the THD coefficient in the steady-state modes is 4%.

7   Conclusions

The combination of several units into a common energy system is a feature of energy complexes based on the micro gas turbine system. Therefore, the On-Grid inverter must be able to work in parallel with the grid. It requires the use of a synchronization system (PLL). The grid inverter control system must work in the coordinate system that rotates synchronously with the voltage vector at the point of connection to the grid. The control system structure is based on the separate control of the inverter output current components along the longitudinal and lateral axes of the coordinate system, which rotates synchronously with the grid voltage vector. The simulation results that the proposed converter structure, the principles of synchronization and control allows us to solve the problems.

Acknowledgements. The work is done by the authors as part of the agreement No 14.604.21.0174 about subsidizing dated 26.09.2017. The topic is "Development of scientific and technical solutions for the design of efficient high-speed generator equipment for a gas micro turbine" by order of the Ministry of Education and Science of the Russian Federation, Federal Targeted Programme (FTP) "Researches and developments in accordance with priority areas of Russian science and technology sector evolution for 2014 – 2020 years". The unique identity code of applied researches (the project) is RFMEFI60417X0174.

References

1. Ahmad, A.: Smart grid as a solution for renewable and efficient energy. In: Advances in Environmental Engineering and Green Technologies. IGI Global (2016). https://books.google.ru/books?id=c6EoDAAAQBAJ
2. Oh, S., Hildreth, A.: Analytics for smart energy management: tools and applications for sustainable manufacturing. Springer Series in Advanced Manufacturing. Springer International Publishing (2016). https://books.google.ru/books?id=p0AGDAAAQBAJ

3. Monti, A., Pesch, D., Ellis, K., Mancarella, P.: Energy positive neighborhoods and smart energy districts: methods, tools, and experiences from the field. Elsevier Science (2016). https://books.google.ru/books?id=KHm0CwAAQBAJ
4. Boicea, A.V., Chicco, G., Mancarella, P.: Optimal operation of a microturbine cluster with partial-load efficiency and emission characterization. In: PowerTech, 2009 IEEE Bucharest, pp. 1–8 (2009)
5. Sioshansi, F.: Smart Grid: Integrating Renewable, Distributed & Efficient Energy. Academic Press (2012). https://books.google.ru/books?id=MQMrLNPjZVcC
6. Giampaolo, T.: Gas Turbine Handbook: Principles and Practice, 5th Edn. Fairmont Press, 15 November 2013. https://www.amazon. com/Botanicals-Phytocosmetic-Frank-DAmelio-Sr/dp/0849321182? SubscriptionId=0JYN1NVW651KCA56C102&tag=techkie-20&linkCode=xm2& camp=2025&creative=165953&creativeASIN=0849321182
7. Gerada, D., Mebarki, A., Brown, N.L., Gerada, C., Cavagnino, A., Boglietti, A.: High-speed electrical machines: technologies, trends, and developments. IEEE Trans. Ind. Electron. **61**(6), 2946–2959 (2014)
8. Soares, C.: Gas Turbines: A Handbook of Air, Land and Sea Applications. Elsevier Science (2014). https://books.google.ru/books?id=ubJZAwAAQBAJ
9. Kolpakhchyan, P.G., Shaikhiev, A.R., Kochin, A.E.: Sensorless control of the high-speed switched-reluctance generator for the steam turbine. In: Abraham, A., Kovalev, S., Tarassov, V., Snasel, V., Vasileva, M., Sukhanov, A. (eds.) Advances in Intelligent Systems and Computing, pp. 349–358. Springer International Publishing (2017). https://doi.org/10.1007/978-3-319-68324-9_38
10. Wang, Q., Fang, F.: Optimal configuration of CCHP system based on energy, economical, and environmental considerations. In: 2011 2nd International Conference on Intelligent Control and Information Processing (ICICIP), vol. 1, pp. 489–494 (2011)
11. Boicea, V.: Essentials of Natural Gas Microturbines. CRC Press (2013). https:// books.google.ru/books?id=cZfNBQAAQBAJ
12. Zhong, Q., Hornik, T.: Control of Power Inverters in Renewable Energy and Smart Grid Integration. Wiley - IEEE. Wiley (2012). https://books.google.ru/books? id=m5kWmDIiuxQC
13. Yamada, K., Higuchi, T., Yamamoto, E., Hara, H., Sawa, T., Swamy, M.M., Kume, T.: Integrated filters and their combined effects in matrix converter. In: Fourtieth IAS Annual Meeting. Conference Record of the 2005 Industry Applications Conference, vol. 2, pp. 1406–1413 (2005)
14. Shepherd, W., Zhang, L.: Power Converter Circuits. Electrical and Computer Engineering. Taylor & Francis (2004). https://books.google.ru/books? id=nF0oL5WInAYC
15. Brown, M.C.: Practical switching power supply design. In: AP Professional and Technical Series. Elsevier Science (2012). https://books.google.ru/books? id=il8ADNvEPygC
16. Surprenant, M., Hiskens, I., Venkataramanan, G.: Phase locked loop control of inverters in a microgrid. In: 2011 IEEE Energy Conversion Congress and Exposition, pp. 667–672 (2011)
17. Arruda, L., Silva, S., Filho, B.: PLL structures for utility connected systems. In: Proceedings of the IEEE Industry Applications Conference, pp. 2655–2660 (2001)

Hybrid Intelligent Multi-agent System Model for Solving Complex Transport-Logistic Problem

Sergey Listopad[✉]

Kaliningrad Branch of the Federal Research Center
"Computer Science and Control" of the Russian Academy of Sciences,
Kaliningrad 236022, Russian Federation
ser-list-post@yandex.ru

Abstract. On the example of a complex transport-logistic problem, computer modeling of problem solving by expert team using one of the divergent thinking techniques namely the brain record pool is considered. Computer modeling of problem solving in a team allows applying a set of dynamically synthesized and modifiable integrated models for a variety of problem solving situations rather than a single tool. The results of a comparison of the probability and magnitude of the synergy effect in hybrid multi-agent intelligent systems with different architectures made it possible to develop rules for a fuzzy knowledge base for selecting an architecture corresponding to different problems.

Keywords: Hybrid intelligent multi-agent system · Divergent thinking
Complex transport-logistic problem

1 Introduction

Management of the complex social and technical systems requires to involve experts of various specialties and to organize their effective interaction in the team. So the issues of organizing collective problem solving in such a system, which have been studied for a long time within social psychology, system analysis, artificial intelligence and other sciences, are topical. Making collective decisions differs significantly from individual decision-making [1]. It's more democratic and takes into account the interests of a multitude of stakeholders. In collective decision-making, the problem can be comprehensively analyzed, since experts in various fields of knowledge are invited. This allows the team to integrate diverse knowledge and dynamically develop a method for solving the problem taking into account its dynamic nature.

The concept of integrating diverse, complementary knowledge for computer support of individual decision-making is embodied in Kolesnikov's concept of hybrid intelligent systems (HIS) [2]. The disadvantage of HIS is the complexity of modeling of the collective processes and effects that arise in expert teams solving complex

The reported study was funded by RFBR according to the research project No. 18-07-00448A.

problems. To overcome this shortcoming a new class of intelligent systems, namely hybrid intelligent multi-agent systems (HIMAS), combining a hybrid approach of professor Kolesnikov and the apparatus of multi-agent systems [3].

2   Process of Problem-Solving by a Team of Experts

Problem solving by a team of experts is a complex process generally consisting of the following stages: definition, formulation and analysis of tasks, data collection and interpretation, solution search, analysis of decision effectiveness and final choice, presentation of results, monitoring and evaluation of results [4]. Periods of collective discussion about common problem can alternate with periods of individual work of experts on their subtasks. The process of the problem-solving is superimposed on the process of formation and development of the team of experts as a single entity, consisting of the following stages [5–7]:

- formation - members of the team get acquainted, exchange official information about each other, make suggestions about the work of the team, adhere to generally accepted points of view, make proposals leading to obvious decisions [6]. If the problem has an obvious solution, then the discussion ends at this stage. Otherwise, the goal of the team is to go beyond the boundaries of established opinions, using, for example, various techniques of divergent thinking;
- "bubbling" is the state of the team, when its members must make efforts to combine the diversity of ideas with their own convictions and values. At this stage there are conflicts and confrontation between members of the team. If differences are too great, some members may leave the team; otherwise they adjust or discuss the contradictions. An important role at this stage is played by facilitator or formal leader, who has to solve conflicts in the team using, for example, parallel thinking techniques [8]. At this stage members of the team adapt to differences in views and cooperate with each other, develop group norms of behavior, the sense of camaraderie and cohesion appears;
- revision of proposals and preparation of alternatives - team members reformulate valuable ideas into concrete proposals and "polish" them until all participants in the discussion come to the final decision embodying the diversity of points of view of the team members. This stage is characterized by "convergent thinking": classification of ideas, their generalization, making of estimates, as opposed to the divergent thinking, inherent to the first stages, within which an open, non-valued discussion and the generation of a large number of decisions are encouraged [7];
- decision making and disbanding - collective decision is being taken, according to the opinions of all experts who took part in the discussion.

This paper is limited to modeling only the first stage of this process, i.e. divergent thinking zone. On the basis of the analysis of methods for organizing divergent thinking in the team, the brainwriting pool was chosen for computer simulation of the problem solving in the team of experts [9]. According to this method the solution of the problem is carried out as follows: (1) the decision-maker passes the conditions of the problem to the experts; (2) each expert develops own version of the problem solution on the basis of his own model and the criterion of optimality and passes it to the decision-maker; (3) the decision-maker passes received solutions for improvement to all experts, except for the source (confidentiality is respected); (4) experts improve received problem solutions and return results to the decision-maker: the third and fourth stages are repeated until each of the experts has processed at least once each solution; (5) the decision-maker evaluates all options, including intermediate solutions, and chooses the best on the basis of its own model of the problem and the criterion of optimality. The functioning of the brainwriting pool model in the team, consisting of a decision maker and two experts, is given schematically in Fig. 1.

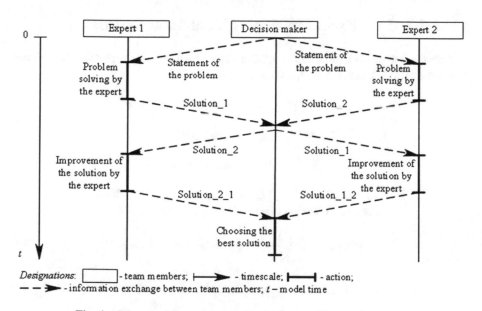

Fig. 1. Scheme of the team work by the brainwriting pool method.

According to Fig. 1, the team decision-making processes alternate with the periods of individual work of experts seeking new solutions or improving existing solutions through the prism of their model of the world. As a result, the problem is reduced to subtasks with an expert-specific criterion, any of which could be solved by the expert with various methods. In such a team there is a synergistic effect when the team common decision is qualitatively better than the decisions obtained by any expert separately.

Synergetic effect in the team, which consists in the high quality of the decisions made, is achieved as a result of the "group compensation of individual disabilities", because the team interaction based on partner cooperation in solving problems "improves efficiency by no less than 10%" [10]. It is effective team interaction that generates a synergetic effect in the team, when the inability of one to perform any work or operation is compensated by the skills and abilities of the other. Thus, the magnitude of the synergetic effect can be used to estimate the effect of team members' interaction. To measure it, the following expression can be used [11]:

$$Syn = SQ_2 - SQ_1,$$ (1)

where SQ1 is the value of the quality criterion of the best problem solution found by experts without interaction, individually; SQ2 is the value of the quality criterion of the best problem solution that the experts received collectively, interacting in team.

3   Model of the Hybrid Intelligent Multi-agent System

To simulate the process of solving problems by a small team of experts, it is proposed to use the HIMAS, defined by the expressions:

$$himas = <AG^*, env, INT^*, ORG, \{so^{goa}\} >$$ (2)

$$AG^* = \{ag_1, \ldots, ag_n, ag^{dm}\}$$ (3)

$$INT^* = \{prot, lang, ont, rcl\}$$ (4)

$$ORG = ORG_{coop} \cup ORG_{neut} \cup ORG_{comp}, \quad ORG_{coop} \cap ORG_{neut} = \emptyset,$$
$$ORG_{coop} \cap ORG_{comp} = \emptyset, \quad ORG_{comp} \cap ORG_{neut} = \emptyset,$$ (5)

$$act_{himas} = (\bigcup_{ag \in AG^*} act_{ag}) \cup act_{ia} \cup act_{as} \cup act_{col}$$ (6)

$$act_{ag} = (MET_{ag}, IT_{ag}), \ ag \in AG^*, \quad | \bigcup_{ag \in AG^*} IT_{ag}| \geq 2$$ (7)

$$ag = ag \vee himas$$ (8)

where AG^* is the set of agents ag, i.e. models of experts, including the decision-maker agent (DMA) ag^{dm}; n is a number of expert agents; env is the conceptual model of the HIMAS environment; INT^* are the elements structuring interactions of agents: $prot$ is interaction protocol; $lang$ is the language of messages; ont is domain model; rcl is the classifier of agent relations (classifies relations between agents depending on their goals into classes: neutrality, competition and cooperation) [1]; ORG is a set of HIMAS architectures (ORG_{coop}, ORG_{neut}, ORG_{comp} are subsets of HIMAS architectures with cooperating, neutral and competing agents, respectively) [1]; $\{so^{goa}\}$ is a set of conceptual models of macro-level processes in HIMAS: so^{goa} is the model of self-organization effect on the basis of agents' goals analysis [1]; act_{himas} is the function of HIMAS as a whole; act_{ag} is the function of the agent from the set AG^*; act_{ia} is DMA's function "analysis of interactions", which is designed to identify the current state of the team problem solving process on the basis of the similarity of the agent's fuzzy goals analysis [1]; act_{as} is the DMA's function "architecture selection", which establishes one of the architectures depending on the problem parameters in order to increase the efficiency of the HIMAS work; act_{col} is the HIMAS collective function, designed dynamically in the process of the system functioning and determined by the current architecture; met_{ag} is the problem solving method; it_{ag} is the intelligent technology, within the framework of which the method met_{ag} is implemented.

4 Functional Structure of the Hybrid Intelligent Multiagent System

The universal functional structure of HIMAS has been developed for computerized implementation of the expert team model (Fig. 2). It can be used to solve a wide range of problems because: (1) common multi-agent ontology is used; (2) the list of solver agents covers five classes of methods from six used in HIS [1]; (3) the order of interaction of agents is determined by the domain model and specific agents' algorithms implemented by the developer.

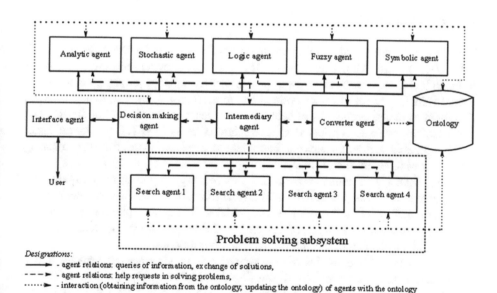

Designations:
⟶ - agent relations: queries of information, exchange of solutions,
- - ▶ - agent relations: help requests in solving problems,
······▶ - interaction (obtaining information from the ontology, updating the ontology) of agents with the ontology

Fig. 2. Universal functional structure of the hybrid intelligent multi-agent system.

Consider the appointment of its elements:

1. interface agent requests input data and outputs the result;
2. DMA dispatches the problem conditions to the search agents, determines the order of their interaction. When the latter solved the problem, it selects an alternative and passes it to the interface agent, or launches a new iteration of the problem solving process, sending each solution to all search agents, except the agent that found it;

3. search agents have knowledge about the subject area and use algorithms specific for each problem to solve the subtasks (ant colony algorithm in case of complex transport-logistic problem);
4. intermediary agent monitors the names, models and capabilities of registered intelligent technologies agents (solvers). Agents address to it to find out which of the solvers can help in solving the subtask set before them;
5. solvers in the upper part of Fig. 2 together with the converter agent implement the hybrid component of HIMAS, combining diverse knowledge, and provide "services" to other agents using the following models and algorithms: algebraic equations for describing the cause-effect relations of the domain's concepts; the Monte Carlo method; production expert system with forward reasoning; the Mamdani's fuzzy inference algorithm, etc.;
6. ontology is the semantic network, the basis for agents' interactions, it's built on the conceptual model of the problem being solved. Agents interpret the meaning of the received messages using this model.

5 Experiments

To evaluate the effectiveness of agent interaction in various architectures of HIMAS and to develop fuzzy knowledge base of DMA, a series of experiments were performed in which it was required to solve a complex transport-logistics problem, i.e. to find for several vehicles the set of routes optimal for four criteria: total cost; total duration of trips for all vehicles; probability of being late to at least one client; reliability (the mean total cost increase was considered as a measure of reliability) [1]. Stochastic factors such as probability of traffic jams and being late to the client, losses from cargo damage and others were taken into account.

Initial data: (1) customer requests for the delivery of goods: name, quantity of goods, time interval for their delivery; (2) information about roads to customers: length, congestion, quality; (3) information about vehicles: fuel consumption, cargo capacity, etc.; (4) information on work schedules and staff salaries: drivers and loaders; (5) information about the cargo: weight, dimensions, fragility, etc.

Output data: a set of routes for the delivery of goods (one per vehicle) and its parameters: cost, duration, reliability and likelihood of late arrival, a composite criterion for the quality of the route. Five problems are developed for test purposes, the quantitative parameters of which are given in Table 1.

Three HIMAS architectures working by the scheme in Fig. 1 were studied: with neutral, cooperating and competing agents. In HIMAS with neutral agents, each of the four search agents minimizes the value of "own" criterion of the solution quality. In HIMAS with cooperating agents, all four search agents minimize all four criteria of the solution quality (similar to DMA). In HIMAS with competing agents, the first agent minimizes the cost and maximizes the duration, the second maximizes the cost and minimizes the duration, the third minimizes the likelihood of delay and maximizes reliability, and the fourth maximizes the probability of delay and minimizes the reliability.

Table 1. Quantitative parameters of tested problems.

Problem	Number of customers	Number of roads	Number of drivers	Number of loaders	Number of vehicles
CTLP_10	10	75	3	3	3
CTLP_15	15	240	5	5	5
CTLP_20	20	420	5	5	5
CTLP_25	25	650	9	9	9
CTLP_30	30	377	6	6	6

The quality of test solutions was evaluated by objective indicators and subjectively by experts. For each problem in Table 1 and each HIMAS architecture 100 computational experiments were carried out. For all the problems and architectures of HIMAS, as well as for architectures without interaction (search agents do not exchange individual solutions), graphical dependencies of the number of situations when the collective solution is better than any individual (Fig. 3), of the synergy effect magnitude (Fig. 4), of average cost, duration, reliability, probability of delay from the number of customers are constructed. The analysis of these dependencies showed the high quality of the routes recommended by HIMAS.

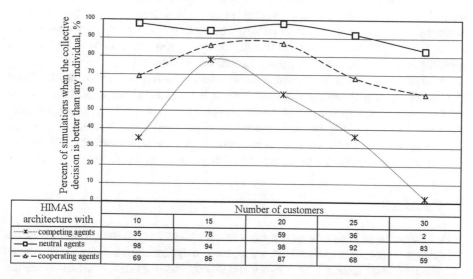

Fig. 3. Dependence of the number of situations when the collective decision is better than any individual from the number of customers in the problem.

In Fig. 3 it can be seen that the quality of decisions taken by HIMAS with neutral agents is higher than that of other architectures. This is a direct consequence of the fact that in HIMAS with neutral agents the probability of a synergetic effect is higher. However, as can be seen in Fig. 4, the smaller the dimension of the problem, the less this effect affects the quality of the solution. As it follows from Fig. 4 synergy arise in

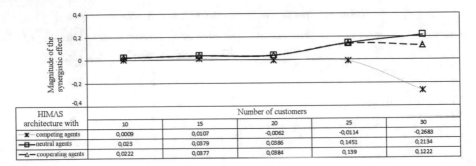

HIMAS architecture with	Number of customers				
	10	15	20	25	30
─✕─ competing agents	0,0009	0,0107	-0,0062	-0,0114	-0,2683
─☐─ neutral agents	0,023	0,0379	0,0386	0,1451	0,2134
─△─ cooperating agents	0,0222	0,0877	0,0384	0,139	0,1222

Fig. 4. The magnitude of the synergistic effect in various types of the HIMAS architecture.

most cases in any HIMAS architecture to some extent. However, it becomes noticeable on problems with high combinatorial complexity: TSLP_20, TSLP_25 and TSLP_30. The HIMAS architecture is particularly important for such problems. In Fig. 4 it can be seen that for problems with more than 20 clients HIMAS with competing agents demonstrates the effect of negative synergy (dysergy), when the collective solution is not better than the solutions of individual agents, which agrees with the studies in [12]. Obviously, this effect is due to the inability of competing agents to "agree" on problems with high combinatorial complexity. Thus, it is shown that with proper organization of interaction in HIMAS there is a synergy effect, which increases the quality of the decisions taken in comparison with HIMAS, in which it is not modeled. Dependencies at the Figs. 3 and 4 were used to develop rules for the fuzzy knowledge base for selecting an architecture corresponding to different problems [13].

Analysis of technical and operational parameters showed that under the conditions of a laboratory experiment, the solution time reduced is from two to 30 min, and the improvement in the quality of solutions compared with those in practice has been confirmed by experts [13]. Based on the results of practical use of HIMAS at two sites, the average total cost and length of cargo delivery per day decreased by 7.2% and by 12.13%, respectively, the average time of building routes per day decreased by 23.14%.

6 Conclusion

In this paper the process of expert team work is considered. Until now it remains the only effective tool for solving complex problems in dynamic environments. The model and the functional structure of a hybrid intelligent multi-agent system for computer simulation of solving problems by such team are proposed. The results of laboratory experiments with the system that implements elements of this model have shown the effectiveness for solving complex problems and the prospects for this direction of research. The study of various models of divergent thinking in teams and the computer simulation of bubbling and convergent thinking stages of the team problem solving process should be highlighted as further research directions on this topic.

References

1. Kolesnikov, A.V., Kirikov, I.A., Listopad, S.V., Rumovskaya, S.B., Domanitskiy, A.A.: Reshenie slozhnykh zadach kommivoyazhera metodami funktsional'nykh gibridnykh intellektual'nykh system [Complex travelling salesman problems solving by the methods of the functional hybrid intelligent systems]. IPI RAN, Moscow (2011). (In Russian)
2. Kolesnikov, A.V.: Gibridnye intellektual'nye sistemy. Teoriya i tekhnologiya razrabotki [Hybrid intelligent systems: theory and technology of development]. SPbGTU Publs., Saint Petersburg (2001). (In Russian)
3. Tarasov, V.B.: Ot mnogoagentnykh sistem k intellektualnym organizatsiyam: filosofiya, psikhologiya, informatika [From multiagent systems to intelligent organizations: philosophy, psychology, and informatics]. Editorial URSS, Moscow (2002). (In Russian)
4. Samsonova, M.V., Efimov, V.V.: Tekhnologiya i metody kollektivnogo resheniya problem: Ucheb. posobie [Technology and methods of group decision. Handbook]. UlSTU, Ulyanovsk (2003). (In Russian)
5. Zankovskiy, A.N.: Organizatsionnaya psikhologiya [Organizational psychology]. Flinta: MPSI, Moscow (2002). (In Russian)
6. Latfullin, G.R., Gromova, O.N. (eds.): Organizatsionnoye povedeniye [Organizational Behavior] Piter Publishing House, Saint Petersburg (2004). (In Russian)
7. Kaner, S.: Facilitator's Guide to Participatory Decision-Making, 2nd edn. Jossey-Bass, San Francisco (2007)
8. De Bono, E.: Parallel Thinking: From Socratic to De Bono Thinking. Penguin Books, London (1994)
9. Kirikov, I.A., Kolesnikov, A.V., Listopad, S.V.: Komp'yuternaya model' sinergii kollektivnogo prinyatiya resheniy [Computer model of synergy of team decision-making]. Inf. Appl. **11**(3), 34–41 (2017). (In Russian)
10. Zimnyaya, I.A.: Pedagogicheskaya psikhologiya [Pedagogical psychology]. Rostov-on-Don, Phoenix (1997). (In Russian)
11. Kirikov, I.A., Kolesnikov, A.V., Listopad, S.V., Rumovskaya, S.B.: Metod izmereniya effekta sinergii v gibridnykh intellektual'nykh mnogoagentnykh sistemakh [Method for measuring synergy effect in hybrid intelligent multiagent systems]. Syst. Means Inf. **27**(3), 99–111 (2017). (In Russian)
12. Johnson, D.W., Maruyama, G., Johnson, R., Nelson, D., Skon, L.: Effects of cooperative, competitive, and individualistic goal structure on achievement: a meta-analysis. Psychol. Bull. **89**, 47–62 (1981)
13. Kolesnikov, A.V., Kirikov, I.A., Listopad, S.V.: Gibridnye intellektual'nye sistemy s samoorganizatsiey: koordinatsiya, soglasovannost', spor [Hybrid intelligent systems with self-organization: coordination, consistency, dispute]. IPI RAN, Moscow (2014). (In Russian)

Decentralized Planning of Intelligent Mobile Robot's Behavior in a Group with Limited Communications

Donat Ivanov[✉]

Southern Federal University Acad. Kalyaev Scientific Research Institute
of Multiprocessor Computer Systems,
2 Chekhov Street, GSP-284, Taganrog 347928, Russia
donat.ivanov@gmail.com

Abstract. In this paper, we consider the integration of intelligent mobile robots by various designs in a coalition that together solve a common group goal, consisting of a set of individual tasks, each of which is available to some single robots of the coalition. A general approach is proposed for control such coalitions of robots, solving the problem of information exchange, integrating information about the environment from onboard sensor systems of robots from the group, planning the actions of the coalition robots. The problem of forming a mobile reconfigured distributed artificial neural network is considered. A generalized algorithm for deploying such a network is proposed. A modification of the "method of spheres" for planning the actions of the coalition robots solving the formation task is proposed. That modification is necessary for the deployment of a mobile re-configurable distributed artificial neural network in a decentralized control strategy and limited communications is proposed. The results of computer simulation of the solution of the formation task in a group of quadrotors for deployment of an artificial neural network are presented.

Keywords: Heterogeneous multi-robot groups · Method of spheres
Formation task · Intelligent mobile robot · Distributed cooperation

1 Introduction

Application area of intelligent mobile robots, especially small-sized, significantly expands when they are combined into groups or coalitions. The term "coalition of robots" can be considered as a synonym for the term "heterogeneous group of robots", that is a group which includes robots with different design, functionality or even a moving environment. For example some ground mobile robots and some air robots (UAVs) can be combined into one heterogeneous group.

The solution of many practical tasks, such as mapping, geological exploration of the shelf, monitoring, the formation of mobile telecommunications networks of rapid deployment, requires a coalition of robots to form a formation with specified configuration. As widely shown in the recent literature [1], robustness and flexibility constitute the main advantages of multiple-robot systems vs. single-robot ones. Therefore, recently active work has been carried out in the field of control coalitions of intelligent

© Springer Nature Switzerland AG 2019
A. Abraham et al. (Eds.): IITI 2018, AISC 875, pp. 418–427, 2019.
https://doi.org/10.1007/978-3-030-01821-4_44

mobile robots united by a single communication network [2, 3]. But in practical use, there are significant limitations on the bandwidth of communication channels and the range of communication between the robots in the coalition.

In this paper, we propose original methods for solving the problem of planning actions in the coalition of intelligent mobile robots with a decentralized control strategy and limited communications with the example of solving the problem of deploying a mobile distributed artificial neural network.

2 Problem Statement

The formalizing of the problem for multi-robot systems was repeatedly cited in the scientific literature, for example in [4] for swarms of robots, and in [2] for heterogeneous groups of mobile robots. We give the statement of the problem with considering limited communications.

There is a coalition R of N intelligent mobile robots $r_i \in R$ $(i = \overline{1,N})$. The current position $p_{r_i}(t)$ of each robot in space is given by the coordinates $p_{r_i}(t) = (x_{r_i}(t), y_{r_i}(t), z_{r_i}(t))$ and the vector of direction $\vec{o}_{r_i}$. Internal state of each $r_i \in R$ $(i = \overline{1,N})$ is given by the state's vector-function $q_i(t) = [q_{i,1}(t), q_{i,2}(t), \ldots, q_{i,h}(t)]^T$. The set $S = \{s_1, s_2, \ldots s_p\}$ of skills of all robots r_i $(i = \overline{1,N})$ in the coalition R. The functionality $S_{r_i} \subseteq S$ of each individual robot r_i of the coalition R is defined as a subset of all the functionality of the coalition robots, i.e. $\cup_{r_i \in R} S_{r_i} = S$. In general, some robots in coalition may have the same functionality, therefore $\cap_{r_i \in R} Sri \neq 0$. The coalition has a group's target T, which could be decomposed into a set of tasks $t_j \in T$. The achievement of the group's target T requires a set B_T of robot's functionality $B_T = \{s_1, s_2, \ldots s_b\}$.

Executing a task t_j requires some set of functionality $b_i = \{s_1, s_2, \ldots s_{b_i}\}$. Thus, the execution of some tasks is not available to some robots in the group, i.e. $S_{r_i} \cap b_i \neq b_i$. The fulfillment of the group goal is impossible for this coalition if. The achievement of the group's target T is impossible for the coalition if $B_T \cap S \neq B_T$.

There is at least one robot in the coalition, which is able to communicate with operator via this communication channel. Information exchange between the robots in the group is carried out through other communication channels, the range l_{ri} of which is so limited that in the coalition there are pairs of robots that do not have a direct connection with each other. If $d_{ri,rj}$ is the distance between the robots r_i and r_j, then in the coalition there is at least one pair r_i and r_j such $l_{ri} < d_{ri,rj}$.

The group's robots need to collect information on environmental parameters, the current robot's state and the planned actions of other robots in coalition. Thus, the planning of cooperative actions in the coalition of intelligent mobile robots needs to divide the group's target into a set of tasks, and distribute the tasks among the robots in the group so as to solve the group's task by a reasonable time, taking into account communication's limitations.

3 The Proposed Approach

3.1 Decomposition of Group's Target

The solution of the group's target implies four components: organization of information exchange; integration of data about state of the environment; change the state of the robots in group; change the state of the environment.

Given the existing restrictions on the radius l_{ri} of communication between the robots of the coalition, as well as the fact that direct communication with the operator is not available to all robots in the coalition, it is necessary to provide a mechanism for routing in the coalition of robots. Unlike stationary telecommunication networks, when the robots move in a coalition, some links between them will disappear, others will appear due to the fact that some robots of the group move away from each other for a distance exceeding the distance of direct communication, while others are converging. These changes in the structure of the telecommunications network are nondeterministic. The analysis of known routing methods in telecommunication networks of the nondeterministic structure showed that it is preferable to use methods of roving intelligence [5, 6] for routing [7–9].

However, the maintenance of the routing system consumes the computing and telecommunication resources of the coalition robots. It is preferable to minimize the amount of information transmitted between robots outside the forward link area. And at the same time, it is necessary to have a mechanism for quickly notifying all coalition workers of important changes, such as changing the group task by the operator or detecting a hazard for the whole group. In this case, it is proposed to use the mechanisms of quick notification of all the robots of the coalition on the basis of the epidemic algorithm [10, 11]. The robot sends an important message to all those coalition robots that fall into its direct communication zone. Those robots transmit this message to all robots that fall into the zone of direct communication. In order to minimize repeated retransmissions of the message, it is suggested to assign the serial number of the robot that transmitted this message at each transmission of the message. This will minimize the transmission of messages to those robots that have already received this message.

To obtain a more detailed map of the environment, we propose a complexing of information. Each robot collects information about the current state of the environment using its on-board sensor devices. With the help of the on-board computer, primary processing of the data obtained, their analysis and structuring is performed, and then the received data indicating the time of their receipt are transferred to the neighboring robots of the group. Each robot, having received data on the state of the environment, produces a refinement of the environmental parameters available on board it.

Changes in the parameters of the robots of the coalition implies, first of all, the execution of movements of robots in space, the formation of the system necessary to solve the group problem.

The change in the environment by a coalition of robots is the performance of those actions that affect the parameters of the environment: the movement of objects, sampling of ground, etc.

3.2   Formation of Mobile Artificial Neural Network of Rapid Deployment

As an example of a group's target, let us consider the problem of forming a distributed artificial neural network (ANN) of rapid deployment in which each robot (its agent) acts as a separate neuron.

Different ANN's configurations requires different formations of the robots in a coalition. In this case, the formation does not define like a set of coordinates of robots, but it defines like a set of distances between neighboring robots, which make it possible direct communication between neurons (robot's agents) in ANN.

Each robot work like a neuron of ANN, until get message from operator or from other robots about new number of target ANN. Then robot take a part of creating new formation, which is corresponding to new ANN. The robot loads parameters of new ANN. ANN can be trained in advance, and the characteristics are stored in memory. The group receives a signal about the construction of the desired ANN, and carries out a rebuild. Then loads the specified INS parameters from memory.

Let us consider in more detail the subtask of formation of the system. There are some well-known methods for solving a formation task in a group of mobile robots, including a behavior based [12, 13], leader-follower approach [14–17], a virtual structure/virtual leader approach [18], based on the game theory [19] etc. However some of the methods make it possible to form a formation by only a certain set of shapes. Some others require considerable computing resources, but there are not such resources on-board of mobile robots.

Many of the methods are aimed at positioning mobile robots in absolute coordinates, but in this task compliance required distances between the mobile robots is more important than the mobile robot's positioning in absolute coordinates.

Thus the formation task in a group of mobile robots needs a computationally simple method, which provide to derive desired target formation of various sharps and precise distances between mobile robots. Such method for solving 2D formation task on a plane is proposed in [20, 21]. The method for creating 3D formations within a group of UAVs is proposed in [22]. But these works do not take into account restrictions on information exchange, and also does not take into account the required direction of the mobile robot in the target system. In this paper, we propose a modification of the "method sphere", taking into account restrictions on communication and robot's orientation.

The target formation is a set V of target positions $v_\mu \in V(\mu = \overline{1,N})$ of single quadrotors. Each target position $v_\mu \in V$ is described by a point $p_\mu(\mu = \overline{1,N})$ with coordinates (x_μ, y_μ, z_μ) and orientation $\vec{o}_\mu$. But there is not information about point's $p_\mu(\mu = \overline{1,N})$ coordinates $x_{p\mu}, y_{p\mu}, z_{p\mu}(\mu = \overline{1,N})$ and about assignments between quadrotors $r_i \in \mathbf{R}$ and target positions $v_\mu \in V$.

The only one available information about the target formation is a set of vectors $\mathbf{O}$ of orientations $\vec{o}_\mu \in \mathbf{O}$ and matrix

$$
D_f = \begin{bmatrix}
0 & d_{1,2} & d_{1,3} & \cdots & d_{1,N} \\
- & 0 & d_{2,3} & \cdots & d_{2,N} \\
- & - & 0 & \ddots & \vdots \\
- & - & - & 0 & d_{N-1,N} \\
- & - & - & - & 0
\end{bmatrix},
\tag{1}
$$

where each variable $d_{i,j}$ of D_f is a distance between points p_i and p_j of target positions v_i and v_j in a target formation.

In the first step, each robot estimates the time $t_{ri,v1}$ which takes to move to the target position, which is closest to the geometric center of the target system. This takes into account the orientation of the robot at the initial time and its orientation in the target position.

Then each robot sends to other robots of the coalition a message with its number and time to reach the target position in question. To send this message, an epidemic algorithm is used.

When the robot r_i received messages with estimates of the time to reach the first (central) target position from all other robots of the coalition, it compares its estimate with the minimum obtained. If his time $t_{ri,v1}$ is less than the minimum received $\min t_{ri,v1}$, he sends a message stating that he is occupying this target position and is moving to this target position. Otherwise, it participates in the further planning of the coalition's actions.

At the second step, robots that have not yet received the target positions make an estimate of the time $t_{ri,v2}$ to reach the second target position v_2, namely, the closest point p_2 to it on the sphere $c_{1,2}$ with center in point p_1 and radius $d_{1,2}$ (from the matrix D_f). Then construct $N - 1$ straight lines, each one passing through the point p_1 and the current position of quadrotors $r_i(i = \overline{2,N})$:

$$
\frac{x_{p2i} - x_1}{x_i - x_1} = \frac{y_{p2i} - y_1}{y_i - y_1} = \frac{z_{p2i} - z_1}{z_i - z_1}, \quad i \in [\overline{2,N}]
\tag{2}
$$

To determine the coordinates of the intersection points of these lines and the sphere $c_{1,2}$ for each quadrotor $r_i(i = \overline{2,N})$ use the system of equations:

$$
\begin{cases}
\frac{x_{p2i} - x_1}{x_i - x_1} = \frac{y_{p2i} - y_1}{y_i - y_1} = \frac{z_{p2i} - z_1}{z_i - z_1} \\
(x_{p2i} - x_1)^2 + (y_{p2i} - y_1)^2 + (z_{p2i} - z_1)^2 = d_{1,2}^2
\end{cases} \quad i \in [\overline{2,N}]
\tag{3}
$$

Get equation's roots $(x_{p2i_1}, y_{p2i_1}, z_{p2i_1})$ and $(x_{p2i_2}, y_{p2i_2}, z_{p2i_2})$ for each $r_i(i = \overline{2,N})$. Then each robot calculate time $t_{ri,v2}$ to achieve the points $(x_{p2i_1}, y_{p2i_1}, z_{p2i_1})$ and $(x_{p2i_2}, y_{p2i_2}, z_{p2i_2})$ with orientation $\vec{o}_2$.

The robot with the minimal time $\min(t_{ri,v2}), q \in [1,2]i \in [\overline{2,N}]$ get an assignment target position v_2 and point p_2 with coordinates (x_{p2}, y_{p2}, z_{p2}) and orientation $\vec{o}_2$.

At the third step the coordinates $p_3(x_{p3}, y_{p3}, z_{p3})$ of the target position v_3 is determined. For this construct two spheres $c_{1,3}$ and $c_{2,3}$. The sphere $c_{1,3}$ has the center in point p_1 and the radius $d_{1,3}$. The sphere $c_{2,3}$ has the center in point p_2 and the radius $d_{2,3}$. Then find the set of points of intersections of spheres $c_{1,3}$ and $c_{2,3}$. It is a circle

$$\begin{cases} (x_{p3} - x_1)^2 + (y_{p3} - y_1)^2 + (z_{p3} - z_1)^2 = d_{1,3}^2; \\ (x_{p3} - x_2)^2 + (y_{p3} - y_2)^2 + (z_{p3} - z_2)^2 = d_{2,3}^2. \end{cases} \tag{4}$$

Then find points $p_{3i}, i \in \overline{[3, N]}$. And each robot $r_i(i = \overline{3, N})$ calculate time $t_{ri,v3}$ to achieve point $(x_{p3,q}, y_{p3,q}), q \in \overline{[1, 2]}$ and orientation $\vec{o}_3$. The quadrotor with the minimal time $\min(t_{ri,v3}), i \in \overline{[3, N]}, q \in \overline{[1, 2]}$ get an assignment target position v_3 and point p_3 with coordinates (x_{p3}, y_{p3}, z_{p3}) and orientation $\vec{o}_3$.

At the fourth step the coordinates $p_4(x_{p4}, y_{p4}, z_{p4})$ of the target position v_4 is determined. For this construct three spheres $c_{1,4}$, $c_{2,4}$ and $c_{3,4}$. The sphere $c_{1,4}$ has the center in point p_1 and the radius $d_{1,4}$. The sphere $c_{2,4}$ has the center in point p_2 and the radius $d_{2,4}$. The sphere $c_{3,4}$ has the center in point p_3 and the radius $d_{3,4}$. Then find points of intersections of spheres $c_{1,4}$, $c_{2,4}$ and $c_{3,4}$:

$$\begin{cases} (x_{p4} - x_1)^2 + (y_{p4} - y_1)^2 + (z_{p4} - z_1)^2 = d_{1,4}^2; \\ (x_{p4} - x_2)^2 + (y_{p4} - y_2)^2 + (z_{p4} - z_2)^2 = d_{2,4}^2; \\ (x_{p4} - x_2)^2 + (y_{p4} - y_2)^2 + (z_{p4} - z_2)^2 = d_{2,4}^2. \end{cases} \tag{5}$$

At the each next step the coordinates $p_\mu(\mu = \overline{4, N})$ of target position v_μ is determined. For this construct 4 spheres $c_{\mu-4,\mu}$, $c_{\mu-3,\mu}$, $c_{\mu-2,\mu}$ and $c_{\mu-1,\mu}$. The first circle $c_{\mu-4,\mu}$ has the center in point $p_{\mu-4}$ and the radius $d_{\mu-5,\mu}$. The second sphere $c_{\mu-3,\mu}$ has the center in point $p_{\mu-3}$ and the radius $d_{\mu-3,\mu}$. The third sphere $c_{\mu-2,\mu}$ has the center in the point $p_{\mu-2}$ and the radius $d_{\mu-2,\mu}$. And the fourth sphere $c_{\mu-1,\mu}$ has the center in the point $p_{\mu-1}$ and the radius $d_{\mu-1,\mu}$.

Then find points of intersections of spheres $c_{\mu-4,\mu}$, $c_{\mu-3,\mu}$, $c_{\mu-2,\mu}$ and $c_{\mu-1,\mu}$. There are two points of intersections in general, but there is only one in some cases.

To determine the coordinates of the intersection points of spheres $c_{\mu-4,\mu}$, $c_{\mu-3,\mu}$, $c_{\mu-2,\mu}$ and $c_{\mu-1,\mu}$ for each quadrotor $r_i(i = \overline{k, N})$ use the system of equations:

$$\begin{cases} (x_{p\mu} - x_{\mu-4})^2 + (y_{p\mu} - y_{\mu-4})^2 + (z_{p\mu} - z_{\mu-4})^2 = d_{\mu-4,\mu}^2; \\ (x_{p\mu} - x_{\mu-3})^2 + (y_{p\mu} - y_{\mu-3})^2 + (z_{p\mu} - z_{\mu-3})^2 = d_{\mu-3,\mu}^2; \\ (x_{p\mu} - x_{\mu-2})^2 + (y_{p\mu} - y_{\mu-2})^2 + (z_{p\mu} - z_{\mu-2})^2 = d_{\mu-2,\mu}^2; \\ (x_{p\mu} - x_{\mu-1})^2 + (y_{p\mu} - y_{\mu-1})^2 + (z_{p\mu} - z_{\mu-1})^2 = d_{\mu-1,\mu}^2. \end{cases} \tag{6}$$

Then robot $r_i(i = \overline{k, N})$ calculate time t_{r_i,v_μ} to achieve point $(x_{p\mu q}, y_{p\mu q}, z_{p\mu q})$. The robot with the minimal time $\min(t_{r_i,v_\mu}), i \in \overline{[\mu, N]}, q \in \overline{[1,2]}$ get an assignment target position v_μ and the point p_μ with coordinates $(x_{p\mu}, y_{p\mu}, z_{p\mu})$ and orientation $\vec{o}_\mu$.

When all assignments and coordinates of all target positions are obtained, each robots moves to its own target position and orientation. After the new formation is created, each robot loads from memory the parameters of the ANN's neuron corresponding to the given formation and its target position and starts working in neuron mode until it receives a message from the operator or from other robots from the coalition that it is necessary to change the formation and its corresponding neural network.

4 Software Simulation

In order to test the efficiency and evaluate the effectiveness of the proposed methods for solving the formation task, the software "Swarm Control" was developed (Fig. 1). The software make it possible to simulate a series of experiments to create formations with different configurations in groups or flight robots - quadrotors (Fig. 2). The averaged results of that experiments are shown at the graph (Fig. 3).

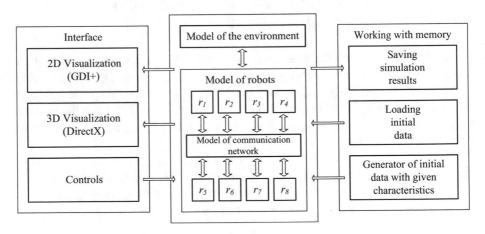

Fig. 1. Structure of the simulation software (arrows show the data flow)

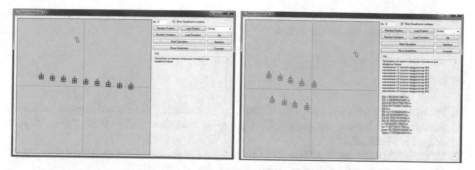

Fig. 2. Screenshots of the simulation of the formation of the system in a group of 9 quadrocopters: the initial position (left) and the formed system (right). The distances are indicated in meters, the average velocity of the quadrocopter horizontally is 5 m/s, the time of formation creation is $T_{penu} \approx 7,8\,c$

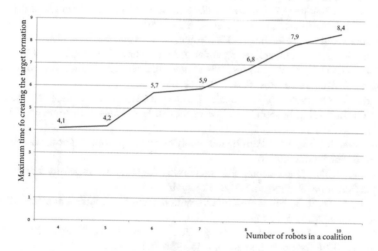

Fig. 3. The graph of the formation time dependence on the number of robots in the group

5   Conclusions and Future Work

This article discusses the planning of actions of intelligent mobile robots in coalitions with limited communications. A general approach to solving this problem is given and an example of the formation of mobile distributed artificial neural networks is considered. A modification of the sphere method is proposed to form the system necessary for creation of mobile distributed artificial neural networks is given.

The results of computer simulation are presented. In the future, it is planned to consider in detail the methods of organizing information exchange, taking into account not only the limitations on the range of communication between the robots of the group, but also on the speed of data transmission and the bandwidth of communication channels.

Acknowledgement. The reported study was funded by RFBR according to the research projects No. 17-29-07054 and No. 16-29-04194.

References

1. Franchi, A., Secchi, C., Ryll, M., Bülthoff, H., Giordano, P.R.: Shared control: balancing autonomy and human assistance with a group of quadrotor UAVs. IEEE Robot. Autom. Mag. **19**, 57–68 (2012)
2. Di Paola, D., Gasparri, A., Naso, D., Lewis, F.L.: Decentralized dynamic task planning for heterogeneous robotic networks. Auton. Robot. **38**, 31–48 (2015)
3. Jones, C.V, Matarić, M.J.: Behavior-based coordination in multi-robot systems. Auton. Mob. Robot. Sens. Control Decis. Appl., 549–569 (2005)
4. Kaliaev, I., Kapustjan, S., Ivanov, D.: Decentralized control strategy within a large group of objects based on swarm intelligence (2011)
5. Dorigo, M., Birattari, M.: Swarm intelligence. Scholarpedia **2**, 1462 (2007)
6. Bonabeau, E., Dorigo, M., Theraulaz, G.: Swarm Intelligence: From Natural to Artificial Systems. Oxford University Press, New York (1999)
7. Schoonderwoerd, R., Holland, O.E., Bruten, J.L., Rothkrantz, L.J.M.: Ant-based load balancing in telecommunications networks. Adapt. Behav. **5**, 169–207 (1997)
8. Di Caro, G., Dorigo, M.: AntNet: distributed stigmergetic control for communications networks. J. Artif. Intell. Res. **9**, 317–365 (1998)
9. Di Caro, G., Ducatelle, F., Gambardella, L.M.: AntHocNet: an adaptive nature-inspired algorithm for routing in mobile ad hoc networks. Eur. Trans. Telecommun. **16**, 443–455 (2005)
10. Kostadinova, R., Adam, C.: Performance analysis of the epidemic algorithms. Intell. Control Autom. **6**, 6675–6679 (2008)
11. Hollerung, T.D., Bleckmann, P.: Epidemic Algorithms. http://my.fit.edu/~gfrederi/ComplexNetworks/09-Epidemic-Algorithms.pdf
12. Balch, T., Arkin, R.C.: Behavior-based formation control for multirobot teams. IEEE Trans. Robot. Autom. **14**, 926–939 (1998)
13. Lawton, J.R.T., Beard, R.W., Young, B.J.: A decentralized approach to formation maneuvers. IEEE Trans. Robot. Autom. **19**, 933–941 (2003)
14. Wang, P.K.C.: Navigation strategies for multiple autonomous mobile robots moving in formation. J. Robot. Syst. **8**, 177–195 (1991)
15. Desai, J.P., Ostrowski, J., Kumar, V.: Controlling formations of multiple mobile robots. In: Proceedings of the 1998 IEEE International Conference on Robotics and Automation (Cat. No. 98CH36146), vol. 4 (1998)
16. Mesbahi, M., Hadaegh, F.Y.: Formation flying control of multiple spacecraft via graphs, matrixinequalities, and switching. In: Proceedings of the 1999 IEEE International Conference on Control Applications (Cat. No. 99CH36328), vol. 2 (1999)
17. Wang, P.K.C., Hadaegh, F.Y.: Coordination and control of multiple microspacecraft moving in formation. J. Astronaut. Sci. **44**, 315–355 (1996)
18. Lewis, M.A., Tan, K.-H.: High precision formation control of mobile robots using virtual structures. Auton. Robot. **4**, 387–403 (1997)
19. Erdoğan, M.E., Innocenti, M., Pollini, L.: Obstacle avoidance for a game theoretically controlled formation of unmanned vehicles. In: 18th IFAC (2011)

20. Ivanov, D., Kalyaev, I., Kapustyan, S.: Method of circles for solving formation task in a group of quadrotor UAVs. In: 2014 2nd International Conference on Systems and Informatics (ICSAI), pp. 236–240 (2014)
21. Ivanov, D., Kalyaev, I., Kapustyan, S.: Formation task in a group of quadrotors. In: Robot Intelligence Technology and Applications, vol. 3. pp. 183–191. Springer (2015)
22. Ivanov, D., Kapustyan, S., Kalyaev, I.: Method of spheres for solving 3D formation task in a group of quadrotors. In: International Conference on Interactive Collaborative Robotics, ICR 2016. Lecture Notes in Computer Science, Budapest, Hungary, 24–26 August 2016, vol. 9812 (2016)

Matrix-Like Structures
for Representation and Processing
of Constraints over Finite Domains

Alexander Zuenko$^{(\boxtimes)}$

Institute for Informatics and Mathematical Modeling – Subdivision of the Federal
Research Centre, Kola Science Centre of the Russian Academy of Sciences,
Apatity Murmansk 184209, Russia
zuenko@iimm.ru
http://www.iimm.ru

Abstract. The paper presents two types of matrix-like structures, the
C-systems and the D-systems, which are proposed to be used in representation and handling the constraints over finite domains. The modifications of the well known constraints propagation algorithms have been
developed by the author. The proposed approach allows representing
the n-ary relations in a compressed form, and accelerating searching the
solution by means of analyzing the specific features of the constraints
matrices.

Keywords: Constraints propagation
Consistency-enforcing algorithms · Constraints over finite domains
Combinatorial search

1 Introduction

Constraint programming is a powerful paradigm for solving combinatorial search
problems that draws on a wide range of techniques from artificial intelligence, computer science, and operations research [1]. Recent applications include computer
graphics, natural language processing (construction of efficient parsers), database
systems (to ensure and/or restore consistency of the data), operations research
problems (like optimization problems), molecular biology (DNA sequencing), business applications (option trading), electrical engineering (to locate faults), circuit
design (to compute layouts), production scheduling, etc. [2].

According to [3], the *constraint satisfaction problem* (CSP) is defined by a set
of *variables* $X = \{x_1, x_2, \ldots, x_n\}$, and a set of *constraints* $C = \{C_1, C_2, \ldots, C_m\}$.
Each variable x_i has a non-empty *domain* D_i. Each constraint C_i includes a
subset of variables that participate in the constraint, setting the *legal* combinations of values that those variables can take on. A CSP is presented as a tuple
$<X, C, D>$, where $X = \{x_i\}, C = \{C_j\}, D = \{D_k\}$.

Supported by Russian Foundation for Basic Research (grants nos. 16-07-00377-a,
16-07-00313-a, 16-07-00273-a, 16-07-00562-a, 18-07-00615-a).

A. Abraham et al. (Eds.): IITI 2018, AISC 875, pp. 428–438, 2019.
https://doi.org/10.1007/978-3-030-01821-4_45

Each state in a CSP is defined as an *assignment* of values to some or to all of these variables, $\{x_i = v_i, x_j = v_j, \ldots\}$. An assignment, which does not violate any constraints, is called a *consistent* or *legal* assignment. An assignment, in which each variable participates, is called a *complete* assignment. The *solution* of the CSP is a complete assignment satisfying all the constraints.

Within the constraint programming paradigm, the integration of various qualitative and quantitative modeling methods allows the combinatorial problems of high theoretical complexity to be solved in practice. The CSP solution methods may differ greatly depending on what we take as CSP variables, what domains we choose for variables, what types of constraints we use. Integers, real numbers, numerical intervals, finite sets, finite multisets, multiintervals, etc., can be used as domains of variables. Constraints can be logical, arithmetic, formulated by the language of some theories, such as the graph theory or the automata theory [4].

Considered below are the CSPs, in which variables have the finite domains. According to Bartak [2], more than 95% of applications of the CSP are likely to use finite domains. If the maximum size of the domain of any variable in a CSP is equal to d, the number of possible complete assignments is $O(d^n)$, i.e. it depends exponentially on the number of variables, which makes us apply the methods of combinatorial search speed-up. Basic methods of a CSP solution fall into three classes [2]. The first class includes the variants of the *backtracking* search algorithms. These algorithms construct a solution by step-by-step extending a partial assignment, relying on different heuristics and using more or less intelligent strategies of backtracking from dead ends. The second class includes the so-called *constraint propagation* algorithms, which eliminate some non-solution elements from the search space, thus resulting in a smaller solution space. In general, these algorithms do not eliminate all the non-solution elements and hence, they do not produce a solution on their own. These are applied either in solution preprocessing before the algorithms of other types are used, or occur intermittently with the steps of the algorithms of another type (for instance, backtracking search) to accelerate the latter. Finally, the *structure-driven* algorithms exploit the information about the structure of the primal or dual graph of the problem. The algorithms of this class decompose the initial CSP into loosely-coupled subproblems, which can be solved by the methods from the previous two classes.

Most constraint programming environments are mainly oriented on processing of numerical constraints, which are specified by means of the base set of arithmetical operations, binary relations equal/unequal, more/less, built-in functions, etc. for which specialized procedures-propagators are developed.

The authors' studies showed that processing of qualitative constraints represented in the form of logical expressions and rules, is not sufficiently effective in the systems like those mentioned above, and cannot be implemented for comprehensible time even at a rather small dimension of problem. To solve the problems within the constraint programming concept, the study proposes an approach based on making use of matrix-like representation of constraints.

Presented in [5,6] are the principles of n-tuple algebra (NTA) as well as its applications are shown to be effective in unifying the various data and knowledge representation and processing and in solution of the problems of logic and logic-and-probability analysis. A similar approach is also applied in [7] to solve the problems of pattern recognition and knowledge base simplification. Unlike the author's techniques given in this paper the methods proposed in the prototype-studies are not oriented at the constraints satisfaction problems solution as these are related to "blind" search techniques, not using advanced strategies of the constraint satisfaction theory.

2   Matrix-Like Structures for Constraints Representation

Many kinds of data and knowledge used in intelligent systems, including production systems, can be expressed in the finite predicates language.

The finite predicates in NTA can be presented in a compressed form by two types of matrix-like structures: C-systems and D-systems [5,6].

By means of C-systems it is convenient to present the disjunctive normal forms (DNF) of finite predicates in a "compressed" form. As an example, let the finite predicate be set as follows:

$$\varphi(x, y, z) = (x = a, b) \wedge (y = a, c) \vee (z = d).$$

Just to make it simple, all the variables are in the same set a, b, c, d. Hereinafter, we use $(x = a, b)$ to designate the expression $(x = a) \vee (x = b)$. Taking into account that the truth set of a one-place predicate $(x = a, b)$ is $\{a, b\}$, the truth set of the predicate $\varphi(x, y, z)$ can be presented as the following C-system:

$$R[XYZ] = \begin{bmatrix} \{a, b\} & \{a, c\} & * \\ * & * & \{d\} \end{bmatrix}. \tag{1}$$

The attributes X, Y, Z of the C-system $R[XYZ]$ correspond to the variables x, y, z of the formula $\varphi(x, y, z)$. Hereinafter, we shall agree to designate the attributes in capital Latin letters and the variables in predicates in small Latin letters. In defining the C- and D-systems, two types of dummy components are used, such as: a full component ("*"), i.e., a set which is equal to the domain of the corresponding attribute; and an empty component ("∅") i.e., a component which does not contain any values. The C-system $R[XYZ]$ is possible to be transformed into a multi-place relation, such as:

$$(\{a, b\} \times \{a, c\} \times \{a, b, c, d\}) \cup (\{a, b, c, d\} \times \{a, b, c, d\} \times \{d\}).$$

The conjunctive normal forms (CNF) of the finite predicates are modeled by means of the D-systems. The D-system is expressed as a matrix of the component-sets put into the reversed direct square brackets.

The D-systems allow the complement of the C-systems to be easily calculated: the complement is calculated for each component-set.

The predicate $\neg\varphi = (\neg(x = a, b) \vee \neg(y = a, c)) \wedge \neg(z = d)$ which is equivalent to $\neg\varphi = ((x = c, d) \vee (y = b, d)) \wedge (z = a, b, c)$, can be expressed as the D-system $\overline{R}[XYZ]$:

$$\overline{R}[XYZ] = \left] \begin{matrix} \{c, d\} & \{b, d\} & \varnothing \\ \varnothing & \varnothing & \{a, b, c\} \end{matrix} \right[. \tag{2}$$

Using the C-systems, it is possible to describe a set of the facts in a "compressed" form, the D-systems allow rules, prohibitions, negation of facts and statements to be formulated.

Given below in a short form are the methods of CSPs solution, based on the matrix representation of constraints.

3   The Constraints Propagation Techniques

The constraint propagation algorithms allow the initial CSP $<X, D, C>$ to be reduced to a simpler problem $<X', D', C'>$, in which each domain of the variables in D' is a subset of the corresponding domain in D, and the constraints C' contain the same set of solutions of the CSP as the constraints C but the constraints C' is naturally simpler. The most popular in the theory of constraints satisfaction are the techniques ensuring node consistency (NC) and arc consistency (AC) [8,9]. It is assumed that the nodes of some graph correspond to the variables and its arcs correspond to the binary constraints. Suggested below are the modifications of these techniques, based on the properties of the matrix structures mentioned above.

Let the problem of constraints satisfaction $<X, D, C>$ be defined as the single D-system, in which each tuple (row of the D-system) corresponds to a certain constraint, where each variable is associated with a certain attribute of the D-system (a column of the D-system), and the variable domain corresponds to the domain of the corresponding attribute.

Consider, as an example, one of the variants of the problem "Magic Square".

Example: Let a square be of 3×3 cell in size. Each cell should contain different numbers. A set of numbers used in the square is $\{1, 2, 3\}$. Initially, only three cells are filled with particular numbers. The values of the rest variables are unknown (see Fig. 1). Let us consider how, by means of the D-system, to express the constraints prohibiting the similar numbers in the first row of "Magic Square". The rest constraints are formulated similarly.

Let each cell be associated with certain variable X_i (the attribute of the D-system). Thus each attribute of the D-system has the domain $\{1, 2, 3\}$.

Consider, at first, the constraint "in two cells X_i and X_j there can not be similar numbers". It is the negation of the statement "in two cells X_i and X_j

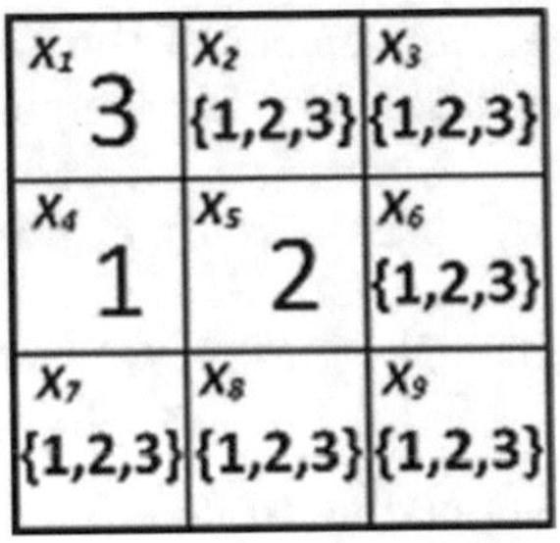

X_1	X_2	X_3
3	{1,2,3}	{1,2,3}
X_4	X_5	X_6
1	2	{1,2,3}
X_7	X_8	X_9
{1,2,3}	{1,2,3}	{1,2,3}

Fig. 1. A variant of the problem "Magic Square"

there are similar numbers", which is formalized by means of the C-system, such as:

$$S[X_i X_j] = \begin{bmatrix} \{1\} & \{1\} \\ \{2\} & \{2\} \\ \{3\} & \{3\} \end{bmatrix}. \tag{3}$$

Then, the required constraint for two cells is written as the D-system, which is an complement to the C-system $S[X_i X_j]$ relative to the scope $\{1,2,3\} \times \{1,2,3\}$:

$$W[X_i X_j] = \overline{S[X_i X_j]} = \overline{\begin{bmatrix} \{1\} & \{1\} \\ \{2\} & \{2\} \\ \{3\} & \{3\} \end{bmatrix}} = \begin{bmatrix} \{2,3\} & \{2,3\} \\ \{1,3\} & \{1,3\} \\ \{1,2\} & \{1,2\} \end{bmatrix}. \tag{4}$$

Now it is possible to formulate the constraint for the first row of the magic square:

$$CS[X_1 X_2 X_3] = W[X_1 X_2] \cap_G W[X_1 X_3] \cap_G W[X_2 X_3].$$

In the resulting D-system there are 9 rows (tuples).

The operation of the generalized intersection of the D-systems is used here. The intersection of two D-systems can be presented as a D-system whose tuples contain only the tuples of the D-systems intersected. Being formed in the different schemes, the D-systems intersected are reduced to a common scheme by means of filling the corresponding positions in the tuples with dummy components.

The fact that in the upper left cell of the magic square there is a particular number "3", is described by means of one more constraint (of the tenth tuple (row) of the D-system).

We have the following D-system:

$$
\begin{array}{c}
\begin{array}{ccc}
X_1 & X_2 & X_3 \\
\{1,2,3\} & \{1,2,3\} & \{1,2,3\}
\end{array} \\
\begin{array}{c}
1 \\ 2 \\ 3 \\ 4 \\ 5 \\ 6 \\ 7 \\ 8 \\ 9 \\ 10
\end{array}
\left[
\begin{array}{ccc}
\{2,3\} & \{2,3\} & \varnothing \\
\{1,3\} & \{1,3\} & \varnothing \\
\{1,2\} & \{1,2\} & \varnothing \\
\{2,3\} & \varnothing & \{2,3\} \\
\{1,3\} & \varnothing & \{1,3\} \\
\{1,2\} & \varnothing & \{1,2\} \\
\varnothing & \{2,3\} & \{2,3\} \\
\varnothing & \{1,3\} & \{1,3\} \\
\varnothing & \{1,2\} & \{1,2\} \\
\{3\} & \varnothing & \varnothing
\end{array}
\right]
\end{array}.
\tag{5}
$$

Now it's time to precise some features of search space pruning (constraints propagation), based on the matrix representation of the finite predicates, namely:

1. How to prune the search space (to simplify the definition of the CSP) without branching?
2. What is the indication of that searching is completed successfully (the constraints propagation, in particular)?
3. What is the indication of that searching is completed unsuccessfully (the indication of failure in constraints propagation, in particular)?

To answer these questions, consider the following affirmations taken here for granted:

Affirmation 1 (**A1**): If, at least, one tuple of the D-system is empty (all the components are empty), the D-system is empty (the corresponding system of constraints is inconsistent).

Affirmation 2 (**A2**): If all the components of an attribute are empty, the attribute can be eliminated from the D-system (all the components in the corresponding column are removed) and the pair "the eliminated attribute – its domain" is included into the partial solution.

Affirmation 3 (**A3**): If in the D-system there is a tuple (row) containing only one nonempty component, all the values not included into this component, are deleted from the corresponding domain. After that the tuple is eliminated from the D-system.

Affirmation 4 (**A4**): If a tuple of the D-system contains, at least, one full component, this tuple is removed (one can remove the corresponding constraint from the system of constraints).

Affirmation 5 (**A5**): If the component of an attribute of the D-system contains the value not belonging to the corresponding domain, this value is deleted from the component.

Affirmation 6 (**A6**): If one tuple of the D-system completely dominates (component-wise) over the other tuple, the dominating tuple is removed from the D-system.

Most affirmations mentioned are the answer to the first question, a part of which allows us to rule out the values from the domains of the attributes (**A3**, **A5**) or even the columns (attributes) (**A2**), and another part enables us to eliminate the redundant tuples (rows) from consideration (**A4**, **A6**).

The *indication of successful finishing* of searching is the elimination of all the rows and columns from the D-system, with no empty rows being formed. In other words, the resulting state in this case is characterized only by a nonempty reduced domains in the vector of the solution.

The answer to the third question is given by **A1**, i.e., *the indication of inconsistency of the CSP* is the D-system's being empty.

Affirmations **A1-A6** are widely used in the constraints propagation algorithms developed by the author, which are given below.

Before these algorithms are presented, let us make some remarks and introduce some term we need below. The value of some attribute of the domain is referred to as "redundant" if there is no solution of the CSP, which would include the given value of the variable (the attribute of the D-system).

The *submatrix* of the D-system is referred to as the D-system generated from the initial D-system by selecting the components which are at the intersection of some rows and columns of the initial D-system.

3.1   Node Consistency

The algorithms establishing node consistency allow the values, which do not satisfy the unary constraints, to be eliminated from the domains of all the variables of the CSP. The typical order of operations in implementation of the algorithm is to test each member of the domain of each variable and conclude whether the value satisfies the unary constraints on the variable. All the values, which violate the unary constraints, are eliminated from the corresponding domains. When the algorithm is finished, the initial problem is reduced to the problem, which is node-consistent.

The algorithm establishing the node-consistency in the matrix-like structures is expressed below:

Algorithm "Node consistency"
For each tuple G of the D-system, to execute:
    If the tuple contains only one nonempty component (B_i), all the values, which do not enter this component, are eliminated from the corresponding domain D_i, that is, $D_i = B_i \cap D_i$.
After that the tuple is eliminated from G.
End.

The constraints propagation procedure described above is iterative, consisting in the cyclical use of affirmations **A3**, **A5** and **A4**.

With the help of the problem "Magic Square", we demonstrate the iterative handling of unary constraints. So, consider again the D-system consisting of 10 rows. It is seen that row 10 contains only one nonempty component corresponding to the attribute X_1. Based on affirmation **A3**, reduce the domain of the attribute X_1 to a one-element set $\{3\}$. Now, eliminate row 10. We have:

$$
\begin{array}{c}
\begin{array}{ccc}
X_1 & X_2 & X_3 \\
\{3\} & \{1,2,3\} & \{1,2,3\}
\end{array} \\
\begin{array}{c}
1 \\ 2 \\ 3 \\ 4 \\ 5 \\ 6 \\ 7 \\ 8 \\ 9
\end{array}
\left[
\begin{array}{ccc}
\{2,3\} & \{2,3\} & \varnothing \\
\{1,3\} & \{1,3\} & \varnothing \\
\{1,2\} & \{1,2\} & \varnothing \\
\{2,3\} & \varnothing & \{2,3\} \\
\{1,3\} & \varnothing & \{1,3\} \\
\{1,2\} & \varnothing & \{1,2\} \\
\varnothing & \{2,3\} & \{2,3\} \\
\varnothing & \{1,3\} & \{1,3\} \\
\varnothing & \{1,2\} & \{1,2\}
\end{array}
\right]
\end{array}
. \qquad (6)
$$

The first column contains the values, which do not belong any longer to the domain of the attribute X_1. We apply therefore **A5** and "customize" the components of the first column to a new domain and eliminate the values not belonging to the new domain. Some components of the first column become empty and some others are full. After rows 1, 2, 4, 5 are eliminated from the D-system, based on affirmation **A4**, the first column contains only empty components. It means that it is possible, based on affirmation **A2**, to remove the first column from the D-system and then extend the partial solution. We have:

A partial solution: $X_1 - \{3\}$.

The remainder of the D-system:

$$
\begin{array}{c}
\begin{array}{cc}
X_2 & X_3 \\
\{1,2,3\} & \{1,2,3\}
\end{array} \\
\begin{array}{c}
3 \\ 6 \\ 7 \\ 8 \\ 9
\end{array}
\left[
\begin{array}{cc}
\{1,2\} & \varnothing \\
\varnothing & \{1,2\} \\
\{2,3\} & \{2,3\} \\
\{1,3\} & \{1,3\} \\
\{1,2\} & \{1,2\}
\end{array}
\right]
\end{array}
. \qquad (7)
$$

In the D-system generated, rows 3 and 6 contain one nonempty component each. It means that, taking **A3** into consideration, the domains of the attributes X_2 and X_3 are reduced to a set $\{1, 2\}$. Then, rows 3 and 6 are deleted from the D-system. After that, based on **A5**, the D-system is being "customized" to new domains. We have:

A partial solution: $X_1 - \{3\}$.

The remainder of the D-system:

$$
\begin{array}{c}
\begin{array}{cc}
X_2 & X_3 \\
\{1,2\} & \{1,2\}
\end{array} \\
\begin{array}{c}
7 \\ 8 \\ 9
\end{array}
\left[
\begin{array}{cc}
\{2\} & \{2\} \\
\{1\} & \{1\} \\
\{1,2\} & \{1,2\}
\end{array}
\right]
\end{array}
. \qquad (8)
$$

According to **A4**, row 9 is eliminated from the D-system as it contains complete components. There is no more propagation to be done. Thus, domains of variables X_1 and X_2 is reduced to set $\{1, 2\}$.

Indeed, analysis of just first row of the "Magic Square" gives the only result that the cells X_1 and X_2 can not be filled with number "3".

3.2   Arc Consistency

Unlike the algorithms of establishing node consistency, the algorithms of this type are more complicated. At present there exist a number of the algorithms of establishing arc consistency. The simplest algorithm is referred to as AC-1 [8], and the most effective one is AC-7 [10]. Of this set, the algorithm AC-3 [9] is used more often, and is also applied in the consistency-enforcing algorithms for the continuous domains.

Below we consider only this algorithm, or to be more exact, its modifications which take into account the advantages of the matrix constraint representation. The basic operation in the arc consistency algorithm is to establish the consistency for each arc.

The procedure of establishing the consistency for one arc can be formulated as follows:

Algorithm "Arc consistency"
Procedure REVICE_DOMAIN(X, Y)
DELETED <–false
For each tuple G of the D-system, to execute:
    If there is no value b in domain D_Y, such that $C[XY] \cap Q[XY] \neq \varnothing$, where
    $Q[XY] = [\{a\}\{b\}]$, $C[XY]$ is binary relation (constraint), $D_X = D_X \setminus \{a\}$
DELETED <–true
return DELETED
End.

Let us precise the notion of "arc". Let us form a submatrix of the initial D-system as follows: select all the rows of the initial D-system, which contain nonempty components only in the positions corresponding to the given attributes X and Y. The submatrix of such kind models a binary relation. This binary relation can be visualized in the graph by two arcs: the arc (X, Y) and the arc (Y, X), where X, Y are some variables. In fact, the arcs set the direction of the binary relation revision: eliminated in the first case are the "redundant" values from the domain of the variable $X(D_X)$, and in the second case – from the domain of the variable $Y(D_Y)$.

The arc (X, Y) can be made consistent by eliminating all the values of D_X, for which there are no the values in D_Y, permitted by the corresponding binary constraint. In other words, eliminated from the D_X are the values, which are not supported in D_Y. In such eliminating of values from D_X, no solution of the initial CSP is removed.

Constraints propagation algorithms developed by the author often allow us to find the solution of the CSP, without branching.

4 Conclusion

The novelty of the research is that the n-ary constraints over finite domains are not divided into simpler (of a lesser arity) ones but are processed as matrix-like structures, namely the D-systems. The D-systems allow, on the one hand, representing the n-ary relations (constraints) in a compressed form, making use of the computer memory more efficient, and, on the other hand, accelerating searching the solution by means of analyzing the specific features of the constraints matrices.

Modifications of the known techniques of the constraints propagation are developed, which, in particular, enforce the node and arc consistency. The modifications allow reducing of the CSP even if it has not been initially represented as a set of the unary and binary constraints only. If the "redundant" values are found in the domains by the modified algorithms, the D-system is "customized" to new reduced domains by elimination of rows (columns) of the D-system and of certain values from the components of the D-system. In "customizing", new unary and binary constraints may appear, i.e. constraints propagation is iterative, allowing some CSPs to be reduced to smaller problems, without branching.

References

1. Baptiste, P., Le Pape, C., Nuijten W.: Constraint-Based Scheduling: Applying Constraint Programming to Scheduling Problems. International Series in Operations Research and Management Science, vol. 39. Springer, New York (2001). https://doi.org/10.1007/978-1-4615-1479-4
2. Bartak, R.: Constraint programming. Pursuit of the Holy Grail. In: Proceedings of the Week of Doctoral Students (WDS 1999), Part IV, pp. 555–564. MathFyzPress, Praque (1999)
3. Russel, S., Norvig, P.: Artificial Intelligence: A Modern Approach, 3rd edn. Prentice Hall, Upper Saddle River (2010)
4. Rossi, F., van Beek, P., Walsh, T.: Constraint programming. In: van Harmelen, F., Lifschitz, V., Porter, B. (eds.) Handbook of Knowledge Representation (Foundations of Artificial Intelligence), vol. 3, pp. 181–211. Elsevier (2008)
5. Kulik, B., Fridman, A., Zuenko, A.: Logical inference and defeasible reasoning in N-tuple algebra. In: Naidenova, X., Ignatov, D. (eds.) Diagnostic Test Approaches to Machine Learning and Commonsense Reasoning Systems, pp. 102–128. IGI Global, Hershey (2013). https://doi.org/10.4018/978-1-4666-1900-5.ch005
6. Kulik, B., Zuenko, A., Fridman, A.: Deductive and defeasible reasoning on the basis of a unified algebraic approach. Sci. Tech. Inf. Process. **42**(6), 8–16 (2015). https://doi.org/10.3103/S0147688215060076
7. Zakrevskij, A.: Integrated model of inductive-deductive inference based on finite predicates and implicative regularities. In: Naidenova, X., Ignatov, D. (eds.) Diagnostic Test Approaches to Machine Learning and Commonsense Reasoning Systems, pp. 1–12. IGI Global, Hershey (2013). https://doi.org/10.4018/978-1-4666-1900-5.ch001

8. Mackworth, A.: Consistency in networks of relations. Artif. Intell. **8**(1), 99–118 (1977). https://doi.org/10.1016/0004-3702(77)90007-8

9. Mackworth, A., Freuder, E.: The complexity of some polynomial network consistency algorithms for constraint satisfaction problems. Artif. Intell. **25**(1), 65–74 (1985). https://doi.org/10.1016/0004-3702(85)90041-4

10. Bessiere, C.: Arc-consistency and arc-consistency again. Artif. Intell. **65**(1), 179–190 (1994). https://doi.org/10.1016/0004-3702(94)90041-8

Programming of Algorithms of Matrix-Represented Constraints Satisfaction by Means of Choco Library

Alexander Zuenko$^{(\boxtimes)}$ and Yirii Oleynik

Institute for Informatics and Mathematical Modeling – Subdivision of the Federal Research Centre, Kola Science Centre of the Russian Academy of Sciences, Apatity Murmansk 184209, Russia
zuenko@iimm.ru
http://www.iimm.ru

Abstract. The paper proposes an original approach to solving the problem of ineffective processing of qualitative constraints of a subject domain in the framework of constraint programming technology. The approach is based on the use of specialized matrix-like structures, providing a "compressed" representation of constraints over finite domains, as well as using author's inference algorithms on these structures. The paper presents practical aspects of implementation of user-developed types of constraints and corresponding algorithms-propagators with the help of constraint programming libraries. The algorithms performance has been assessed to clearly demonstrate the advantages of representation and processing of qualitative constraints of a subject domain by means of the above matrix structures.

Keywords: Constraint satisfaction problem
Constraint programming · Matrix-like representation of constraints
Qualitative constraints

1 Introduction

The paper continues the article also presented at the conference "Intelligent Information Technologies for Industry" 2018 [1] and presents practical aspects of implementation of original techniques of qualitative constraint propagation with the help of specialized constraint programming libraries (the Choco library taken as an example). The algorithms performance has been assessed to clearly demonstrate the advantages of representation and processing of qualitative constraints of a subject domain by means of the matrix-like structures. Briefly recall the basic concepts needed for further discussion.

According to [2] the *constraint satisfaction problem* (CSP) consists of three components: X, D, C.

Supported by Russian Foundation for Basic Research (grants nos. 16-07-00562-a, 16-07-00377-a, 16-07-00313-a, 16-07-00273-a, 18-07-00615-a).

X – a set of variables $\{X_1, X_2, \ldots, X_n\}$.

D – a set of domains $\{D_1, D_2, \ldots, D_n\}$ where D_i is the domain of variable X_i.

C – a set of constraints $\{C_1, C_2, \ldots, C_m\}$ that specify allowable combinations of the values of variables.

Each domain D_i describes a set of the admissible values $\{v_1, \ldots, v_k\}$ for variable X_i. Each constraint is a pair $< scope, rel >$ where $scope$ is a set of variables which participate in the constraint and rel – is the relation defining admissible combinations of values, which the variables from $scope$ can take on.

Constraints can be presented either explicitly, i.e. by enumeration of all the admissible combinations of the values for a set of variables specified, or implicitly, i.e. as an abstract relation supporting two operations: checking if a tuple is an element of the given relation, and enumeration of all the elements of the relation. The second way, in fact, requires specifying the characteristic function of the given relation.

Each $state$ in a CSP is defined by an assignment of values to some (partial assignment) or to all the variables (complete assignment):$\{X_i = v_i, X_j = v_j, \ldots\}$. The $solution$ of a CSP is complete assignment which satisfies all the constraints.

As an example, consider a CSP. Let $X = \{X_1, X_2\}$. We assume that $D_1 = D_2 = \{a, b, c\}$. Let a set C consists of an only constraint, that is, $C = \{C_1\}$. Constraint C_1 describes that fact that the values X_1 and X_2 must have different values.

The given constraint can be expressed implicitly, that is:

$$C_1 = << X_1, X_2 >, X_1 \neq X_2 >. \qquad (1)$$

The same constraint can be expressed explicitly, that is:

$$\begin{aligned} C_1 &= <<X_1, X_2 >, \{< a, b >, < a, c >, \\ &\quad < b, a >, < b, c >, < c, a >, < c, b >\} >. \end{aligned} \qquad (2)$$

Note, that constraint (2) can be expressed in a more compressed way:

$$C_1 = << X_1, X_2 >, \{a\} \times \{b, c\} \cup \{b\} \times \{a, c\} \cup \{c\} \times \{a, b\} >. \qquad (3)$$

There is a Table in Fig. 1a, which vividly represents expression (2). Figure 1b shows a matrix corresponding to expression (3). In fact, in case of a matrix representation (Fig. 1b), the sign of operation $\times$ (Cartesian product) between the components of one row is omitted, and the sign of operation $\cup$ between rows (union of sets) is not written explicitly. In [3], the similar representation is referred to as an "compressed" representation of the relation.

Unlike the article mentioned above, this paper concerns two types of matrix structures to represent constraints: C-systems and D-systems. In [4], the set-theoretical operations with the given structures are introduced. The similar structures are also used in [5] to solve pattern recognition and knowledge base compression problems.

X_1	X_2
a	b
a	c
b	a
b	c
c	a
c	b

X_1	X_2
{a}	{b, c}
{b}	{a,c}
{c}	{a,b}

a)   b)

Fig. 1. The tabular constraint representation (a); the constraint representation in the form of specialized matrix (b).

Expression (3) can be represented in the form of the C-system:

$$C_1[X_1 X_2] = \begin{bmatrix} \{a\} & \{b,c\} \\ \{b\} & \{a,c\} \\ \{c\} & \{a,b\} \end{bmatrix}. \tag{4}$$

D-systems allow calculating the complement of the C-systems: a complement is taken for each component-set.

Let's assume that we have a constraint $C_1[X_1 X_2]$ meaning that $X_1 \neq X_2$. It is necessary to express the constraint $X_1 = X_2$. Then it is possible to represent D-system as follows:

$$\overline{C_1}[X_1 X_2] = \begin{bmatrix} \{b,c\} & \{a\} \\ \{a,c\} & \{b\} \\ \{a,b\} & \{c\} \end{bmatrix}. \tag{5}$$

The D-system representation is equivalent to the expression:

$$\overline{C_1} = <<X_1, X_2>, \{[(D_1\backslash\{a\}) \times D_2 \cup D_1$$
$$\times (D_2\backslash\{b,c\})] \cap [(D_1\backslash\{b\}) \times D_2 \cup D_1 \times (D_2\backslash\{a,c\})]$$
$$\cap [(D_1\backslash\{c\}) \times D_2 \cup D_1 \times (D_2\backslash\{a,b\})] > . \tag{6}$$

The D-system represented can also be expressed as an intersection of three C-systems of the same scheme $< X_1, X_2 >$ namely $\overline{C_1} = K_1[X_1 X_2] \cap K_2[X_1 X_2] \cap K_3[X_1 X_2]$:

$$\begin{bmatrix} \{b,c\} & * \\ * & \{a\} \end{bmatrix} \cap \begin{bmatrix} \{a,c\} & * \\ * & \{b\} \end{bmatrix} \cap \begin{bmatrix} \{a,b\} & * \\ * & \{c\} \end{bmatrix}. \tag{7}$$

In specifying C-systems, a designation "$*$" (a complete component) may be used, which is equivalent to the indication of domain of the corresponding variable.

There is one more type of dummy components – a empty component (designated as "$\varnothing$"), that is a component containing no value.

Before comparing the effectiveness of the qualitative constraint satisfaction techniques developed by the authors and built-in methods of Choco library used for processing such constraints, we'll describe the main components of this library necessary to create user-developed types of constraints and corresponding satis-faction algorithms. Based on these components of the Choco library, the author's techniques of satisfaction of qualitative constraints represented in the matrix form were developed.

2 Defining CSPs and Algorithms of Their Solution by Means of Modern Tools of Constraint Programming (The Choco Library Taken as an Example)

Constraints programming environments allow usage of built-in types of con-straints and the algorithms of their satisfaction and make it possible to develop original types of constraints, methods of their propagation, as well as construct original search strategies.

The following libraries are most widely used in constraints programming: Choco and JaCoP for Java, as well as GeCode and Z3 for C++.

To implement the original algorithms of inference on constraints that are presented as specialized matrix structures, choice was made of the Choco library. Choco library is the open source software created to define and solve constraint satisfaction problems [6].

To describe CSPs by means of the Choco library, the following basic abstrac-tions are used: Model, Variable, Constraint, Propagator, Search Strategy, Solver, Solution.

2.1 Model

Abstraction "*Model*" is presented by a special class of the Choco library, on the basis of which all the further defining of the CSP is formed.

2.2 Variable

In the library Choco, variables are represented by specialized classes, depending on their type. There are determined four different types of variables in the library:

- **Boolean variable** is a variable with two possible values: true/false (0/1).
- **Integer variable** is a variable taking values from a set of integers. The domain of an integer variable can be specified as an interval $[a, b]$ (bounded domain). Such a representation consumes a small amount of memory, but does not allow processing the gaps in domains. In other words, it is impossible to eliminate an inadmissible value being inside the interval. Another way to specify domains of integer variables is to explicitly enumerate all possible values of a variable (enumerated domain). In so doing, the values should be linearly ordered.
- **Set variable** is a variable whose values are sets of integers. The variable of this type is specified by two sets, i.e. by the upper and lower boundary. It can take the values which are the subsets of the upper boundary and necessarily includes the lower boundary.

- **Real variable** is a variable taking values from the specified interval with the specified precision. At present this type of variables is supported rather poorly in the library.

The variables are added into the defining of a CSP either by means of the methods of class *"Model"*, which associates the variables with the corresponding model, or by means of classes implementing the interfaces of creating the variables of corresponding types. The interfaces associate the variables with the model specified in the class constructor. Thus, variables cannot exist by themselves and should be necessarily associated with the model.

2.3   Constraint

A certain logical formula which specifies admissible combinations of values of variables, is referred as a constraint. In the Choco library, *a constraint* is defined by a set of variables and by propagators (filtering algorithms) which delete the values from the specified variables domains, which do not correspond to the legal assignments. The library contains a set of built-in constraints, for example, global ones, *Alldifferent* in particular, i.e. a constraint meaning that the values of variables in a solution, should differ. Also presented in it are various arithmetical constraints; constraints presented as logical formulae; as rules, etc. There are standard methods-propagators for each built-in type of constraints.

An example of an arithmetical constraint:

model.arithm(var, "=",5).

This constraint means that variable var should take the value equal to 5.

The methods of class *"Model"* allow constructing more complicated constraints. The library also allows constructing original constraints and propagators. To create original program class of constraint, it is necessary to specify propagator (one or several) and set of variables, over which the constraint is set.

An example of user-developed constraint creation:

Constraint c=new Constraint("My",new Dpropogator(vars));

"My" – string-name of the constraint;

new Dpropogator(vars) – a propagator for a constraint over variables vars (there is a possibility to specify several of these).

2.4   Propagator

Abstraction *"Propagator"* is specified in the Choco library as class. Constructing class of propagator, it is necessary to define two main methods: a method *propagate* and a method *isEntailed*. The method *propagate* implements the logic of the constraint propagation. During the propagation, the method will be called repeatedly till it causes changes in variable domains. The method *isEntailed* is called at the end of propagation. The method *isEntailed* can return three parameters:

- *ESat.TRUE* – the constraint is completely satisfied;
- *ESat.UNDEFINED* – the constraint status failed to be determined (the propagation ends but the constraint has not been satisfied);
- *ESat.FALSE* – the constraint cannot be satisfied, backtrack is necessary.

2.5   Search Strategies

If the final solution has not been reached with the help of propagators, the search space is further studied in accordance with a certain search strategy. In fact, the search strategy defines the way the CSP solution should be constructed.

The Choco Version 4.0.0 constructs a binary search tree (for example, if assignment "$x = 5$" cannot be extended to a solution, then "$x \neq 5$" is considered). The search strategy like this, which implements backtracking search, is typical for CSPs. However, other search strategies shouldn't be neglected.

The strategy correctly selected for a certain problem allows the solution to be generated much faster. The types of strategies are related to the types of variables, i.e. each type is supplied with a specific set of strategies. Like in case with constraints, the Choco library provides built-in program classes for popular strategies. However, if necessary, user-developed search strategy may be constructed. To describe a user-developed strategy, it is necessary to additionally define two (three, optionally) classes: a strategy to select a particular variable from a set of the CSP variables (Class 1), a strategy to select a particular value of the specified variable from its domain (Class 2), and, optionally, a class implementing the strategy to select the branch of a search tree (Class 3).

Let's characterize Class 3 separately. It is necessary to define two basic methods here. The first one, *apply(IntVar, int)*, is the method for processing a pair *<a variable, its value>* obtained as a result of the variable selection strategy application and the variable value selection strategy application. The second, *unapply(IntVar,int)*, is the method specifying how modifications introduced by the method *apply* should be droped. If the search-tree branch selection strategy is not determined, the method *apply* assigns the selected value to the specified variable, trying to extend the partial assignment to a complete one. Having worked out this variant, the method *unapply* eliminates the considered value from the domain of the corresponding variable. The search proceeds in alternative directions.

2.6   Solver

"*Solver*" is an abstraction presented by the class, which keeps the stages of solution process, search strategies and the CSP solutions if those have been reached. The Solver type object is the field of class "*Model*". After the necessary solver parameters have been specified, its method *solve()* is called to start searching.

2.7   Solution

A "*Solution*" is an abstraction presented by the Choco library class, which serves as a storage of a complete or partial assignment.

Now we shall try to answer the question: "When is the representation and handling of a CSP in a kind of C- and D-systems capable to ensure the highest computing performance". In other words: "In what cases should CSPs be represented and processed as the C- and D-systems?"

3   Comparison Between the Algorithms of Choco Library and the Procedures Developed

To determine the effectiveness of the classes and algorithms developed we used the problem of placing n chess queens on an $n \times n$ chessboard so that no two queens threaten each other ("N-Queens problem"), due to the simplicity of its scaling.

The aim of the given example was not to generate solutions of the "N-Queens problem" (one or all possible) as fast as possible. Moreover, the authors realized that, by means of standard numerical constraints, it is possible to define the given problem more implicitly. Taking into account of the chessboard symmetry and constraints propagation techniques based on the interval analysis, allow the solutions to be generated faster than in the procedure described in the study presented. The aim of the analysis made is to demonstrate that qualitative dependencies processing by means of modern constraint programming environments, by the Choco library in particular, is less effective than that by the matrix representation proposed in this paper.

That is why the "N-Queens problem" will be described in the form of a set of qualitative constraints and a comparison will be made between the propagation algorithms performance for the two cases. In the first case, the problem is formulated in the form of logical expressions in the Choco language and standard classes-propagators are used. In the second case, the constraints are formalized in the form of a set of the matrix-like structures suggested. Original constraint propagation algorithms are proposed, on the basis of which own classes- propagators extending the base functional of the Choco library are developed.

Thus, the algorithms compared differ in the qualitative constraints representation and the propagators used. After the propagation has stopped, the search tree branching strategy, typical for Choco, is applied.

Consider a simplified variant of "N-Queens" problem. In this example, the chessboard size is 4×4. It is necessary to find possible variants of four queens placing.

Let's associate the i-th horizontal with variable X_i. Then each variable (attribute) will be defined in the domain as $\{a, b, c, d\}$, where a, b, c, d are the labels of the verticals. As an example, let's formulate the constraint "two queens placed on horizontals 1 and 2 are not threatened by each other" in the form of the D-system:

$$
\begin{array}{c}
X_1 \qquad X_2 \\
\{a,b,c,d\} \; \{a,b,c,d\} \\
\begin{array}{c}
1 \\ 2 \\ 3 \\ 4
\end{array}
\left[
\begin{array}{cc}
\{b,c,d\} & \{c,d\} \\
\{a,c,d\} & \{d\} \\
\{a,b,d\} & \{a\} \\
\{a,b,c\} & \{a,b\}
\end{array}
\right.
\end{array}
\qquad (8)
$$

In particular, the first row of the given D-system shows that if a queen is on the field a_1(the intersection of the first horizontal and the first vertical), then, in the second horizontal, other queen can occupy fields c_2 and d_2 only. In the language of logic, it is expressed as follows:

$$
\begin{aligned}
&(x_1 = a) \rightarrow ((x_2 = c) \vee (x_2 = d)) \\
&\text{or } \neg(x_1 = a) \vee (x_2 = c) \vee (x_2 = d) \\
&\text{or } \neg(x_1 = b,c,d) \vee (x_2 = c) \vee (x_2 = d).
\end{aligned}
\qquad (9)
$$

Comparing different pairs of horizontals, it is possible to write out all the constraints on the inter-relative positioning of all the 4 queens. In our case, the number of constraints is calculated by the formula: $4\,C_4^2$ (for each pair of horizontals, four constraints are formed, the total number of pairs is $-\,C_4^2$), that is $\frac{4^2 \cdot (4-1)}{2}$. For a board of $n \times n$ in dimension, the number of constraints is calculated by the formula: $\frac{n^2 \cdot (n-1)}{2}$. So, the CSP considered can be expressed by a D-systems set (like the D-system shown above) describing to the admissible positions of pairs of queens relative to each other.

In the Choco library, the D-system can be represented as a set of built-in logical constraints of the library. For example, the D-system

$$
\begin{array}{c}
X_1 \qquad X_2 \\
\{b,c,d\} \; \{a,b,c,d\} \\
\begin{array}{c}
2 \\ 3 \\ 4
\end{array}
\left[
\begin{array}{cc}
\{c,d\} & \{d\} \\
\{b,d\} & \{a\} \\
\{b,c\} & \{a,b\}
\end{array}
\right.
\end{array}
\qquad (10)
$$

can be described, using the Choco library by encoding the values of variables by integers (Fig. 2). It is clear from Fig. 2 that in order to describe even such a small D-system, a significant number of constraints is required: one constraint per each value of the component (cell) of the D-system, one constraint per each component, constraints uniting components of a row, and one common constraint uniting rows of a system. As the system increases in size in such a representation, the number of constraint also increases essentially, which will take much more time to reach the solution. That is why an original representation of the D-system has been developed to specify it by only one constraint.

Presented below are comparative plots of time required for solutions of "N-Queens problems" for both representations specified. The units of measure for a vertical scale of plots are milliseconds, i.e. the time to solve the problem. However, the time assigned to the solution, has been limited to 2 min (120000 ms).

```
Model testmodel=new Model ("Test");
int []x = (2,3,4),y = (1,2,3,4) ;
IntVar X = testmodel.intVar("X",x);
IntVar Y = testmodel.intVar("Y",y);
Constraint node11=testmodel.or(testmodel.arithm (X,"=",3),
                               testmodel.arithm(X,"=",4));
Constraint node12=testmodel.arithm(Y,"=",4);
Constraint node21=testmodel.or(testmodel.arithm(X,"=",2),
                               testmodel.arithm(X,"=",4));
Constraint node22=testmodel.arithm (Y,"=",1);
Constraint node31=testmodel.or(testmodel.arithm(X,"=",2),
                               testmodel.arithm(X,"=",3));
Constraint node32=testmodel.or(testmodel.arithm (Y,"=",1),
                               testmodel.arithm(Y,"=",2));
Constraint firstrow=testmodel.or(node11, node12);
Constraint secondrow=testmodel.or(node11, node12);
Constraint thirdrow=testmodel.or(node31, node32);
testmodel.and(firstrow,secondrow,thirdrow).post();
```

Fig. 2. *D*-system described by the means of Choco library.

Plots in Fig. 3 show the time spent for reaching the first solution of the "*N*-Queens problem". The plots are broken when the testing computer memory is exhausted. The plots in Fig. 4 show the time spent for reaching all the solutions of the "*N*-Queens problem". These are broken in the points in which the program was not able to reach all the solutions at the time assigned.

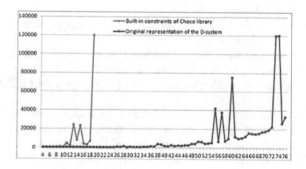

Fig. 3. Reaching the first solution of the "*N*-Queens problem" (ms/*N*).

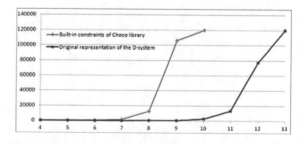

Fig. 4. Reaching all the solutions of the "*N*-Queens problem" (ms/*N*).

4 Conclusion

The studies have demonstrated that the matrix constraint representation proposed by the authors, as well as the original methods of their propagation are quite suitable for practical application. Moreover, in case of qualitative constraint modelling, the application of the approach proposed gives an essential gain in time against the algorithms of qualitative constraint propagation, which are built in the Choco library. In particular, solving the "N-Queens problem", when it was required to reach all the solutions under the time limit of two minutes, the approach proposed permitted processing the search space of 12^{12} in dimension. In doing so, the standard tools of the Choco library did not give a possibility to study the search space of greater than 9^9 in dimension. When the task was to achieve even one solution for the same two minutes, the time for the standard tools of the Choco library to reach a solution was not enough even when the search space was of 19^{19} in dimension. The methods proposed allowed the authors to reach a solution even for the search space of 76^{76} in dimension.

References

1. Zuenko, A.: Matrix-like structures for representation and processing of constraints over finite domains. In: Proceedings of the Intelligent Information Technologies for Industry 2018 (IITI 2018) (2018)
2. Russel, S., Norvig, P.: Artificial Intelligence: A Modern Approach, 3rd edn. Prentice Hall, Upper Saddle River (2010)
3. More, T.: Axioms and theorems for a theory of arrays. IBM J. Res. Dev. **17**(2), 135–175 (1973). https://doi.org/10.1147/rd.172.0135
4. Kulik, B.: A logic programming system based on cortege algebra. J. Comput. Syst. Sci. Int. **33**(2), 159–170 (1995)
5. Zakrevskij, A.: Integrated model of inductive-deductive inference based on finite predicates and implicative regularities. In: Naidenova, X., Ignatov, D.(eds.) Diagnostic Test Approaches to Machine Learning and Commonsense Reasoning Systems, pp. 1–12. IGI Global, Hershey (2013). https://doi.org/10.4018/978-1-4666-1900-5.ch001
6. Choco Documentation. http://www.choco-solver.org. Accessed 20 Feb 2018

Author Index

© Springer Nature Switzerland AG 2019
A. Abraham et al. (Eds.): IITI 2018, AISC 875, pp. 449–450, 2019.
https://doi.org/10.1007/978-3-030-01821-4

Printed in the United States
By Bookmasters